Emerging Infectious Diseases of the 21st Century

Series Editor: I.W. Fong
Professor of Medicine, University of Toronto

More information about this series at http://www.springer.com/series/5903

I.W. Fong

Emerging Zoonoses

A Worldwide Perspective

I.W. Fong
Professor of Medicine
University of Toronto
Toronto, ON, Canada

Emerging Infectious Diseases of the 21st Century
ISBN 978-3-319-84513-5 ISBN 978-3-319-50890-0 (eBook)
DOI 10.1007/978-3-319-50890-0

Softcover reprint of the hardcover 1st edition 2017

Printed on acid-free paper

This Springer imprint is published by Springer Nature
The registered company is Springer International Publishing AG
The registered company address is: Gewerbestrasse 11, 6330 Cham, Switzerland

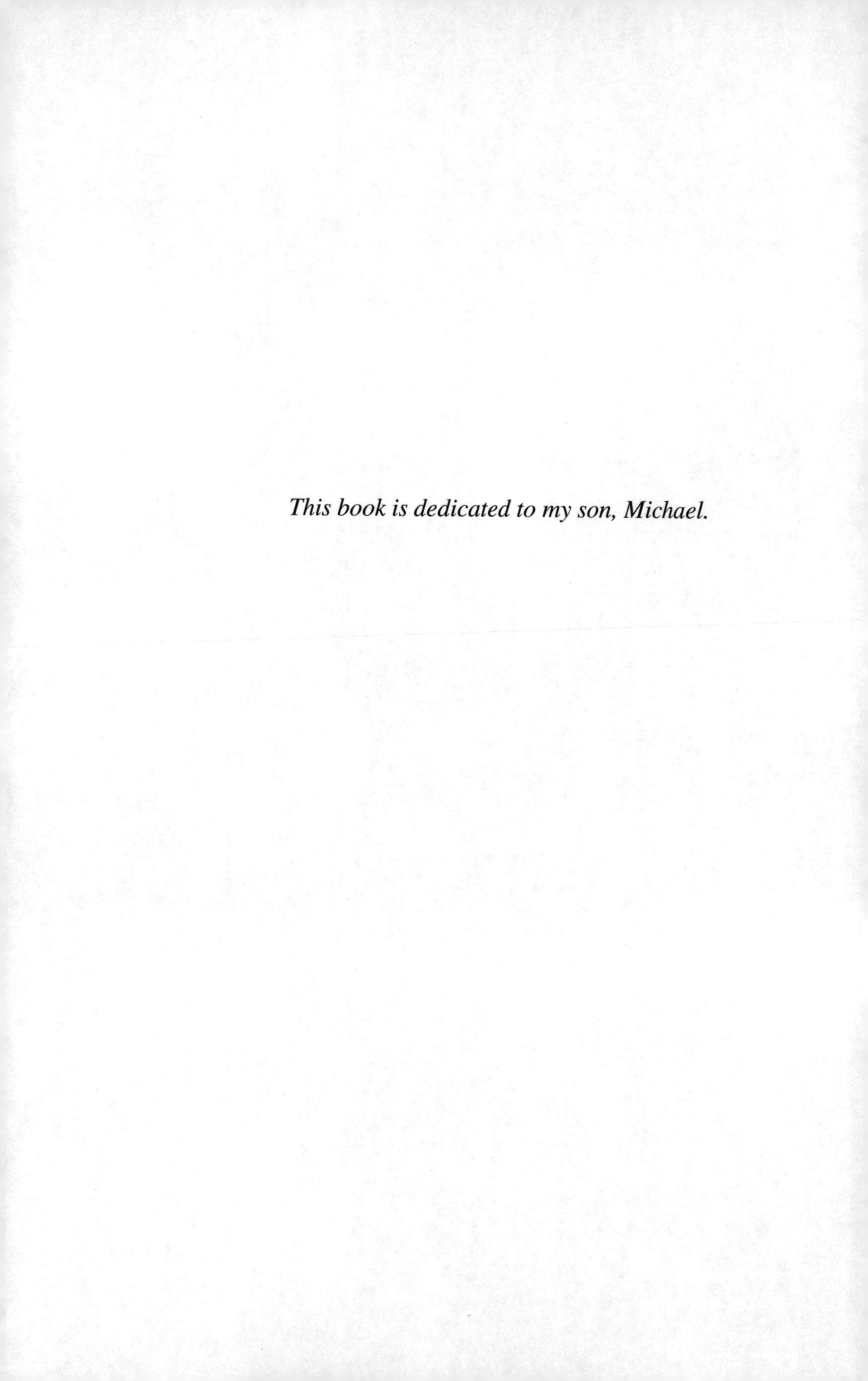

This book is dedicated to my son, Michael.

Preface

Zoonoses, diseases of animals that may spread to humans under natural conditions, have been recognized as causing pandemics of infectious disease from the Middle Ages and even before, and continue to be a threat to global health. Many endemic and sporadic diseases in all countries of the world are still transmitted by animals directly or indirectly. Zoonoses are estimated to account for about 75% of all new and emerging infectious diseases. It is predicted by public health experts that the next major pandemic or future pandemics of infectious disease will be of animal origin. Thus, it is important for all specialists in the health-care field—public health, health-care planners, infectious diseases, infection control/microbiology, veterinary medicine, community epidemiology, and clinical health-care first responders—to be familiar with emerging zoonoses. In this edition the first chapter reviews ancient pandemics of the Middle Ages caused by the "Black Death" or bubonic plague, the Spanish Influenza pandemic [derived from avian influenza], and the more modern pandemic of AIDS/HIV infection, which originated in Africa from primates. However, the main chapters are focused on zoonoses that have been recognized recently since the late twentieth century to the present, such as SARS and MERS coronaviruses, new avian influenza viruses, new tick-borne viruses [Henan fever virus] from China, and recently from the United States [Heartland virus], and recently recognized bacterial pathogens such as *Streptococcus suis* from pigs. In addition, the reemergence of established zoonoses that have expanded their niche is reviewed, such as the spread of West Nile, Chikungunya, and Zika viruses to the Western Hemisphere, the emergence and spread of Ebola virus infection in Africa, and others. New and emerging parasitic zoonoses that were previously considered to be only animal pathogens but have started appearing with increased frequency in human populations are also reviewed. A chapter is also devoted to an overview of the mechanisms and various types of animals involved in the transmission of diseases to humans, as well as to potential means of control and prevention.

Toronto, ON, Canada I. W. Fong

Acknowledgements

I am indebted to Carolyn Ziegler and Mark Naccarato for their invaluable assistance in literature searches, and to Debbie Reid-Marsden for her administrative assistance.

The updated original online version for this book can be found at
DOI 10.1007/978-3-319-50890-0_12

Contents

Chapter 1
Pandemic Zoonoses from the Middle Ages to the Twentieth Century

1.1 Historical Aspects

Zoonoses are diseases of animals transmitted to humans which have been in existence for thousands of years [1]. It has been surmised that zoonoses had afflicted ancestors of humans [Neanderthals, *Homo erectus*, and early *Homo sapiens*] from the beginning of our evolution. Ancient humans living in caves depended on their source of food from hunting and gathering of wild fruits and plants. Thus, exposure to animal pathogens and cross infection was likely a constant threat of daily living. Examination of preserved bones of Neanderthals [more than 100,000 years old] and cavemen by paleontologists has revealed evidence of tuberculosis [likely bovine origin] and chronic brucellosis [1].

Outbreaks or epidemics of zoonoses most likely were never experienced until after the Agriculture Revolution about 10,000 years ago, when social communities and farming developed. Metal tools and domestication of wild animals for food began in Eurasia [the Bronze Age] 6000 years ago [1], and small epidemics of animal-related infections may have started to appear. The original description of a disease outbreak, an epidemic, had been attributed to Hippocrates between 480 and 380 BC [2]. Although epidemics of infectious diseases were recorded by historians from the start of the Bronze and Iron Ages, across Europe and Middle East to Asia, their microbial etiology remains speculative. The plague of Athens in 430 BC and the plague of Antoninus in Rome [second century AD] were probably not secondary to the plague bacillus [3]. Recurrent outbreaks of infectious diseases originating in animals in this era have been attributed to overcrowding with the population explosion in cities, domestication of animals, unsanitary conditions, and traveling between cities and countries.

I.W. Fong, *Emerging Zoonoses*, Emerging Infectious Diseases of the 21st Century,
DOI 10.1007/978-3-319-50890-0_1

1.2 Plague Pandemics

1.2.1 History Revisited

A possible epidemic of bubonic plague was first described in the Old Testament, The First Book of Samuels circa 1000 BC, when the Philistines were afflicted by a plague of rodents. The two great epidemics of the Middle Ages included the Justinian plague in the sixth century and the Black Death of the fourteenth century. The first epidemic of 523 AD killed 10,000 inhabitants of Constantinople alone, and it is the first documented epidemic in history, in which the etiology is almost certainly due to the bubonic plague [4]. The description of the clinical symptoms with the characteristic plague sore in the groin [buboes] and axilla, followed by sloughing from necrosis, and subsequent toxemia and death were well documented. The Justinian plague epidemic in the sixth century was confined to the Mediterranean basin and subsequently subsided with occasional flares of isolated cases.

The plague pandemic of the fourteenth century was more terrifying and extended from Europe across Southwest Russia and to India. Historians believe that the plague originated in India and reached the Crimea in 1347 and was imported into Venice and Genoa by shipborne rats and then spread across Europe [5]. It has been surmised that the practice of biological warfare occurred around this time. Tartars besieged the seaport of Caffa near Crimea and catapulted plague corpses into the walled town of Christians, resulting in a plague outbreak that decimated the community [4]. Rapid spread of the disease occurred in nearby populations. Seaports were subsequently identified as foci for reintroduction and perpetuation of plague epidemics, and this gave birth to introduction of quarantines to curb outbreaks. Ports and cities free of plague instituted quarantine of passengers and foreigners arriving by ships to be held in guarded barracks outside of town for 40 days [quaranta giorni], before allowing freedom of movements [4].

It is estimated that between 1348 and 1720 there were at least ten plague epidemics or pandemics [5]. The mortality estimates varied, but it is believed that at least 25 million Europeans perished, and up to 40 million people worldwide died from plague [4]. Regulations governing public health came into being after the ravages of the Black Death in the fourteenth century, and controls for foods in markets were instituted just before. In the early years of the fourteenth century, it was suspected that plague was spread by invasion of rats, and measures were instituted to prevent ship rats getting on shore from vessels carrying sick passengers. Affected ships had to hoist a quarantine flag on the foremast and remain offshore for the duration of the quarantine [5]. Strict hygiene regulations were also enforced, such as proper disposal of excrement and refuse, and clearing of rubbish and filth from streets and passage-ways to discourage attraction of rats. Although the fleas or rats and other rodents [ground squirrels, chipmunks, prairie dogs, and hamsters] were considered vectors of the plague from the Middle Ages, it was not until 1894 that the etiology agent was identified in Hong Kong [5]. The smaller pandemic that preceded the discovery of the plague bacillus began in China in the 1860s and spread to Hong Kong by 1894. The disease was subsequently spread by rats transported on ships to

California and port cities of South America, Africa, and other parts of Asia [6]. The plague bacillus, *Yersinia pestis*, was discovered by the Swiss-born Alexandre Yersin and the Japanese Shibasaburo Kitasato. The main vectors are fleas [*Xenopsylla cheopis*] which transmit the non-motile, gram-negative bacillus by biting [5]. Measures used to control the rat and flea population included various poisons such as hydrocyanic acid, thallium, and phosphorus and later DDT derivatives.

1.2.2 Current Epidemiology

Plague still exists endemically worldwide in many countries, and several cases are reported each year in the United States [US], but most outbreaks are reported from developing countries in Africa, Southeast Asia, and Asia. Sporadic cases in the US occur predominantly in the Four Corners area where Colorado, Utah, Arizona, and New Mexico conjoin, and prairie dogs and rabbits are the primary hosts of infected fleas [6]. When the infected animal dies, the fleas spread the disease to other hosts, such as pet dogs and humans. Although most outbreaks of plague are transmitted by rodent fleas, the "human flea" [*Pulex irritans*] have been implicated in local epidemics and intermittent transmission in Africa [7]. Plague can also be transmitted by direct contact or ingestion of infected animal tissues and by inhalation of droplets from patients with pneumonic plague or infected pets such as a cat [8]. Pneumonic plague is highly contagious and lethal if not treated early. Hence, aerosolization of the plague bacillus is considered a weapon of bioterrorism [9].

1.2.3 Microbiological and Clinical Aspects of Plague

Yersinia pestis has been divided into three biovars on the basis of glycerol fermentation and reduction of nitrate. Antigua biovar from East Africa was considered to have descended from bacteria that caused the first pandemic, biovar Medievalis from Central Asia may have descended from bacteria that caused the second pandemic, and the third pandemic was linked to the Orientalis biovar [10]. Subsequent studies, however, on *Y. pestis* DNA from ancient human remains in Europe indicate that only the Orientalis biovar caused the three pandemics [11].

Plague usually presents clinically in one of three forms: bubonic plague presents with swollen, tender lymph glands in the groin or axilla, with severe prostration, fever, chills, myalgia, rash, and headaches, which can progress to confusion, delirium, and death if untreated [this is the main form of previous pandemics]; primary septic plague without signs of localized infection but presenting with overwhelming sepsis; and pneumonic plague with usual symptoms of cough, pleurisy, hemoptysis, and respiratory distress with rapid progression and high fatality rate and radiological evidence of lobar or diffuse bronchopneumonia pattern [9]. Septicemia and pneumonia can also occur from the other clinical forms. Severe pharyngitis and gastrointestinal disease can sometimes occur from ingestion of infected meat or liver [9, 12], and in general the incubation period is short [2–6 days] but occasionally longer.

The prognosis of plague in the modern era is very good with early administration of effective therapy. Up to 50% of patients with bubonic plague can survive without antimicrobials, but untreated pneumonic or septic plague is nearly always fatal without treatment. Early diagnosis and institution of effective antibiotics is essential for a good outcome. Several antimicrobials have in vitro activity against *Y. pestis*, including quinolones, ceftriaxone, and sulfamethoxazole-trimethoprim, but guidelines recommend one of the following [based on past experience]: streptomycin or gentamicin, tetracycline or doxycycline, or chloramphenicol for 7–10 days or after 3 days of clinical recovery [9]. With adequate treatment a fatality rate of 17% still occurs but this is often related to delayed treatment.

1.3 Pandemic Zoonotic Influenza

The word "influenza" was derived from the Italian word meaning influence in 1743, from the belief at the time that environmental miasma or the stars could influence disease outbreaks. The initial source of influenza is surmised to be from animal origin. Communities for centuries had noted human outbreaks of influenza coincided with epidemics in ducks, pigs, and horses [13]. Thus human influenza outbreaks probably dates back to 2000–5000 BC at the time of domestication of animals. However, the earliest documented accounts of influenza epidemics were in Europe around 1387 [13]. It is estimated that there have been at least ten zoonotic influenza A pandemics in the past 300 years. The first influenza pandemic occurred in the sixteenth century and 5–10 more in the eighteenth and nineteenth centuries [13].

The influenza virus frequently undergoes minor genetic drifts in its surface proteins on a regular basis since its existence. However, it is believed that there are major antigenic drifts nearly every 60 years, which may explain cyclical pandemics at 50–100 years interval in populations immunologically naïve to new circulating viruses with novel proteins and epitopes. These new strains of influenza virus with novel antigens usually arise from avian strains with coinfection of pigs with porcine strains, or humans with circulating endemic strains, resulting in novel viruses with major genetic changes but adapted to human receptors. The pandemics of 1833 and 1889 originated from Asia and possibly from Eastern Russia with strains of influenza A of bird or swine origin. The epidemics spread across Central Asia and to the Pacific islands by land and ships. The 1833 epidemic infected nearly half the population of some European cities and caused hundreds of thousands of death [13]. In 1889 a more rapidly expanding pandemic occurred, aided by rapid transit of trains and steamships, and killed over a quarter million of people in Europe alone [13].

In 1918–1919, the Spanish influenza pandemic caused the worst ever medical disaster in the history of mankind. It is estimated that one-third of the world's population [about 500 million] was afflicted, resulting in 20–100 million deaths [13, 14]. The 1918 influenza virus was a H1N1 subtype, based on its hemagglutinin [H] and neuraminidase

[N] surface proteins. The evolutionary history of the 1918 influenza virus and the relationship with the preceding viruses of the 1833 and 1889 pandemics have been debated by scientists for decades. Archaeovirologists proposed that an H2 strain caused the pandemic in 1889 and that an H3 subtype was responsible for a smaller epidemic in 1900 [15]. This was supported by RNA found in archival tissue samples from subjects with pneumonia that died in 1915. The influenza H1N1 strain of 1918 is believed to have expressed novel antigens, possibly derived from avian influenza virus, to which most humans and swine were immunologically naïve at the time [15]. It was observed during the 1918–1919 pandemic that there were almost simultaneous outbreaks in humans and pigs.

There have been three influenza pandemics in the twentieth century, the deadly H1N1 1918 epidemic and subsequent less severe outbreaks in 1957 [H2N2 virus] and in 1968, caused by a H3N2 strain [16]. These pandemics were the result of major antigenic shifts with introduction of novel hemagglutinin subtypes from animal source to humans [17, 18]. Recent investigations indicate that genetic elements of the 1918 virus circulated in swine and humans even by 1911 [H1N2 strain] and were not a novel avian influenza virus [19]. It appears most likely that reassortment of an avian strain occurred in swine before emergence of the Spanish pandemic. Results of phylogenetic analysis suggest that novel gene segments of H2N2/1957 and H3N2/1968 pandemic viruses originated from avian influenza viruses, and the offending strains resulted by reassortment with circulating human viruses [19].

1.3.1 Unique Features of the 1918 Pandemic

Most influenza pandemics originate in Asia and then spread across the rest of the world. Historical and epidemiological data indicate that the 1918 pandemic occurred in three distinct waves over a year. The first spring wave started in March 1918 and expanded unevenly throughout Asia, Europe, and North America over 6 months, with high rates of clinical illness but low mortality rates [20]. A second wave in the fall of 1918 [September–November] spread globally with high fatality rates; and a third wave occurred in early 1919 in many countries with high frequency of complicated cases [20]. The reason for the unprecedented rapid successive waves, with brief quiescent periods, still remains a mystery. Some historians believe that the second wave may have been facilitated by troops and supply ships that spread the virus worldwide, starting around August 1918 [13].

Another unique feature of the 1918–1919 pandemic was that the absolute risk of influenza death was highest in those younger than age 65 compared to older people [19]. Nearly half of all influenza-related deaths were in young adults 20–40 years of age, unprecedented before and since this pandemic. Comparative influenza-related deaths in the three pandemics of the twentieth century were in 1918–1919 that was associated with excess mortality of 99% in adults under age 65 years, but this age group accounted for influenza deaths of only 36% in the1957 outbreak and 48% in the 1968 epidemic [20]. The explanation for the high mortality in healthy young adults

[5–20-fold greater than expected] in the 1918 pandemic is controversial. Some experts believe that the H1N1 strain was hypervirulent but others disagree. The entire genome sequence of the 1918/H1N1 virus was completed in 2005 from archival tissues [21], and mice infected with the reconstructed virus demonstrated marked bronchial and alveolar epithelial necrosis, inflammation and upregulation of genes involved in apoptosis, and oxidative damage [22, 23]. In contrast control mice infected with modern human influenza virus produced little lung disease and limited viral replication. It is also postulated that the high fatality rate in young adults was related to partial protection of older adults born before 1889 from previous exposure to a then circulating precursor virus [20]. Others opined that the Spanish influenza virus was not particularly more virulent than usual seasonal influenza virus, as the first wave of the pandemic was mild. Instead they postulate a sequential infection hypothesis and suggest that most patients who succumb in 1918–1919 pandemic died from bacterial pneumonia rather than from primary influenza infection [24]. Historical records and animal studies suggest that prior influenza from the 1890 A/H3Nx pandemic strain resulted in dysregulated immunopathological effects that increased susceptibility to lethal bacterial pneumonia [25]. Moreover, the 1918 influenza A/H1N1 is genetically similar to the novel H1N1 pandemic strain of 2009, with a hundred fold less mortality rate [26]. Seasonal influenza can predispose to secondary bacterial pneumonia by alteration of the innate immunity and causing respiratory epithelium damage and dysfunction. High levels of interferon-Y in response to influenza can impair alveolar macrophage function and decrease bacterial phagocytosis [27].

Recently, investigators using a highly accurate molecular clock technique capable of determining rates of molecular evolution, combined with serological and epidemiological data, have proposed a new theory. Their results suggest that the pandemic 1918 virus originated shortly before 1918 when a human virus emerged around 1907 acquired avian N1 neuraminidase and internal proteins jumped directly to swine then to the human population, but was displaced in humans around 1922 by a reassortment with an antigenically distinct H1 hemagglutinin [28]. Previous pandemics of 1820–1833, 1847–1850, and 1889–1893, based on seroarcheological analysis, were surmised to be associated with H1N, H1N8, and H3N8, respectively. The elderly born before 1834 may have been exposed to a H1 strain that emerged in 1830, provided some protection against the 1918 H1N1 influenza virus, which resulted in lower mortality and morbidity in this age group during the Spanish pandemic. Similarly, children exposed during 1834–1893 probably had antibodies to the H1 antigen, whereas children exposed to the H3N8 virus during 1889–1893 had no significant protection in 1918–1919, with resulting severe disease in the 20–40-year age group [28].

1.3.2 Other Pandemics of the Twentieth Century

After the 1918–1919 pandemic, regular influenza returns to the usual pattern, with less virulent regional epidemics in the 1930s, 1940s, and 1950s [16]. In the spring of 1957, a rapidly evolving epidemic in Hong Kong heralded the Asian influenza

[H2N2] pandemic. The novel H and N antigens did not induce cross-reactive protection from previous influenza A exposure; hence the population at large was unprotected. Animal studies with the new H2N2 virus did not reveal any difference in virulence compared to other influenza A subtypes [16]. Many of the deaths were attributable to primary influenza pneumonia and secondary bacterial pneumonia, with the greatest at-risk population similar to regular seasonal influenza, the elderly, presence of chronic diseases, and pregnant women in the third trimester of pregnancy [29]. Mortality rates were lower than the 1918 pandemic with improved medical care and availability of antibiotics. Over the next several years, the population gradually developed protective antibodies and the virus disappeared from the human population only after 11 years.

A decade after, in 1968, a new pandemic arose from Southeast Asia, the Hong Kong influenza [H3N2] pandemic. The epidemic started in Hong Kong and spread through mainland China and Asia and then the rest of the world. The severity of illness and extent of regional outbreaks were spotty and variable, with relative mild illness in Japan and Western Europe, but with more severe disease and higher mortality on the West Coast of the US [16]. This smoldering pandemic was probably relative mild because, although the H3 antigen was novel, the N2 antigen was similar to that of the previous pandemic virus. The Asian influenza of the 1957 pandemic had acquired three genes, H2N2, and the polymerase subunit protein [PB1], from avian viruses infecting wild ducks, in a backbone of the circulating human influenza strain [30], whereas the Hong Kong influenza H3N2 strain of 1968 acquired two genes from a duck virus [H3 and PB1] in a background of the circulating human H2N2 virus [30, 31]. It has been speculated that the variability in severity of illness in different regions of the world during the Hong Kong pandemic was related to differences in previous exposure and immunity to the N2 antigen [32]. Even after more than 37 years the H3N2 virus continued to circulate among the human population [16].

1.4 The AIDS/HIV Pandemic

The acquired immunodeficiency syndrome [AIDS], due to the human immunodeficiency virus [HIV], is the largest and deadliest infectious disease pandemic ever experienced by mankind. It evolved slowly and insidiously with clinical appearance in the early 1980s, presenting with unusual opportunistic infections and Kaposi sarcoma in apparently healthy gay men in California and New York [33]. Initially AIDS was considered a "gay-related disease," and by 1982 it was found in Haiti, Europe, and Central Africa, where it was recognized to afflict mostly heterosexuals and prostitutes and their male contacts. Later AIDS was recognized to be an affliction of intravenous drug abusers [IVDAs] and hemophiliacs, from blood and blood product venous exposures. Following the development of a diagnostic test for HIV infection and clinical criteria for AIDS in 1985, the full extent of the epidemic became evident. By 1990 all countries reported AIDS cases, with universal fatality in a few years until the advent of highly active antiretroviral therapy [ART].

HIV belonged to the family of retroviruses first discovered in1970 and later cultured in the laboratory in 1978 [34], just before recognition of AIDS. The retroviruses invade cells as an RNA virus, are converted to DNA by reverse transcriptase enzyme in the cytoplasm, and enter the nucleus to become integrated [by the integrase enzyme] into the host's genes. There is a long latency period before activation of the virus, which is transmitted to daughter cells and eventually produces disease. The first retrovirus discovered was the human T-cell lymphotropic virus [HTLV-1] which causes adult T-cell leukemia/lymphoma, first described in Japan in 1977, and later recognized as the cause of tropical spastic paraparesis. A close relative HTLV-2 was found in 1982 that causes hairy cell leukemia, and the HIV was discovered by French scientists in 1983, and they were later awarded the Nobel Prize in medicine.

HIV and the other retroviruses are believed to have originated in Africa from primates for immeasurable millennia [34], and at some unknown time crossed the species barrier to infect humans. Stored blood samples from the 1950s and 1960s have found evidence of HIV infection in Central Africa and neighboring countries, with increased frequency over the years and in the US in the early 1980s [34]. HIV-1 group M is the main culprit of the AIDS pandemic, and it is surmised to have evolved from the closely related simian immunodeficiency virus [SIV] of the chimpanzee [*Pan troglodytes*], with cross-species transmission in Central Africa early in the twentieth century [35]. The human exposure to simian viruses may have been from hunting and eating "bush meat." HIV-2, which causes a less rapidly progressive disease and rarely found outside Western Africa, is believed to have originated in sooty mangabeys, a primate [35]. Recent phylogeographic analysis of archived samples collected in the Democratic Republic of Congo [DRC] and neighboring countries, using molecular clock techniques for estimates of viral mutation rate, reconstructed the spread of HIV in Africa. HIV-1 group M first emerged in Kinshasa around 1920 and spread to surrounding regions connected to Kinshasa by railways [36]. It is surmised that around 1960 the growth rate of HIV-1 increased and spread by commercial sex workers and by IVDAs and contaminated blood products. HIV-1 [gp M, subtype B], which predominates in the Americas and Europe, may have been carried out of Africa by Haitians who worked in the DRC in the 1960s [36]. Subtype C, which accounts for about 50% of HIV infection worldwide, may have developed in the mining regions of the DRC and spread throughout Africa by migrant workers.

Using molecular sequencing, investigation had also suggested that HIV have been slowly spreading in Africa for 100–200 years or longer [36]. Due to the long incubation period of 8–10 years before developing clinical AIDS, the HIV epidemic largely remains unrecognized for decades. The rapid advance of the AIDS pandemic in the 1980s–1990s can be partly explained by several factors, increased clinical recognition and the availability of diagnostic tests, ease of global travel, change in social attitudes toward sexual freedom and adventurism, and a large core of HIV-infected asymptomatic population. The recent rapid growth of the HIV epidemic over the past three decades may reflect normal development of an epidemic, especially if the doubling time were slower in the earlier times. It is postulated that with a doubling time of three years and exponential growth, it would take 30 years for the prevalence of HIV infection to change from a thousandth of a percent to a noticeable 1%, but only 3 years to change from 10% to 20% [36].

1.4.1 Progress in AIDS

Exceptional progress has been made in the last 35 years since the recognition of AIDS. The advancements made first in the discovery of the HIV as the cause, defining the life cycle and key enzymes in the process of development, and discoveries of highly active treatment in a relatively short period of time are unprecedented in medical science.

Highly effective antiretroviral therapy [ART] became available in developed rich countries since 1996, and this has made a dramatic impact on the life-span and quality of life of AIDS-/HIV-infected people. Just 20 years ago, AIDS diagnosis was considered a "death sentence" with universal fatality in a few years. Now AIDS/HIV infection is considered a chronic manageable disease with life expectancy similar to that of diabetes mellitus. The development of new drugs over the years to meet challenge of drug-resistant strains and the ability to detect resistant mutants and monitor viral loads and CD4 T-cells have made the management of HIV/AIDS an artful science available to most infected subjects in resource-rich countries. The list of licensed drugs to treat HIV in North America and Europe is now over 30, from five classes aimed at four viral targets: site of entry into the cell, reverse transcriptase, integrase, and protease enzymes. Effective therapy includes a combination of at least three drugs from at least two classes, and there are now four triple ART combinations in a single pill [Atripla, Complera, Stribild, and Triumeq] that improve compliance and ease of administration.

However, there are marked disparities in the world between rich and poorer developing countries in the access and availability of ARTs and medical facilities for optimal management of HIV infection. Indeed, the countries with the greatest burden of the pandemic with the greatest needs are the ones with the least resources available to counter the AIDS/HIV epidemic. In North America recent analysis of data collected in 2009 indicates that 71% of participants were retained in medical care, 82–89% were receiving ART, and 72–78% had viral loads <200 copies [37, 38]. The disadvantaged groups such as the homeless, IVDAs, alcoholics and drug abusers, and blacks were less likely retained in medical care or received ART. In developing countries the situation is drastically different, where millions of people with HIV infection that need effective therapy are not receiving ART or regular medical care. In the past decade, there have been progress made in developing countries to provide affordable ART to those in need through national programs; availability of generic, cheaper drugs made in India and Brazil; and funding through the PEPFAR and Global Fund. Although a greater proportion of HIV-infected patients are now receiving ART in developing countries, there is still a large gap between the well-to-do and the have-not countries.

1.4.2 Present State and Remaining Challenges

It is estimated that since the start of the HIV pandemic about 78 million people have become infected and 39 million people have died from AIDS-related illness [39]. Worldwide 2.1 million people became newly infected in 2013 and 1.5 died from AIDS complications. At the peak of the epidemic in 2005, HIV caused 1.7 million deaths

Table 1.1 Global statistics of HIV: 1990–2013

	1990	2000	2013
People with HIV	8.5 million	28.6 million	35 million
New HIV infection	1.9 million	3.5 million	2.1 million
AIDS-related deaths	350,000	1.8 million	1.5 million

Data derived from UNAIDS, December 2014

worldwide [40], but modern treatment has saved about 200,000 lives annually. Since the introduction of ART and institution of therapy in pregnant women to prevent mother-to-child transmission, recent analysis indicates that 19.1 million life years have been saved, 70% in developing countries [40]. Although access to HIV care has markedly improved in recent years in developing countries, with 12.9 million people living with HIV that had access to ART in 2013, this represents only 37% of all HIV-infected population, and 24% of children with HIV received life-saving therapy [40]. The global status of the HIV pandemic over the last 14 years is summarized in Table 1.1, derived from UNAIDS [39]. The greater number of people living with HIV over these years is due to improved survival despite yearly reduction of newly infected cases. In the US the estimated annual rates of HIV transmission were highest in the early 1980s [>30%] and has been steadily declining since then to 5% from 2006 [41].

The current HIV epidemic has been classified by UNAIDS as generalized epidemic in the population in 46 developing countries and concentrated epidemics in 114 middle-/high-income countries, affecting mainly male homosexuals, prostitutes and their customers, IVDAs, and immigrants from countries with high endemic rates [40]. Although the rate of spread of the HIV pandemic has been slowing, due to multiple factors such as campaigns for safe sex, prevention measures in pregnancy and after childbirth, earlier institution of ART, and needle exchange programs, huge hurdles still remain. This is quite evident from the statistics released from the UNAIDS World AIDS Day data of 2014 [39]:

1. HIV is the leading cause of death in women of reproductive age.
2. 54% of pregnant women in low- and middle-income countries did not receive an HIV test in 2013.
3. Nearly 60% of new HIV infection in 2013 of young people occurred among adolescent girls and young women.
4. AIDS-related diseases are the leading causes of death among adolescents in Africa.
5. Globally the prevalence of HIV infection in special groups is still much higher than the general population, 19-fold higher in gay men, 12-fold greater in sex workers, 49-fold higher in transgender women, and 28 times greater in IVDAs.

1.4.3 Prospects of Meeting the Challenges

Limiting the spread of the HIV pandemic is a key goal of the WHO and the United Nations, as well as the health planners of all countries. The current measures for preventing the ongoing epidemic, including starting ART at any CD4

T-cell count and ART for preexposure or postexposure prophylaxis, will only have limited success in curbing the pandemic. The cornerstone for prevention of the HIV epidemic globally has been the exhortation of safe sex with condoms; however, there is evidence that the populations of many countries have resorted to increasing practice of unsafe sex in adolescents, young adults, and MSM. This has been attributed to "consumer fatigue" from hearing the same message repeatedly. In the US where MSM account for nearly half of all new HIV infection the practice of unprotected anal sex have been increasing, from 48% in 2005 to 57% in 2011 [42, 43]. Moreover, 38% of HIV-infected MSM unaware of their HIV infection reported unprotected sex. In 2006 the Center for Disease Control and Prevention [CDC] recommended universal opt-out HIV screening in all heath care setting, to help limit the spread of HIV by people unaware of their infection. However, universal testing is still not practiced in North America or anywhere else in the world.

Researchers from the WHO, using data from South Africa to apply a mathematical model, claim that it is theoretically possible to halt the generalized HIV epidemic with scale-up universal testing for HIV and institution of ART soon after the diagnosis [43]. However, this is not feasible or practical due to high unaffordable cost and lack of resources and need for yearly HIV testing. Also finding a cure for HIV infection is unlikely to occur in the foreseeable future. Hence an effective vaccine is greatly needed to terminate the ongoing HIV pandemic. Intensive research to find an effective vaccine has been in progress for the past two decades, but preliminary results of vaccine candidates have been disappointing until recently. The only effective vaccine developed for viruses that persist in the host latently indefinitely is the varicella zoster virus vaccine to prevent chickenpox, and this is a live attenuated virus vaccine. The risk of a whole-virus HIV vaccine that is attenuated or even killed is too great to develop, because of the fear of inadequate killing or integration in the host genome and subsequent reversion to a virulent form [44].

Development of vaccines based on HIV envelope proteins have failed to provide protection due to inadequate production of neutralizing antibodies [45, 46]. Destroying cells with replicating HIV by specific CD8 T-lymphocytes responses was considered feasible with vaccine in development [47]. However, phase 3 trials using adenovirus vectors to express HIV proteins had failed to show any protection but possible increased risk [48, 49]. However, a subsequent large trial in Thailand using the prime-boost dual vaccines [ALVAC-HIV and AIDSVAX B/E] showed promise of achieving modest protection [50]. After three years the vaccine combination showed a protective efficacy of 31.2%, only after a modified intention to treat analysis and mainly in the lower risk group of subjects. The partial protection has been postulated to be related to antibodies produced against epitopes in the V1V2 loop of the envelope [44]. Recent research suggest that it may be possible to develop a more effective combination vaccine, a prime-boost dual vaccine to generate broadly neutralizing antibodies, with a cytomegalovirus vector vaccine with SIV inserted genes to induce CD8 T-cell responses [44].

References

1. Karlen A (1995) Revolution. In: Man and microbes: disease and plagues in history and modern times. GP Putnam's Sons, New York, pp 29–46
2. Martin PMV, Martin-Granel E (2006) 2500-year evolution of the term epidemic. Emerging Infect Dis 12:976–980
3. Karlsen A (1995) Ruthless curse. In: Man and microbes: disease and plagues in history and modern times. GP Putnam's Sons, New York, pp 65–78
4. Venzmer G (1972) The great epidemics of the Middle Ages. In: Five thousand years of medicine. Macdonald & Co. Ltd, London, pp 122–131
5. Schreiber W, Mathys FK (1987) Plague. In: Infection: infectious diseases in the history of medicine. Hoffmann-La Roche & Co. Ltd, Basle, pp 11–36
6. Coffman K (2014) Three more cases of rare human plague found in Colorado. Reuters, 2014. Kay D. Editorial Comment. Clin Infect Dis 59: News i.
7. Laudisoit A, Leirs H, Mankundi RH et al (2007) Plague and the human flea, Tanzania. Emerging Infect Dis 13:687–693
8. Butler T (2000) Yersinia species including plague. In: Mandel GL, Bennett JE, Dolin R (eds) Principles and practice of infectious diseases, 5th edn. Churchill-Livingstone, Philadelphia, pp 2406–2413
9. Denis DT (2005) Plague as a biological weapon. A new dilemma for the 21st Century. In: Fong IW, Alibek K (eds) Bioterrorism and infectious agents. Springer, New York, pp 37–70
10. Drancourt M, Roux V, Dang LV et al (2004) Genotyping, orientalis-like *Yersinia pestis*, and plague pandemics. Emerging Infect Dis 10:1585–1592
11. Drancourt M, Signoli M, Dang LV et al (2005) *Yersinia pestis* orientalis in remains of ancient plague patients. Emerging Infect Dis 11:332–333
12. Bin Saeed AA, Al-Hamdan NA, Fontaine RE (2005) Plague from eating raw camel liver. Emerging Infect Dis 11:1456–1457
13. Karlen A (1995) Victory it seems. In: Man and microbes. Disease and plagues in history and modern times. GP Putnam's Sons, New York, pp 129–147
14. Osterholm M (2005) Preparing for the next pandemic. N Engl J Med 352:1839–1842
15. Enserink M (2006) Influenza. What came before 1918? Archeovirologist offers a first glimpse. Science 312:1725
16. Kilbourne ED (2006) Influenza pandemics of the 20th century. Emerging Infect Dis 12:9–14
17. Gething MJ, Bye J, Skehel J, Waterfield M (1980) Cloning and DNA sequence of double-stranded copies of haemagglutinin genes from H2 and H3 strains elucidates antigenic shift and drift in human influenza virus. Nature 287:301–306
18. Fang R, Jou WM, Huylebroeck D, Devos R, Fiers W (1981) Complete structure of A/duck/Ukraine/63 influenza hemagglutinin gene: animal virus as progenitor of human H3 Hong Kong 1968 influenza hemagglutinin. Cell 25:314–323
19. Smith GJD, Bahl J, Vijaykrishna D et al (2009) Dating the emergence of pandemic influenza viruses. PNAS 106:11709–11712
20. Taubenberger JK, Morens DM (2006) 1918 influenza: the mother of all pandemics. Emerging Infect Dis 12:15–22
21. Taubenberger JK, Reid AH, Lourens RM, Wang R, Jin G, Fanning TG (2005) Characterization of the 1918 influenza virus polymerase genes. Nature 437:889–893
22. Kash JC, Basler CF, Garcia-Sastre A et al (2004) Global host immune response: pathogenesis and transcriptional profiling of type A influenza virus expressing the hemagglutinin and neuraminidase genes from the 1918 pandemic virus. J Virol 78:9499–9511
23. Kash JC, Tumpey TM, Proll SC et al (2006) Genomic analysis of increased host immune and cell death responses induced by 1918 influenza virus. Nature 443:578–581
24. Brundage JF, Shanks GD (2008) Deaths from bacterial pneumonia during 1918–19 influenza pandemic. Emerging Infect Dis 14:1193–1199
25. Shanks GD, Brundage JF (2012) Pathogenic responses among young adults during the 1918 influenza pandemic. Emerging Infect Dis 18:201–207

26. Louie JK, Acosta M, Winter K et al (2009) Factors associated with death or hospitalization due to the pandemic 2009 influenza A [H1N1] infection in California. JAMA 302:1891–1902
27. Sun K, Metzger DW (2008) Inhibition of pulmonary antibacterial defense by interferon-Y during recovery from influenza infection. Nat Med 14:558–564
28. Worobey M, Han G-Z, Rambaut A (2014) Genesis and pathogenesis of the 1918 pandemic H1N1 influenza virus. PNAS 111:8107–8112
29. Louria DD, Blumenfeld HL, Ello JT, Kilbourne ED, Rogers DE (1959) Studies on influenza in the pandemic of 1957–58. Pulmonary complications of influenza. J Clin Invest 38:213–215
30. Fang R, Jou WM, Huylebroeck D, Devos R, Fiers W (1981) Complete structure of A/duck/Ukraine/63 influenza hemagglutinin gene: animal virus as progenitor of human H3 Hong Kong 1968 influenza hemagglutinin. Cell 25:315–323
31. Monto AS, Kendal AP (1973) Effect of neuramidase antibody on Hong Kong influenza. Lancet 1:623–625
32. Vibound C, Grais RF, Lafont BA, Miller MA, Simonson L, Multinational Influenza Seasonal Morbidity Study Group (2005) Multinational impact of the 1968 Hong Kong pandemic: evidence for a smoldering pandemic. J Infect Dis 192:223–248
33. Karlen A (1995) An old thread, new twists. In: Man and microbes. Diseases and plagues in history and modern times. GP Putnam's Sons, New York, pp 175–194
34. De Cook KM, Jaffe HW, Curran JW (2011) Reflections on 30 years of AIDS. Emerging Infect Dis 17:58–66
35. Faria NR, Rambaut A, Suchard MA et al (2014) HIV epidemiology. The early spread and epidemic ignition of HIV-1 in human populations. Science 346:56–61
36. Anderson RM, May RM (1992) Understanding the AIDS pandemic. Sci Am 2660:58–66
37. Althoff KN, Rebeiro P, Brooks JT et al (2014) Disparities in the quality of HIV care when using US Department of Health and Human Services indicators. Clin Infect Dis 58:1185–1189
38. Blair JM, Fagan JL, Frazier EM et al (2014) Behavioral and clinical characteristics of persons receiving medical care for HIV infection medical monitoring project, United States, 2009. MMWR, Surveillance Summaries 63:1–28
39. Fact Sheet: World AIDS Day 2014.http://www.unaids.org/en/WAD2014factssheet.
40. Murray CJ, Ortblad KF, Guinovart C et al (2014) Global, regional, and national incidence and mortality for HIV, tuberculosis, and malaria during 1990–2013: a systemic analysis for the Global Burden of Disease Study 2013. Lancet 384:1005–1070. doi:10.1016/S0140-6736(14)60844-8
41. Holtgrave DR, Hall HI, Rhodes PH, Wolitski RJ (2009) Updated annual HIV transmission rates in the United States, 1977–2006. J Acquired Immune Defic Syndr 50:236–238
42. Centers for Disease Control and Prevention [CDC]. HIV testing and risk behaviors among gay, bisexual and other men who have sex with men united States. MMWR Morb Mortal Wkly Rep 2013; 62: 958.
43. Granich RM, Gilks CF, Dye C, De Cock KM, Williams BG (2009) Universal voluntary HIV testing with immediate antiretroviral therapy as a strategy for elimination of HIV transmission: a mathematical model. Lancet 373:48–57
44. Fauci AS, Folkers GK, Marston HD (2014) Ending the global HIV/AIDS pandemic: the critical role of an HIV vaccine. Clin Infect Dis 59(S2):S80–S84
45. Flynn NM, Forthil DN, Harro CD et al (2005) Placebo controlled phase 3 trial of a recombinant glucoprotein 120 vaccine to prevent HIV-1 infection. J Infect Dis 191:654–665
46. Pitisuttithum P, Gilbert P, Gurwith M, Bangkok Vaccine Evaluation Group et al (2006) Randomized, double-blind placebo controlled efficacy trial of a bivalent recombinant glycoprotein 120 HIV-1 vaccine among injection drug users in Bangkok. Thailand J Infect Dis 194:1661–1671
47. Korber BT, Levin NL, Haynes BF (2009) T-cell vaccine strategies for human immunodeficiency virus, the virus with a thousand faces. J Virol 83:8300–8314
48. Cohen J (2007) Promising AIDS vaccine's failure leaves field reeling. Science 318:28–29
49. Bushbinder SP, Mehrotra DV, Duerr A et al (2008) Efficacy assessment of a cell-mediated immunity HIV-1 vaccine [the Step Study]: a double-blind, randomized, placebo-controlled, test-of-concept trial. Lancet 372:1881–1893
50. Rerks-Ngarm S, Pitisuttithum P, Nitayaphan S et al (2009) Vaccination with ALVAC and AIDSVAX to prevent HIV-1 infection in Thailand. N Engl J Med 361:2209–2220

Chapter 2
Animals and Mechanisms of Disease Transmission

2.1 Introduction

A diverse range of microbial pathogens can be transmitted from domestic and wild animals to human populations. These include viruses, bacteria, parasites, fungi, and prions. Strictly defined, not all infectious diseases common to animals and humans are zoonotic, and they can contract pathogens from the same sources, i.e., soil, water, plants, and invertebrates. For instance, acquisition of common human pathogens [*Escherichia coli*, *Enterococci* spp.] from animal food source with multiple antimicrobial resistance profile due to the liberal usage of antibiotics in animal feed may not be considered as zoonoses. Zoonotic diseases are due to transmissible infectious agents that affect more than one animal species, including humans, and cause clinical or subclinical infections. The resurgence of zoonotic infectious diseases in the past two decades globally is of major concern. These diseases have resulted in significant morbidity and mortality on a large segment of the global human population, with over a billion afflicted resulting in millions of annual deaths [1]. Moreover, outbreaks of zoonoses have had adverse impact on regional economies as a result of huge financial burden on the affected communities, but also indirectly affecting commerce or trade with more affluent consumer countries from the fear of spreading the affliction to their population. A systematic survey in 2007 estimated that in 1399 species of human pathogens, 87 of which were first reported in humans since 1980, most were viruses of animal reservoirs [1]. The World Organization for Animal Health estimates that 75% of emerging infectious diseases in humans originate from domestic or wild animals, appealing for collaboration between animal health and human public health organizations and authorities [www.oie.int/int/edito/en-avr09.htm]. A new strategy to incorporate animal health data to assist public health authorities was adopted by the European Union under the One Health concept [2]. The socioecology of zoonoses is complex and dynamic, and the epidemiology and resurgence of these diseases are influenced by various

I.W. Fong, *Emerging Zoonoses*, Emerging Infectious Diseases of the 21st Century,
DOI 10.1007/978-3-319-50890-0_2

conditions that can be classified as human related, pathogen related, animal/host related, and climate/environment related [3]. Significant interaction and combination of these factors are usually present.

2.2 Various Means of Transmission

2.2.1 Socioecology Factors

Human-related factors include the results of industrialization and expansion of communities to accommodate the global population explosion. Developmental progress with clearing of forests for new roads, residences, towns, and farmlands impinge on wildlife ecology. Moreover, human intrusion on animal ecosystem is influenced by globalization of trade, alteration of farming and food chain practices, increased hunting and pet ownership, ecotourism, and expansion of culinary practice [3]. Human exposures to animals and wildlife may be affected by changes in political regimes, conflict and wars, famine, mass migration and loosening of border controls, and breakdown of public health infrastructure. Furthermore, starvation and nutritional deficiencies usually result in a highly susceptible population to various diseases, including infections transmitted directly or indirectly from animals.

Pathogen-related factors are influenced by changes in the ecosystem and biodiversity that affect local fauna composition and quantity that may result in greater numbers of vectors and disease reservoirs/hosts, selection pressures for development of greater microbial resistance and virulence, and genetic variability [3]. Climate and environmental factors are of increased concern in recent years, as global climate change with abnormal north-south hemisphere alteration of climate pattern can influence host-vector life cycles and change in fauna and ecology of animals and vectors.

2.2.2 Mechanisms of Transmission

There are several ways by which animals can transmit infectious agents to humans, and these include: direct contact with live animals or carcasses, indirect contact through animal products such as milk or eggs, intermediary transmission through vectors [fleas, mites, ticks, mosquitoes, etc.], and remote contact from exposure to contaminated waters, soil, and air. Direct contact with animals or carcasses can result in transmission of disease by a few different ways, most commonly by oral ingestion as in foodborne zoonoses, or accidentally from handling pets without proper hand sanitization. Transmission of disease by animal bites, scratches, or mucosal exposure to infected secretions occurs relatively infrequently, such as in rabies and animal-related wound infections. Exposure to animal pathogens can be through inhalation of droplets of infected secretions, as believed to have occurred in China with local outbreaks of avian influenza arising from exposures to infected fowls or ducks in open markets. On rare occasions inhalation of microbial spores or contamination of the abraded or broken skin from exposure to contaminated animal hide can result in infection such as anthrax.

Exposures to animal excretions are common causes of unsuspected animal infectious pathogens and in sporadic cases the origin is usually undetected. Only in local outbreaks of diseases can the source of infection be detected by epidemiological investigations, as sometimes found from ingestion of contaminated vegetables from infected manure used for fertilization, as in listeria infection after ingestion of contaminated cold-slaw. Inhalation of aerosolized excreta from infected animals is another means of transmission, i.e., in "cave disease" that occurs in spelunkers after exploration of bat-infested caverns contaminated with *Histoplasma capsulatum* spores or sporadic cases of hantavirus pulmonary disease from unwitting inhalation of rat excreta aerosolized with dust. Infected animal excretions may also transmit diseases through contamination of water used for drinking or bathing or from accidental skin exposure to puddles of water containing rat urine to produce leptospirosis.

Intermediary transmission by vectors represents a major mode of transmitting animal pathogens to humans. In most instances humans are incidental hosts and are not essential for maintaining the life cycle of the parasite or pathogen. Vector-transmitted zoonoses have been a scourge to humanity since antiquity and continue to plague human populations at present and for the foreseeable future. Chagas disease, transmitted by the "kissing bug" [triatomine], afflicted humans as early as 9000 years ago [4]. The causative *parasite Trypanosoma cruzi* [discovered in 1909] can be found in at least 150 species of domestic and wild animals; and paleoparasitologists detected *T. cruzi* DNA from human mummies of ancient times [5]. Vector-borne zoonoses continue to emerge in the modern era, for example, Lyme disease, and expand with greater resurgence as exemplified by the global expansion of dengue fever and West Nile virus disease.

2.3 Animal Disease via the Food Chain

Accurate estimates of the global burden of foodborne infectious diseases are difficult to attain and are grossly underestimated in countries with active surveillance. Even in developed affluent countries such as the United States [US], most foodborne diseases are unreported, and only a fraction of these cases have a microbiological etiology confirmation. Estimates of foodborne infections are even more difficult to derive in developing countries, where there are inadequate facilities and healthcare infrastructure and the global burden of foodborne zoonoses is the greatest. It is estimated that each year at least one-third of world's population is afflicted with foodborne infections and a large fraction from animal pathogens. International collaboration spearheaded by the World Health Organization [WHO] launched surveillance programs since 2000 to determine the burden of foodborne diseases in resource-poor countries and globally [6]. Previous estimates from the Center for Disease Control and Prevention [CDC] in 2011 determined that 31 major pathogens acquired in the US caused at least 9.4 million episode of foodborne illness each year, but could be >48 million cases [7]. Common animal-derived pathogens such as nontyphoidal *Salmonella* spp., *Campylobacter* spp., and *Toxoplasma gondii* were the leading causes of hospitalization [58%]. It is surmised that about 30% of all emerging infectious diseases in the past 60 years were caused by pathogens commonly transmitted through food and were of animal origin [8].

There are more than 250 different foodborne diseases, the majority due to transmissible microbes [bacteria, viruses, parasites, and prions] and the remainder by chemicals and toxins [9]. Although foodborne zoonoses can be asymptomatic and the majority present with acute self-limited disease, occasionally severe disease and fatality occur and chronic disability can be present. Infection may result from eating raw or undercooked animal products, or contaminated vegetables and fruits, or drinking contaminated water or milk [9]. Moreover, infections and outbreaks can occur from prepared food contaminated by an ill person or asymptomatic carrier. It is often difficult to trace animal-mediated infectious diseases from non-meat products as shown by recent experiences. This is exemplified by the outbreak in Europe of a novel Shiga toxin-producing *E. coli* 0104:H4 from contaminated sprouts originating in Germany, resulting in 50 deaths, and the outbreak in the US with *Listeria*-contaminated cantaloupes which killed 29 people [10–12]. Foodborne zoonoses are usually classified according to the microbial etiology, and a list of various pathogens is summarized in Table 2.1 [this may not be totally inclusive].

2.3.1 Bacterial Foodborne Zoonoses

Campylobacter sp. and *Salmonella* sp. are the leading causes of bacterial foodborne zoonoses in the US and European Union from domestic acquisition [7, 13, 14]. Although the majority of outbreaks of these bacterial zoonoses are related to

Table 2.1 (a) Bacterial foodborne zoonoses and (b) Foodborne parasitic zoonoses

Microbes	Animals	Frequency	Distribution
(a)			
Campylobacter spp.	Poultry, pigs, cattle, sheep, rabbits, and pets	>1 million/year in US	Worldwide
Salmonella spp.	Same as above	>1 million/year in US	Worldwide
Listeria monocytogenes	Cattle, ruminants	>16,000/year in US	Worldwide
E. coli—STEC	Cattle, sheep, goat	>265,600/year in US	Worldwide
Yersinia enterocolitica	Pigs, other farm animals	N/A	Worldwide

(continued)

Table 2.1 (continued)

Microbes	Animals	Frequency	Distribution
Vibrio species			
V. parahaemolyticus	Sea crabs, oysters	N/A	Southeast Asia, Eastern Coast US
Brucella abortus	Cattle	500,000/year	Western Europe
B. mellitensis	Sheep and goats	N/A	Mediterranean, worldwide
B. suis	Pigs, cattle	N/A	Asia, Africa, S. America
Coxiella burnetii	Sheep, goats	N/A	Worldwide except New Zealand
Mycobacteria bovis	Cattle, buffaloes	N/A	Worldwide
(b)			
Toxoplasma gondii	Pigs, sheep, cats, and rabbits	½ billion globally	Worldwide
Clonorchis sinensis	Freshwater fish, pigs, cats, rats	35 million worldwide	Asia
Fasciola hepatica	Wild and domestic ruminants	N/A, variable	Worldwide, high rates in S. America, Nile Delta, Asia, Northern Europe
Opisthorchis sp.	Canines, felines, freshwater fish	N/A, variable	SE. Europe, Siberia, SE. Asia, Northern Thailand
Paragonimus westermani [lung fluke]	Freshwater shellfish, crabs, pigs	20 million global	Asia, Africa, Central and South America
Diphyllobothrium latum [fish tapeworm]	Freshwater fish	N/A	N. hemisphere, Baltic region, America, Northern Japan, China
Taenia saginata [beef tapeworm]	Cattle, deer, rare beef	4 million global	Worldwide, cosmopolitan, high in E. Africa, SE. Asia, China
Taenia solium [pork tapeworm]	Pigs, undercooked pork	N/A	Central and S. America, Africa, SE. Asia, South and Eastern Europe
Visceral cysticercosis	Fecal-oral	10–20% in endemic areas population	Central and S. America, Africa, SE. Asia, South and Eastern Europe
Angiostrongylus cantonensis	Freshwater crabs, crayfish, rats—lung worm	N/A	SE. Asia, India, Australia, Caribbean, Southern US
Anisakis simplex [Herring worm]	Sea—fish, squids, marine mammals	N/A	Cosmopolitan, Spain, France, Japan, Netherlands, rarely in US
Gnathostoma spinigerum [Gnathostomiasis]	Freshwater fish, frogs, snails, chickens, cats, dogs	N/A	SE. Asia, Central America, Australia

Data obtained from Kauss h, Weber A, Appel M et al. [eds]. Zoonoses: Infections transmissible from animals to humans, 3rd Edition, Appendix E, 2003, ASM Press, Washington, 423–37; Alturi VL, et al. Ann.Rev.Microbiol. 2011; 65: 523–41; Durr S, et al. PLOS Neglected Trop. Dis. 2013; 7: e2399, PMID: 24009789

HUS hemolytic uremic syndrome, *TTP* thrombotic thrombocytopenic purpura, *STEC* Shiga toxin *E. coli*, *N/A* not available "high-risk" foods commonly associated with illness, for instance, pink hamburger, raw oysters, unpasteurized milk or milk products [cheese], runny eggs, and alfalfa sprouts [15], outbreaks have occurred with vegetables, fruits, peanut butter, and raw nuts [16]. The majority of resulting illnesses are usually self-limited gastroenteritis, but chronic complications may include irritable bowel syndrome, reactive arthritis/spondylitis, and rarely Guillain-Barre syndrome. These bacteria are widespread in nature and often colonize the intestines of many domestic and wild animals, i.e., poultry, pigs, cattle, sheep, etc. Hence, raw meat from these animals are frequently contaminated with pathogenic bacteria during the preparation of carcasses for wholesale or retail in broilers and supermarkets, with poultry more frequently colonized with disease-producing bacteria than pork or beef. The European Food Safety Authority reported that poultry meats from broilers were contaminated by *Campylobacter* in 75.8% and by *Salmonella* in 15.6% [17]. However, even higher rates were reported in China with the presence of *Campylobacter* in carcasses of chicken in 94%, duck in 96%, rabbit in 97%, pork in 31%, and beef in 35% [18]. *Salmonella* contamination of meats for retail in the marketplace of China was somewhat less, with rates in chicken of 54%, in pork of 31%, in lamb of 20%, and in beef of 17% [19].

Recent reports in the US ranked the disease burden of food source infections and listed ten pathogen-food combinations, which included five pathogens [*Campylobacter*, *Salmonella*, norovirus, *Listeria monocytogenes*, and *T. gondii*], detected mainly from eight food categories, poultry, pork, deli meats, dairy, beef, eggs, farm produce and complex foods. Poultry was ranked as the number one cause of significant disease burden [mainly *Campylobacter* and *Salmonella*], followed by complex foods [83% caused by norovirus], and thirdly by pork [largely from *T. gondii*]. The five pathogens listed account for more than 50% of the total food-related illnesses due to 14 pathogens, responsible for $8 billion overall cost and 36,000 quality-adjusted life years [20]. The global burden of nontyphoidal *Salmonella* infections [mainly from foodborne zoonosis] is estimated to be 80.3 million cases, with 155,000 deaths [21].

Listeria monocytogenes, a nonsporulating, gram-positive bacillus, found mainly in ruminants but can affect all species of animals, is an infrequent and serious infection that can lead to meningitis with a mortality of 20–30% for at-risk persons [pregnant, neonates, elderly, and the immunocompromised] [12]. *Listeria* poses special risk in current trend of culinary habits, as 1–10% of ready-to-eat food may be contaminated with *Listeria* [22]; and the bacteria can survive and grow at low temperatures, high salt level, and low pH used in food processing [23]. Contaminated food products with *Listeria* more commonly include unpasteurized milk and dairy products [soft and feta cheese] and meat from ruminants and sometimes poultry [24]. Listeriosis has a variable incubation period of 2–70 days and the food source is often difficult to ascertain. The Active Surveillance Network in the US reported 762 listeriosis cases over a 6-year period [2004–2009], with an overall fatality rate of 18% [25]. However, most cases are unreported, and it is estimated that as many as 1600 cases of invasive listeriosis and 260 related deaths occur each year in the US [7]. In China *L. monocytogenes* had been recovered from about 4% of various types of prepared food samples and 6.28% of raw meat [26].

Enterohemorrhagic *E. coli* or Shiga toxin-producing *E. coli* [STEC] is a foodborne zoonosis transmitted to humans from contaminated food and water or contact with infected animals or persons. Infections are caused predominantly by *E. coli* serotype 0157:H7, but novel serotypes are emerging that produces the toxin, as exemplified by the outbreak in Europe with *E. coli* 0104:H4 [10, 11]. Infections can result in acute uncomplicated diarrhea or with severe hemorrhagic colitis and hemolytic uremic syndrome with acute renal failure and death [mainly children under 10 years of age], and thrombotic thrombocytopenic purpura [mostly in adults]. These severe complications occur in about 10% of infected subjects [27]. Infections with STEC are estimated to occur in about 265,630 persons with 31 deaths annually in the US [7]. Outbreaks of STEC infections occur worldwide but the global burden of disease is unknown. There are over 100 STEC serotypes, but most outbreaks and sporadic diseases are attributed to the 0157 serotype. The diagnosis is probably frequently missed as many laboratories only test for the 0157 STEC, without a test to detect Shiga toxin as recommended [28]. Shiga toxins [STXs] or verotoxins are potent cytotoxins, found mainly in zoonotic pathogenic *E. coli*, and are produced through lysogenic bacteriophages inserted into the chromosomes to encode the genes for six or more STXs [27]. Most STEC disease is the result of STX1 or STX2. Transmission of STEC readily occurs as the infectious dose is low [10–100 colony-forming units], and the bacteria are resistant to gastric acid [27]. The STECs are mainly found in cattle, sheep, and goats, but other animals can be carriers; and infection in humans mainly results from ingestion of undercooked beef or contaminated milk, yogurt, and cottage cheese and rarely from salads, vegetables, and apple juice [27]. The prevalence of STEC in cattle bowels ranges from 0.1 to 16%, and shedding of the pathogen in the excreta is intermittent [29].

The large outbreak of the hypervirulent STEC 0104:H4 that started in northern Germany and spread to 15 European countries in 2011 is unique. The epidemic affected over 4000 people and caused 54 deaths, and 22% of those afflicted developed the hemolytic uremic syndrome [10]. Several unusual features of this outbreak included median incubation period of 8 days which usually is 3–4 days with STEC 0157; 88% of the cases with hemolytic uremic syndrome occurred in adults [men age 42 years], which generally occurs in children. The pathobiologic mechanisms differed from the STEC 0157 infections in the absence of the usual virulence enterocyte effacement locus, characteristic for enteropathogenic *E. coli* used for epithelial adherence. However, this strain possessed several genes and virulence plasmids typical for enteroaggregative *E. coli*, carried primarily by humans [10]. This suggest that the STEC 0104 strain resulted from mixing of human and animal [cattle] STEC to produce a recombination of genes in a novel bacteria probably of a zoonotic source.

Some foodborne bacterial zoonoses including *Yersinia enterocolitica* infection have declined significantly in the US since 1996 [30]. Outbreaks are rarely reported and sporadic cases that occur are mainly associated with consumption of undercooked pork. In China two outbreaks of *Y. enterocolitica* occurred in the early 1980s which affected more than 500 people [31]. The main animal reservoir is the pig, as a pharyngeal commensal [32], but the organism has been isolated in more than ten different types of animals in China, and 32% of the strains were considered pathogenic [33]. Foodborne vibriosis due to non-cholera species is mainly from consumption of raw or undercooked seafood, especially oysters, appears to be increasing in the US, and is estimated to cause about 800,000 illnesses, 500 hospitalizations, and 100 deaths each year [7, 34]. Vibrios are natural commensals in marine and estuarine seawater, and *Vibrio parahaemolyticus* is the most common cause of foodborne infection, presenting mainly as a self-limited gastroenteritis. *Vibrio vulnificus*, which is associated with soft tissue infection and septicemia, is rarely foodborne [34]. Other bacterial foodborne zoonoses which are rare in North America but more common in the Mediterranean basin and developing countries include Q-fever and brucellosis. Persistence of these pathogens in these regions may largely be related to unregulated animal husbandry. *Streptococcus suis* infection is an emerging foodborne zoonosis in Asia and is reviewed in a separate chapter.

2.3.2 Foodborne Parasitic Zoonoses

Foodborne animal-related parasitic infections are globally distributed, and the burden of disease is underestimated in developed countries. In the US it is estimated that only 2% of foodborne illnesses are caused by parasites annually [7]. However, *T. gondii* which is the most common foodborne parasite is frequently subclinical, and only 15% of recently infected subjects are clinically ill [35]. *T. gondii* is the most widespread parasitic infection in the world and infects half a billion people or one-third of the global population [27, 29]. The parasite infects a wide variety of domestic and wild animals, and nearly all warm-bloodied animals can be infected. The seroprevalence of toxoplasmosis varies greatly in different countries, which may be related to culinary habits and hygienic standards. Although toxoplasmosis can be transmitted from contamination of the hands while changing cat [kitten] liter from sporulated oocysts, most infection results from ingestion of raw or undercooked meat with tissue cysts. The prevalence of infection in cat ranges between 10 and 80%, and oocysts are shed only in the first 14 days of primary infection [36]. Consumption of raw or undercooked pork and lamb is most commonly implicated in the transmission of toxoplasmosis, less commonly from beef due to a short developmental stage in cattle, but infection can occur from eating raw/undercooked liver, caribou, and seal meat and rarely from drinking contaminated water [36]. The seroprevalence of toxoplasmosis in Central Europe ranges between 37 and 58%, in the US 3 and 35%, in Latin America 51 and 72%, and in West Africa 54 and 77% [36]. Infection in healthy persons is usually subclinical or sometimes present with the mononuclear syndrome, but infection in pregnancy and immunosuppressed hosts can produce serious consequences. Congenital toxoplasmosis in the first trimester occurs in about 15% from primary infection and may cause severe complications [abortion, microcephaly, hydrocephalus, and mental retardation]; and infection in the third trimester can result in less severe abnormalities [chorioretinitis, epilepsy, deafness, jaundice at birth, learning difficulties later in life] but more frequently about 65% [36].

Giardiasis caused *by Giardia lamblia* [*G. intestinalis*] is not usually a foodborne parasitic infection but is carried by dogs, cats, cattle, sheep, pigs, and rodents and indirectly causes human infection from contamination of drinking water and food. It is worldwide in distribution and causes millions of infection each year. Other foodborne parasitic zoonoses are rare in well-developed countries, but cryptosporidiosis was fairly common in AIDS patients before the advent of highly active antiretroviral agents. *Cryptosporidium parvum* is an intracellular protozoa that affect over 150 mammalian species and is worldwide in distribution; it causes disease in calves and less commonly in lambs, cats, and dogs [36]. Transmission is by contaminated food or water, and outbreaks can occur from contaminated municipal drinking water as the organism is chlorine resistant. It is estimated that >56,000 domestically acquired infection occurs annually in the US [7]. In healthy subjects *C. parvum* causes a self-limited gastroenteritis, but the immunosuppressed hosts are prone to severe, protracted diarrhea with dehydration, weight loss, or wasting, and it is an AIDS-defining disease.

Several parasitic foodborne diseases are common in developing countries but are rare in North America and Europe, largely due to differences in standards of hygiene and governments regulations, and culinary practices. Consumption of undercooked infected meat and fish poses a major health risk to people of Asia and other tropical and subtropical regions of the world. Foodborne parasitic zoonoses are estimated to affect about 150 million people in China alone [31]. National surveys in China between 2001 and 2004 indicated that the number of people affected with clonorchiasis, trichinellosis, paragonimiasis, and angiostrongyloidiasis had increased compared to previous surveys in 1988–1992 and are of public health concern [31, 37]. *Clonorchis sinensis* is a trematode that can be asymptomatic or cause liver/biliary tract disease in all Asian countries from consumption of raw freshwater fish or shrimps; and it is estimated to that 35 million people who are infected globally, with 15 million residing in China [31]. The cycle of the liver fluke is perpetuated by fecal contamination of freshwater by animal hosts [humans, pigs, cats, and rats], with ingestion of larvae by snails and release of cercariae to infect fish. Eradication and control has been difficult as rice paddies, swamps, ponds, and streams commonly have infected fish [mainly carps], which is a main source of food supply. In some provinces of China, about 17% of freshwater fish are infected with *C. sinensis*, and rates of infection in the population vary from 4.75% to 31.6% [31, 38]. Despite the common practice of eating raw fish [sushi] in Japan, clonorchiasis is very rare, as the fish is mainly from the sea. The bulk of the infected people with clonorchiasis are asymptomatic, but heavy burden of infection for many years can result in gallbladder and liver disease, cholangitis, and cholangiocarcinoma.

Viral zoonoses transmitted by food are few in number and include mainly hepatitis E and severe acute respiratory syndrome [SARS] which are reviewed in separate chapters. Prion, a transmissible altered protein, was first recognized to cause disease in sheep [scrapie] for more than 50 years, but more recent transmission to humans from consumption of beef from cattle with bovine encephalopathy ["mad cow disease"] had arisen in Britain and spread to Europe almost two decades ago. A summary of some foodborne zoonoses is shown in Table 2.1.

2.4 Pets as a Source of Zoonoses

The domestication of animals for companionship in human societies probably exists for over 10,000 years. Ancient Egyptian writings in the Kahun Papyri [1900 BC] acknowledged the recognition of animal diseases; and ancient Mesopotamians passed laws [Eshuna Code of 2300] for containment of rabid dogs [39]. The number of households or the proportion of families in the world with pets is unknown but this is quite substantial. In affluent countries pet ownership is increasing, not only for cats and dogs but also for more exotic animals. In the US it was estimated in 2006 that 37% of households owned 72 million pet dogs and 32% owned 81 million cats [2007 US Pet Ownership and Demographics Sourcebook]. Similar figures were available for the United Kingdom [UK] with 8 million pet dogs in 23% of households and 8 million cats in 19% of households [http://www.pfma.org.uk].

Companion animals can cause infectious diseases by several means of transmission such as by direct contact, bites scratches, incidentally by fecal-oral route, and indirectly by vectors, i.e., fleas, ticks, sandflies, and mosquitoes. Pet ownership rarely results in disease transmission overall, and most infections that result are usually asymptomatic or subclinical [e.g., toxoplasmosis from handling infected cat litter]. The majority of zoonoses reported from pets are related to dogs and cats, but there is a wide variety of animals and pathogens that can be involved such as reptiles/amphibians, birds, rodents, monkeys, and even pet-household pigs. The risk of infection of potential zoonoses depends on the animal, local zoonoses endemic in the region, hygienic practice, laws governing animal vaccinations and license, and restrictions of exotic animals in households. The major pet-associated zoonoses are shown in Tables 2.2, 2.3, 2.4, and 2.5.

Table 2.2 Pet animal-associated zoonoses: diseases by bites and scratches

Microbes	Animals	Disease	Distribution
Rabies virus	Dogs and cats	Rabies encephalitis	Worldwide, esp. Africa and Asia
Pasteurella multocida	Cats and dogs	Wound infection	Worldwide
Pasteurella canis	Dogs	Soft tissue infection	Worldwide
Capnocytophaga canimorsus	Dogs	Cellulitis, sepsis in splenic dysfunction	Worldwide

Table 2.3 Pet animal-associated zoonoses: vector-borne disease

Microbes	Animals	Disease	Vector	Distribution
Leishmania infantum	Dogs	Visceral leishmaniasis	Sandflies	Southern Europe, Central and S. America, Asia, North Africa
Bartonella henselae	Cats	Cat scratch disease, bacillary angiomatosis	Cat fleas	Worldwide
Rickettsia rickettsiae	Dogs	RMSF	Dog tick	US, Central and northern South America
Rickettsia conorii	Dogs	MSF	Dog tick	Southern Europe, Middle East, North Africa, India, Pakistan
Ehrlichia canis	Dogs	Ehrlichiosis	Dog tick	Venezuela [rare]

Table 2.4 Pet animal-associated zoonoses: parasitic zoonoses from pets

Microbes	Animals/transmission	Disease	Distribution
Toxoplasma gondii	Cat/kitten via litter	Toxoplasmosis	Worldwide
Echinococcus granulosus	Dogs via feces	Hydatid cyst [cystic echinococcosis]	Sheep-rearing regions worldwide
Echinococcus multilocularis	Foxes, wolves, dogs, cats	Alveolar hydatid cyst	Central and South Europe, Turkey, Northern China, N. Canada and Japan
Toxocara canis	Dogs—feces on playground	Visceral larva	Worldwide
Toxocara cati	Cats—feces in sandbox	Migrans	

Table 2.5 Pet animal-associated zoonoses: pet birds

Microbes	Animals/transmission	Disease	Distribution
Chlamydophila psittaci	Parrots, parakeets, budgies, canaries, finches, lovebirds, aerosol of excreta	Psittacosis	Worldwide
Mycobacteria avium	All birds by aerosol of excreta	Scrofula, disseminated infection in AIDS	Worldwide
Mycobacteria genavense			
West Nile virus	Birds in open-air aviary and zoos via mosquitoes	Meningitis, polio-like paralysis	Americas, Caribbean, Asia, SE. Asia, Africa

In most developed countries, pet-associated clinically recognized infections are usually secondary to accidental bites, scratches, or contact of open wounds with saliva of cats or dogs. Dog bites most commonly affect children, and in the US 4.7 million people a year suffer from dog bites, with 368,345 persons requiring emergency room treatment for dog-bite injuries [40]. Rabies, a fatal zoonosis transmitted by dog or cat bite, is rare in North America and most well-developed countries, and most cases occur in Africa and Asia with 55,000 deaths globally each year [41]. The majority of cases result from dog bites [99%] and most frequently from free-roaming animals. In North America rabies is mainly associated with bats in human cases. Animal rabies has declined in all domestic vertebrates in the US since 2010, with only 69 affected dogs but 303 infected cats, 63% of all domestic animals [42]. Free-roaming cats account for most of the human exposure incidents and need for postexposure rabies prophylaxis in the US [43].

The most common clinical infection reported in humans associated with pets is wound infection after puncture wounds, crush injuries, or soft tissue tears. Mixed infection with oral flora of the animal, including *Pasteurella* species, streptococcus, *Staphylococcus aureus*, and anaerobes, is mainly found. In cat-bite wounds, *Pasteurella* sp. [mainly *P. multocida*] is found in 75% to 90% and in dog-bite wounds in 20–50%, but mainly *Pasteurella canis* [44]. Anaerobes, *Fusobacterium* species, and *Bacteroides* sp. are present in 30–40% of animal-bite wounds; and a fastidious gram-negative bacteria, *Capnocytophaga canimorsus or C. cymodemgi*, most commonly from dog bites, can cause serious infection in the elderly, asplenic subjects, alcoholics, and the immunosuppressed who are prone to severe sepsis and metastatic infection with mortality >30% [45].

2.4.1 Vector-Borne Zoonoses from Pets

Vectors are important transmitters of companion animal zoonoses worldwide; see Table 2.3. The most important disease globally in this category is visceral leishmaniasis which is caused *by Leishmania infantum* [*L. chagasi*], transmitted by sandflies from the major reservoir of domestic dogs [39]. The disease is endemic in many countries of southern Europe, South and Central America, northern Africa, and Asia; and a high proportion of dogs in these areas are infected. It was estimated by the WHO in 2007 that about 12 million people were infected worldwide and 2 million new cases occurred annually; also there was increased risk of severe disease in the HIV-infected population [46]. Control of leishmaniasis in the endemic areas is a great challenge and extension to non-endemic regions is of great concern, due to increasing travel access of pets across borders. Control measures instituted in Brazil consisted of serologically testing dogs and culling positive animals [47], and vaccination of dogs with a commercial vaccine [Leishmune] has resulted in decreased prevalence in canine and human leishmaniasis from reduced transmission [48].

Cat scratch disease is the most common zoonotic infection caused by *Bartonella* bacteria. It is of worldwide distribution but more common in temperate zones where pets are kept indoors. The cat although asymptomatic is *a large* reservoir for human infections with *Bartonella henselae*, *Bartonella clarridgeiae*, *and Bartonella koehlerae*. Cat scratch disease occurs mainly in children and is transmitted by the infected flea feces via the claws of kittens from scratching. Thus it is a vector-borne infection transmitted from cats with bacteremia to other cats and humans by fleas. Dogs can be infected with various *Bartonella* species, but their role in human disease is unclear [49]. The infection is usually manifested by fever and regional lymphadenopathy, but rare severe complications can occur such as encephalitis, granulomatous hepatitis, retinitis, choroiditis, arthritis, and osteomyelitis [27]. In patients with AIDS, *Bartonella henselae* can rarely cause a rash of bacillary angiomatosis with secondary lesions in the liver [bacillary peliosis], spleen, and lymph nodes. In developing countries the prevalence of Bartonella seropositivity in cats is about 27% and bacteremia approximately 10% [39].

There are several tickborne diseases common to both humans and companion animals that are of concern with respect to increased transmission and diseases in human communities. These include borreliosis [Lyme disease], ehrlichiosis, babesiosis, rickettsiosis, anaplasmosis, Q-fever, tularemia, and tickborne encephalitis, found in Europe and Northwestern Asia [39]. However, only a few of these conditions have been documented to be transmitted via contact with pets. In some of these diseases, the pets [dogs and cats] are not natural hosts for the vectors [i.e., black-legged deer ticks], but the animals may facilitate human exposure from outdoor activities, such as Lyme disease, ehrlichiosis, babesiosis, and anaplasmosis.

Hence, the pets pose a minimal risk to humans to predispose to acquiring these conditions. Infections in the pets as diagnosed by a veterinarian may be used as a sentinel for monitoring the risk of disease in the community of an endemic area [50]. Dog ticks can transmit rickettsia infection to humans, and dog ownership is a risk factor for acquiring rickettsiosis in endemic regions. Rocky Mountain spotted fever [RMSF] is endemic in many states of the US, greatest in the mid-south Atlantic states and west south-central region than the Rocky Mountain States, but also present in Central America [Mexico, Costa Rica, and Panama] and northern South America [Columbia and Brazil] [27, 51]. The main vector for RMSF is the American dog tick, *Dermacentor variabilis*, in the Eastern US, but in the Western US, it is the Rocky Mountain wood tick, *Dermacentor andersoni*. Only the adult *ticks* accidentally feed on humans, but all stages of the tick are infected and the organism is transmitted to the offspring transovarially [51]. The brown dog tick, *Rhipicephalus sanguineus*, is the vector for RMSF in Mexico, Arizona, California, Texas, and Brazil [52, 53]. Field mice, other rodents in the wild, rabbits, and their ticks are the natural reservoir *for R. Rickettsiae*, and the dog is an incidental host. Mediterranean spotted fever [MSF], caused by *Rickettsia conorii,* is also transmitted by the brown dog tick from bites or from mucus membrane contact with crushed tick [53]. The endemic areas are found across southern Europe, northern Africa, the Middle East, the Indian subcontinent, and parts of Asia. Dogs are subclinically infected and are important biological hosts. The prevalence of MSF in dogs in endemic areas can be high, 26–60%, and the cycle of the rickettsia is maintained between rodents, ticks, and dogs, with humans as accidental hosts. The disease often starts with an eschar at the site of bite; then progresses to maculopapular, erythematous rash with fever; and usually lasts for about 10 days [27*]. Ehrlichia canis* [a rickettsia-like organism] causes severe disease in dogs and rarely can be transmitted to humans. The vector is the brown dog tick, and human disease *with E. canis* has been reported primarily from Venezuela [54].

2.4.2 Parasitic Zoonoses from Pets

Besides toxoplasmosis pets can transmit other parasites to humans. Transmission of giardiasis and cryptosporidiosis by pets has been proposed but not documented to be related to pet contact [55]. Helminthic zoonoses can be indirectly transmitted to humans from cats and dogs, mainly in children. There is concern that dog tapeworm disease [echinococcosis] is reemerging in some countries of Europe and that toxocariasis, from cat and dog roundworms, is still persisting in large endemic areas [56]. Echinococcosis is a cystic [hydatid] disease produced by tapeworms of canids and humans become accidentally infected by ingestion of fertile eggs or proglottids. *Echinococcus granulosus* causes cystic hydatid disease which is a global zoonosis transmitted within a dog-sheep cycle in sheep-rearing pastoral regions. *Echinococcus multilocularis* is a less common infection but produces a more invasive and aggressive disease, alveolar echinococcosis, and the red or arctic fox is the natural final host [36]. *E. granulosus* infection is fairly common in the Mediterranean basin and is reemerging in southern Europe endemic areas [57]. The sheep strain *of E. granulosus* prevalence in farm or shepherd dogs in Italy and Spain varies from 0% to 31%, in Lithuania 14.2%, and in Wales it increased from 0% in 1993 to 10.6% in 2008 [56]. The pig strain *of E. granulosus,* subspecies intermedius, is present in the Baltic countries, Poland, Austria, and Romania. Home slaughter of farm animals such as sheep and pigs in several of these countries may be responsible for maintaining the cycle of the parasite between farm animals and the dogs by feeding slaughtered animal parts such as offal to the pets. *E. multilocularis* prevalence is lower in pets and dogs and cats are incidental hosts. However, there is evidence that pet and human infections are present in many areas of Eastern, Central-Eastern, and Southeastern Europe [58]. In Switzerland alveolar echinococcosis has been slowly increasing from 0.10–0.16 cases per 100,000 to 0.24 per 100,000 people, and increased prevalence in the fox population of 30–60% [59]. It has been estimated that 490,000 dogs in Switzerland are infected with *Echinococcus multilocularis* and up to 13,000 dogs in Germany; and in the cat population, the prevalence ranges from 0% to 5.5% in various endemic areas [56].

Toxocariasis [visceral larva migrans] from roundworms of cats [*Toxocara cati*] and dogs [*Toxocara canis*] are present worldwide in carnivorous animals and humans, predominantly affecting children. The prevalence of *T. canis* in dogs of Western Europe varies from 3.5% to 34%, lowest for household pets and highest for rural animals; and in cats *T. cati* is present in 8–76%, depending on the environment [56, 60]. The prevalence of *Toxocara* infection and worm burden is highest in puppies and kittens less than six months of age. Children become infected from ingestion of embryonated eggs from contaminated soil while playing and without adequate hand sanitization. Playing with pets is usually not a source of the infection as the eggs need several weeks to become infectious [56]. High rates of soil contamination [10–30%] with *Toxocara* eggs have been found all over the world in areas where children commonly play, backyards, sandpits, parks, and lake beaches, and patent eggs can survive for up to a year; and dog roundworm eggs are most commonly found in public parks, while cat roundworm eggs most commonly contaminate sandboxes [56]. *Toxocara* infection in the global population is common with 80% in children and 50% under 3 years of age, but most infections are subclinical or misdiagnosed. In human communities the seroprevalence of toxocariasis varies from 2.5% in Germany, 4.6–7.3% in the US, 19% in the Netherlands, and up to 83% in children of the Caribbean [56]. Clinical syndromes such as visceral larva migrans with liver, small intestine, lung, and heart involvement are very rare, and ocular and brain involvement with seizures may be seen [61].

2.5 Birds and Bats in Zoonoses

The most important avian-related zoonosis is the avian influenza which is reviewed in Chap. 3. Birds and bats can play a role in dissemination of some pathogens via environmental contamination with colonized excreta which may not be considered as zoonoses, such as histoplasmosis and cryptococcosis. Birds are involved in the transmission of infectious diseases by several mechanisms, as a food source [poultry], as pets, and as wild birds. Birds in the wild play a major role in the amplification and life cycle of several zoonotic viruses, which are transmitted by vectors such as mosquitoes. These include West Nile virus, western equine and eastern equine encephalitis viruses which are endemic in North America, and Venezuelan equine encephalitis virus which is endemic in Central and South America; in the latter disease, horses and not birds are the amplification hosts. Bats are being recognized as important hosts for several emerging novel vial zoonoses, but their role in transmission of infections is rarely through direct contact, and in most cases the association with bats is obscure.

2.5.1 *Pet Birds*

Infectious diseases from pet birds are uncommonly recognized or reported but may be from direct contact, droplet, fomites, and from vector transmission. Areas of transmission of infectious disease can be in the homes, pet shops, bird fairs, and markets and through international trade. Pet birds are primarily songbirds [Passeriformes] such as canaries, finches, and sparrows and Psittaciformes parrots, parakeets, budgeries, and lovebirds. A previous study in 2007 estimated that there were 11–16 million pet or exotic birds in the US [62]. One of the most important bird zoonosis is psittacosis, also known as chlamydophilosis, ornithosis, or parrot fever, caused *by Chlamydophila psittaci*. Although psittacines are highly susceptible to infection and are the primary hosts, other birds can be infected including canaries, finches, etc. Most recognized human infections occur in veterinarians and bird breeders, and owners represent only about 40% of the cases [63]. Clinical diseases vary from mild respiratory infection to severe interstitial pneumonia, and infection occurs from inhalation of contaminated dust or excretion of infected birds. Asymptomatic psittacines and pigeons are the most important source of infection [27]. In a study from Belgium, involving 39 breeding facilities, infection *with C. psittaci* was detected in 19.2% of birds [mainly psittacines] and 13% of owners and workers. Of the persons with viable organisms isolated, 66% had mild respiratory symptoms and 25.6% of bird owners reported a history of pneumonia shortly after ownership. Despite antibiotic therapy of the birds in 18 breeding facilities, 66.6% were still positive for *viable C. psittaci* [64]. It is unclear from this report whether this was due to the development of antibiotic resistance or due to reinfection.

Pet birds, especially psittacines, are sources and excretors of mycobacteria species, most *commonly Mycobacterium avium* and *Mycobacterium genavense,* but they are rarely ill from the infection [64]. These bacteria are ubiquitously present in the environment and probably disseminated widely by birds via their feces [pigeons and birds in the wild], and there is no evidence that pet birds play a significant role in human infections. On rare occasion *Mycobacteria tuberculosis* have been transmitted from an infected owner to the pet bird [green-winged macaws], and this may have resulted in acute infection [Mantoux skin test conversion] in veterinarians handling the sick birds [65, 66]. *M. tuberculosis* infection has also been reported in a canary and Amazon parrot [67]. Incidental infections that could arise from pet birds include *Salmonella* and *Campylobacter* gastroenteritis [63]. Vector-borne infections such as West Nile virus could affect captured birds in an open-air aviary and zoos and thus act as reservoir for infections to humans via urban mosquitoes. Ixodes ticks that transmit Lyme disease can also infect birds, and although migrating birds in the wild may play a role in spreading the vectors to new regions [68], captured or pet birds are not likely to facilitate the transmission of Lyme disease to humans.

2.5.2 Bats

Bats [Chiroptera] are found on all continents except Antarctica, and they represent about 20% of mammalian species and are important in maintaining a balanced ecology and a striving agriculture industry [69]. However, bats are reservoirs for several zoonotic pathogens and may represent amplification hosts for emerging viruses [70]. In the past two decades, it has become evident that bats are important in the life cycle of novel viral pathogens, arising from tropical and subtropical countries and with the potential for global expansion. There are several features of these creatures that facilitate the spread of infectious diseases and novel pathogens: the large diversity of species with different feeding habits, the ease of migration by flying, their nocturnal activity that allows contact and spread of diseases to other animals [domestic or in the wild], and the ecology and social activity in forming large communities that facilitate the potential to harbor and spread multiple viral pathogens [71]. Bats form the largest aggregate of mammals worldwide with over 900 species, and their diversity of diet is unparalleled among mammals [72]. Various species feed on insects, fruits, leaves, flowers, nectar, pollen, fish, blood, and other vertebrates. The ability to fly may in part explain their feeding, roosting, and social behavior. Insectivorous bats are found at all latitudes and bats living above 38°N and below 40°S are all insectivorous. Few species of bats are carnivorous or sanguinivorous [blood sucking], and they are confined to tropical and subtropical regions [72]. The nocturnal feeding habit, highly specialized sense of hearing, and ability to capture their prey by echolocation [sonar] facilitate transmission of infection among various mammals.

It has been recognized for many decades that bats can be reservoirs for rabies and transmit the disease to humans by bites [vampire bats], scratches, or direct contact and by aerosol of the virus in bat caves [73]. Bat rabies occurs worldwide and can be found in islands free of canine rabies. In North America there are 40 species of insectivorous bats that can be infected with rabies. Transmissions of bat rabies have been reported in North America, Central and South America, and Europe [73]. In most cases of bat rabies, there is no definite history of a bite, and in North America indigenously acquired rabies in humans are primarily bat rabies variant associated with the silver-haired bat [73].

There are several features of the bat life cycle and ecology that are favorable for maintaining and transmission of zoonoses. Bats have a long life-span of 24–35 years, and their crowded roosting behavior predisposes to intra- and interspecies transmission of viruses and persistence of infection, which allows for cross infection of other vertebrates [74]. There are a large number of viruses that have been isolated from bats, but

Table 2.6 Zoonoses associated with bats

Disease	Type of bats	Means of transmission	Distribution
Diseases transmitted by bats directly or indirectly			
Lyssavirus-1-rabies virus	Vampire & insectivorous	Bite, aerosol, direct contact	South and Central America, North America
Lyssavirus-2-7 [rabies like]			
Lagos bat virus	Fructivorous	Aerosol, contact?	Africa
Mokola virus	Fructivorous	Aerosol, contact?	Africa
Duvenhage virus	Fructivorous	Aerosol, contact?	South Africa
European bat virus	Insectivorous	Aerosol, contact?	Europe
Australian bat virus	"Flying foxes"	Contact	Australia
Henipaviruses			
Hendra virus	"Flying foxes"	Contact with horses	Australia
Nipah virus	"Flying foxes"	Contact with pigs	Malaysia, Bangladesh, India, Singapore
Coronaviruses			
SARS-CoV	Fructivorous	Contact with palm civets, raccoon dogs	China
MERS-CoV	Fructivorous	Contact with camels?	Middle East
Rubulavirus:			
Menangle	"Flying foxes" [fructivorous]	Contact with pigs	Australia

Data obtained from references [71, 74], Krauss H et al., Zoonoses. Infectious Diseases transmissible from animals to humans. 3rd Edition, 2003, ASM Press, Washington, Appendix E, p 423–37

most of them have not been shown to cause human disease [74]. The viral zoonoses associated with bats besides rabies include Nipah and Hendra viruses, and SARS-coronavirus (SARS-CoV)-like virus of bats have all arisen in Southeast Asia or Asia; and the newly recognized MERS-coronavirus (MERS-CoV) from the Middle East may also have a reservoir in bats; see Table 2.6. There are other viruses found in bats that are arthropod-borne diseases of the families alphavirus, flavivirus, and bunyavirus, but their role in maintaining the life cycle and as reservoir of these agents is unclear [74].

2.6 Animals in the Wild

Wildlife poses increasing challenges with respect to emerging and existent zoonoses. Available evidence suggests that ecological changes in the environment ecosystems, such as nutrient enrichment, often exacerbate infectious diseases caused by parasites with simple life cycles [75]. The mechanisms include changes in host and vector density, host distribution, resistance, and virulence/toxicity of pathogens. Many emerging zoonoses occur as a result of increasing exposures of domestic animals and humans to wildlife. Predicting the emergence of novel zoonoses will require tracking the general trends of emerging infectious diseases, analyzing the risk factors for their emergence, and examining the environmental changes that are influential [76]. Wolfe et al. [77] estimated that the three main factors that govern the emergence of new wildlife zoonoses are the diversity of wildlife pathogens in a region, the zoonotic pool; environmental changes that affect the presence of pathogens in the wildlife population; and the frequency of contacts of domestic animals and humans with wildlife harboring potential zoonoses. Emergence of a new infectious disease usually results from a change in ecology of the hosts or pathogens or both [76]. The frequency and density of wildlife infectious agents with the potential to cross over to humans appears to be greatest in tropical and subtropical regions of the world, where humans are encroaching on large areas of previously virgin forests, i.e., in Africa, Asia, and the Amazon basin. Environmental factors such as global climate change are predicted to result in expansion of wildlife zoonoses, especially vector-borne diseases. The recent emergence of chikungunya fever in the western hemisphere with rapid spread throughout the Caribbean, Southern US, and Central and northern South America may be related to recent climate change.

Human factors that contribute to the emerging wildlife zoonoses include population expansion, with urbanization of previous forested areas that encroach on wildlife habitat. Deforestation of tropical forests for commercial log industry and agriculture expansion and hunting of wildlife are major sources of increased human contact with wildlife zoonoses. Deforestation and road construction in virgin forests also displace small wildlife reservoir of infectious pathogens, result in dispersal to new habitat, and expose livestock and domestic animals to these microbes and indirectly to humans. This may have occurred in the US where new road construction caused dispersal of the white-footed mouse [*Peromyscus leucopus*] which is a reservoir *for Borrelia burgdorferi* and resulted in expansion of Lyme disease to new areas [78]. Hunting of wildlife is an ancient tradition that increases the risk of exposure to multiple zoonotic pathogens. Hunting for bushmeat involves contact with potential infectious pathogens at several stages: tracking, capturing and handling the animals, then butchering and transporting the carcass, and trading and consumption of the meat. Moreover, there is increased exposure to potential infected vectors while hunting. Africa poses the greatest risk for emergence of novel wildlife zoonoses which may threaten the health of the global population, by adaptation and transmission from humans to humans. Bushmeat trade and consumption and demand in West and Central Africa appear to be highest in the world, four times greater than in the Amazon basin, with about 4.5 million tons of bushmeat marketed for consumption annually or >287 g per person per day [79, 80]. Furthermore, the practice of hunting primates for bushmeat is still a common practice in Africa, which is associated with a particular high risk for crossover infection to humans, especially chimpanzees which are phylogenetically the closest to humans [77]. This is commonly believed to be a major factor in the emergence of HIV and Ebola virus from Africa. Other factors driving the emergence of novel zoonoses from Africa may include the ecologically diverse wildlife in large areas of tropical rainforests, new and changing patterns of land utilization, and the increased demand for bushmeat in urban communities [77]. The reverse pattern of transmission of domestic animal pathogens to wildlife to maintain a large natural reservoir of zoonoses is also possible, but transmission to humans would still be more likely to occur from contact with domestic livestock. Table 2.7 shows some of the major zoonoses associated with wildlife.

Table 2.7 Some zoonoses associated with wildlife

Microbes	Diseases	Transmission	Animals	Location
Arenaviruses				
Hemorrhagic fevers [HF] Junin virus	Argentinian HF	Aerosol of excreta	Field mice	Argentina, Paraguay
Machupo virus	Bolivian HF	Aerosol of excreta	Field mice	Bolivia
Guanarito virus	Venezuelan HF	Aerosol of excreta	Field mice	Venezuela
Lassa virus	Lassa fever	Aerosol of excreta, food contamination	Mice	West Africa
Bunyaviruses				
California E. viruses	Encephalitis	Mosquitoes	Rabbits, squirrels	N. America
Crimean-Congo viruses	Crimean-Congo HF	Ixodes ticks	Rodents, cattle hedgehogs	Congo, Asia, E. Europe
Hantaviruses				
Hantaan virus	Korean HF	Excreta in water and aerosol	Field mice, bats	East Asia
Seoul virus	HFRS	Via excreta	Norwegian rats	Worldwide, seaports
Puumala virus	Nephropathia epidemica	Via excreta	Bank voles	Europe
Dobrava virus	Balkan hemorrhagic fever	Via excreta	Field mice	S. Europe
Sin Nombre virus	HPS	Via excreta	Deer mice	US, North Mexico
Laguna Negra virus	HPS	Via excreta	Field mice	Argentina
Andes virus	HPS	Via excreta	Rice rats	S. America
Filoviruses				
Marburg virus	Hemorrhagic fever	Blood contact	Monkeys	East Africa
Ebola virus	Hemorrhagic fever	Blood, meat contact	Chimpanzees, gorillas	West Africa, Congo
Flaviviruses				
Japanese E. virus	Encephalitis	Culex mosquitoes	Boars, birds, snakes, farm animals, bats	Asia, SE. Asia
St. Louis E. virus	Encephalitis	Culex mosquitoes	Wild and domestic birds, bats	N. America, Mexico

(continued)

Table 2.7 (continued)

Microbes	Diseases	Transmission	Animals	Location
West Nile virus	Encephalitis	Aedes and culex mosquitoes	Wild birds	Africa, Asia, Americas, Australia
Dengue virus	Dengue fever	Aedes mosquitoes	Monkeys	Asia, Africa Americas, Caribbean
Yellow fever virus	Hepatitis	Several mosquitoes	Monkeys	Africa, South and Central America
Powassan virus	Encephalitis	Ixodes ticks	Rabbits, squirrels, skunks	North America, Mexico, Russia
Tickborne E. virus	Encephalitis	Ixodes ticks	Feral and domestic animals	Europe, Asia
Bacteria				
Leptospira interrogans	Leptospirosis	Rat urine via water	Rodents, foxes, hares	Worldwide, mainly tropics
Yersinia pestis	Plague	Rodent fleas	Rodents, squirrels, marmosets	Africa, Asia, S. America, SW. US
Francisella tularensis	Tularemia	Ticks, fleas, lice, direct contact	Hares, rodents, squirrels	US, Russia, Europe
Rickettsia typhi	Murine typhus	Fleas	Rats	Tropics, subtropics, Europe, US
Orientia tsutsugamushi	Scrub typhus	Mites	Rodents, rabbits	Eastern Asia, Australia, Pacific islands
Spirillum minus	Rat-bite fever	Bite of rodents	Rats and mice	Mainly Asia
S. moniliformis	Rat-bite fever	Rat bites	Rats, mice, squirrels, ferrets	Worldwide
Borrelia species				
B. burgdorferi	Lyme disease	Ixodes ticks	White footed-mice deer, roe, hedgehog	Worldwide, esp. US, Cen. Europe, Asia
B. duttoni	Relapsing fever	Soft ticks	Wild rodent, domestic animals	Africa
Ehrlichia species	Ehrlichiosis	Ixodes ticks	White-tailed deer, rodents, raccoons, rabbit	US, Europe, Mali
Protozoa				
Plasmodium knowlesi	Malaria	Anopheles mosquitoes	Wild macaques	Southeast Asia
Babesia microti, *B. divergens*	Babesiosis	Ixodes ticks	Rodents, rabbits, cattle	Worldwide, all continents
Trypanosoma Cruzi	Chagas disease	Triatomine bugs	Wild and domestic animals	Central and South America
Trypanosoma brucei	Sleeping sickness	Tsetse flies	Antelopes, monkeys, lions, hyenas, domestic animals	Africa

Data obtained from Refs. [27, 36, 71], Krauss H et al. [eds.], Zoonoses: Infectious Disease Transmissible from animals to humans. 3rd Edition, 2003, ASM Press, Washington, Appendix E, p 423–37

HPS hantavirus pulmonary syndrome, *HFRS* hemorrhagic fever with renal syndrome, *E* encephalitis

References

1. Woolhouse M, Gaunt E (2007) Ecological origins of novel human pathogens. Crit Rev Microbiol 33:231–242
2. European Commission. Health and Consumer Protection Directorate General, 2007. A new animal health strategy for the European Union [2007–2013]. Where "prevention is better than cure", Communication to the Council, the European Parliament, the European Economic and Social Committee, and the Committee of the Regions [COM 539 final]. http://ec.europa.eu/food/animal/disease/strategy/docs/animal-health-strategy-en-pdf
3. Cascio A, Bosilkovski M, Rodriguez-Morales AJ, Pappas G (2011) The socio-ecology of zoonotic infections. Clin Microbiol Infect 17:336–342
4. Rassi A Jr, Rassi A, Marin-Neto JA (2010) Chagas disease. Lancet 375:1388–1402
5. Aufderheide AC, Salo W, Madden M et al (2004) A 9,000-year record of Chagas disease. Proc Natl Acad Sci U S A 101:2034–2039
6. Flint JA, van Duynhoven YT, Anmgulo FJ et al (2005) Estimating the burden of acute gastroenteritis, food borne disease, and pathogens commonly transmitted by food: an international review. Clin Infect Dis 41:698–704
7. Scallen E, Hosktra RM, Angula FJ et al (2011) Foodborne illness acquired in the United States—major pathogens. Emerg Infect Dis 17:7–15
8. Jonos KE, Patel NG, Levy MA et al (2008) Global trends in emerging infectious diseases. Nature 451:990–993
9. Militosis MD, Bier JW (2003) International handbook of foodborne pathogens. Marcel Dekker, New York
10. Frank C, Werber D, Cramer JP et al (2011) Epidemic profile of Shiga-toxin producing *Escherichia coli* 0104: H4 outbreak in Germany. N Engl J Med 365:1771–1780
11. Blaser MJ (2011) Deconstructing a lethal foodborne epidemic. N Engl J Med 365:1835–1836
12. Kahn RE, Morozov I, Feldman H, Richt JA (2012) 6th International Conference on Emerging Zoonoses. Zoonoses Public Health 59(Suppl 2):2–31
13. Havelaar AH, Ivason S, Lofdahl M, Nauta MJ (2012) Estimating the true incidence of campylobacterosis and salmonellosis in the European Union, 2009. Epidemiol Infect 141:293–302
14. Kendall M, Crim S, Fullerton K et al (2012) Travel-associated enteric infections diagnosed after return to the United States, Foodborne Disease Active Surveillance Network [Food Net], 2004-2009. Clin Infect Dis 54(S5):S4807
15. Shiferaw B, Verrill L, Booth H, Zansky SM, Norton DM, Crim S, Henao OL (2012) Sex-based differences in food consumption: Foodborne Diseases Active Surveillance Network [Food Net] Population Survey, 2006-2007. Clin Infect Dis 54(S5):S453–S457
16. Centers for Disease Control and Prevention (2004) Outbreak of Salmonella serotype Enteritidis infections associated with raw almonds—United States and Canada, 2003–2004. MMWR Morb Mortal Wkly Rep 53:484–487
17. Anonymous (2010) Analysis of baseline survey on the prevalence of Campylobacter in broiler batches and of Campylobacter and Salmonella on broiler carcases in the EU. 2008-part A: Campylobacter and Salmonella prevalence estimates. Eur Food Saf Auth J 8:1503
18. Sun Y, Wu Q, Zhou Y, Zhou R (2005) Investigation on contamination by Campylobacter of retail raw meat in Shenyang. Chin J Public Health 21:985–987
19. Yang B, Qu D, Zhang X (2010) Prevalence and characterization of Salmonella serovars in retail meats of marketplace in Shanxi. China Int J Food Micobiol 141:63–72
20. Batz MB, Hoffmann S, Morris JG Jr (2012) Ranking the disease burden of 14 pathogens in food sources in the United States using attribution data from outbreak investigation and expert elicitation. J Food Prot 75:1278–1291
21. Majowicz SE, Musto J, Scallan E et al (2010) The global burden of nontyphoidal Salmonella gastroenteritis. Clin Infect Dis 50:882–889
22. Public Health Agency of Canada, 2011: Policy on *Listeria monocytogenes* in ready to-eat-foods. http//www.hc.gc.ca/fu-an/legislation/pol/policy-listeria-monocytogenes-2011-eng.php
23. Bertolsi R (2008) Listeriosis: a primer. CMAJ 8:795–797

24. Ghandi M, Chikindas ML (2007) Listeria: a foodborne pathogen that knows how to survive. Int J Food Microbiol 113:1–15
25. Silk BJ, Date KA, Jackson KA et al (2012) Invasive listeriosis in the Foodborne Diseases Active Surveillance Network [Food Net], 2004-2009: further targeted prevention needed for higher-risk groups. Clin Infect Dis 5(S5):S396–S404
26. Yan H, SB N, Guan W et al (2010) Prevalence and characterization of antimicrobial resistance of foodborne *Listeria monocytogenes* isolates in Hebei province of Northern China, 2005-2007. Int J Food Microbiol 144:310–316
27. Krauss H, Weber A, Appel M et al (2003) Bacterial zoonoes. In: Zoonoses: infectious diseases transmissible from animals to humans, 3rd edn. ASM Press, Washington, DC, pp p173–p252
28. Centers of Disease Control and Prevention (2009) Recommendations for diagnosis of Shiga toxin- producing *Escherichia coli* infections by clinical laboratories. MMWR Recomm Rep 58:1–11
29. Dharma K, Rajagunalan S, Chakrabarty S, Verma AK, Kumar A, Tiwari R, Kapoor S (2013) Foodborne pathogens of animal origin—diagnosis, prevention, control and their zoonotic significance: a review. Pak J Biol Sci 16:1076–1085
30. Ong KL, Gould LH, Chen DL et al (2012) Changing epidemiology of *Yersinia enterocolitica* infections: Markedly decreased rates in young black children, foodborne diseases active Surveillance Network [Food Net], 1996-2009. Clin Infect Dis 54(S5):S385–S389
31. Shao D, Shi Z, Wei J, Ma Z (2011) A brief review of foodborne zoonoses in China. Epidemiol Infect 139:1497–1504
32. Veerhagen J, Charlier J, Lemmens P et al (1998) Surveillance of human *Yerinia enterocolitica* infections in Belgium: 1967–1996. Clin Infect Dis 27:59–64
33. Wang X, Cui Z, Jin D et al (2009) Distribution of pathogenic *Yersinia enterocolitica* in China. Eur J Clin Microbiol Infect Dis 28:1237–1244
34. Newton A, Kendall M, Vugia DJ, Henoa OL, Mahon BE (2012) Increasing rates of vibriosis in the United States, 1996-2010: Review of surveillance data from 2 systems. Clin Infect Dis 54(S5):S391–S395
35. World Health Organization (1967) Toxoplasmosis. Tehnical report series no. 431. WHO, Geneva
36. Krauss H, Weber A, Appel M et al (2003) Parasitic zoonoses. In: Infectious diseases transmissible from animals to humans, 3rd edn. ASM Press, Washington DC, pp p261–p403
37. Zhou P, Chen N, Zhang RL, Lin RQ, Zhu XQ (2008) Foodborne parasitic zoonoses in China: perspective for control. Trends Parasitol 24:190–196
38. Lun ZR, Gassert R, Lai D-H, Li AX, Zhu X-Q, Yu X-B, Fang Y-Y (2005) Clonorchiasis: a key foodborne zoonoses in China. Lancet Infect Dis 5:31–41
39. Day MJ (2011) One health: the importance of companion animal vector-borne diseases. Parasit Vectors 4:49
40. Center for Disease Control and Prevention (2003) Non-fatal dog bite-related injuries treated in hospital emergencies departments—United States, 2001. MMWR Morb Mortal Wkly Rep 61(23):436
41. Dachenx L, Delmas O, Bourthy H (2011) Human rabies encephalitis prevention and treatment: progress since Pasteur's discovery. Infect Disord Drug Targets 11:251–299
42. Blanton J, Palmer D, Dyer J, Rupprecht C (2011) Rabies surveillance in the United States during 2010. J Am Vet Med Assoc 239:773–783
43. Gerhold RW, Jessup DA (2013) Zoonotic diseases associated with free-roaming cats. Zoonoses Public Health 60:189–195
44. Krauss H, Weber A, Appel M et al (2003) Animal bite infections. In: Zoonoses. Infectious Disease transmissible from animals to humans, 3rd edn. ASM Press, Washington DC, pp p405–p410
45. Lion S, Escande F, Burdin JC (1996) *Canocytophaga canimorsus* infection in humans: review of the literature and cases report. Eur J Epidemiol 12:521–533
46. WHO: Report of the 5th Consultative Meeting on Leishmania/HIV co-infection. Addis Ababa, Ethiopia, 2007.

47. Nunes CM, Pires MM, Marques da Silva K, Assis FD, Filho JG, SHV P (2010) Relationship between dog culling and incidence of human visceral leishmaniasis in an endemic area. Vet Parasitol 170:131–133
48. Palatnik-de-Sousa CB, Silva-Antunes I, de Aguiar MA, Menz I, Palatnik M, Lavor C (2009) Decrease of the incidence of human and canine visceral leishmaniasis after dog immunization with Leishmune in Brazilian endemic areas. Vaccine 27:3505–3512
49. Chomel BB, Karsten RW (2010) Bartonellosis, an increasingly recognized zoonosis. J Appl Microbiol 109:743–750
50. Harner SA, Tsao JI, Walker ED, Mansfield LS, Foster ES, Hickling JG (2009) Use of tick surveys and serosurveys to evaluate pet dogs as sentinel species for emerging Lyme disease. Am J Vet Res 70:49–56
51. Walker DH, Raoult D (2000) *Rickettsia ricketsii* and other Spotted Fever group Rickettsiae [Rocky Mountain Spotted Fever and other Spotted Fever]. In: Mandel GL, Bennett JE, Dolin R (eds) Principles and Practice of Infectious Diseases, 5th edn. Churchill Livingstone, Philadelphia, pp 2035–2056
52. Demma LJ, Traeger MS, Nicholson WL et al (2005) Rocky Mountain spotted fever from an unexpected tick vector in Arizona. N Engl J Med 353:587–594
53. Nicholson WL, Allen KE, McQuiston JH, Breitschwerdt EB, Little SE (2010) The increasing recognition of rickettsial pathogens in dogs and people. Trends Parasitol 26:205–212
54. Perez M, Bodor M, Zhang C, Xiong Q, Rikihisa Y (2006) Human infection with *Erlichia canis* accompanied by clinical signs in Venezuela. Ann N Y Acad Sci 1078:110–117
55. Esch KJ, Peterson CA (2013) Transmission and epidemiology of protozoal diseases of companion animals. Clin Microbiol Rev 26:58–85
56. Desplases P, van Krepen F, Schweiger A, Overgauuw PAM (2011) Role of pet dogs and cats in the transmission of helminthic zoonoses in Europe, with a focus on echinococcosis and toxocariasis. Vet Parasitol 182:41–53
57. Jenkins DJ, Romig T, Thompson RCA (2005) Emergence/reemergence of Echinococcus spp.—a global update. Int J Parasitol 35:1205–1214
58. Siko SB, Desplazes P, Ceica C, Tivadar CS, Bogolin I, Popescu S, Cozma V (2011) *Echinococcus multilocularis* in south-eastern Europe [Romania]. Parasitol Res 108:1093–1097
59. Schweiger A, Ammann RW, Candinas D et al (2007) Human alveolar echinococcosis after fox population increase. Switz Emerg Infect Dis 13:878–882
60. Lee CY, Schantz PM, Kazacos KR, Montgomery SP, Bowman DD (2010) Epidemiological and zoonotic aspects of ascarid infections in dogs and cats. Trends Prasitol 26:155–161
61. Magnaval JF, Glickman LT (2006) Management and treatment option for human toxocariasis. In: Holland CV, Smith HV (eds) Toxacara, the enigmatic parasite. CABI Publishing, CAB International, Oxfordshire, pp 113–126
62. American Veterinary Medical Association. US pet ownership & demographic sourcebook. 2007 Edition. https//www.avma.org//KB/Resouces/Statistics/Pages/Market-research-statistics-US-Pet-Ownetrship-Demographics-Sourcebookaspx.
63. Boseret G, Losson B, Mainil JG, Thiry E, Saegerman C (2013) Zoonoses in pet birds: review and perspectives. Vet Res 44:31
64. Vanrompay D, Harkinezhad T, Van de Walle M et al (2007) *Chlamydophila psittaci* transmission from pet birds to humans. Emerg Infect Dis 13:1108–1110
65. FM W, Hoefer H, Kiehn TE, Grest P, Bley CR, Hatt JM (1998) Possible human-avian transmission of *Mycobacterium tuberculosis* infection a green-winged macaw [Arachloroptera]: report with public health implication. J Clin Microbiol 36:1101–1102
66. Steinmetz HW, Rutz C, Hoop RK, Grestr P, Bley CR, Hatt JM (2006) Possible human-avian transmission of *Mycobacterium tuberculosis* in a green-winged macaw [Arachloroptera]. Avian Dis 50:641–645
67. Hoop RK (2002) *Mycobacterium tuberculosis* infection in a canary [*Serius canaria* L] and a blue-fronted Amazon parrot [*Amazona amazon aestiva*]. Avian Dis 46:502–504
68. Mather SA, Smith RP, Cahill B et al (2011) Strain diversity of *Borrelia burgdorferi* in tick dispersed in North America by flying birds. J Vector Ecol 36:24–29

69. Simmons NB (2005) Order Chiroptera. In: Wilson DE, Reeder DM (eds) Mammal species of the world: a taxonomic and geographical reference. John Hopkins University Press, Baltimore, pp 312–529
70. Drexler JE, Corman VM, Wegner T et al (2011) Amplification of emerging viruses in a bat colony. Emerg Infect Dis 17:449–456
71. Krauss H, Weber A, Appel M et al (2003) Viral Zoonoses: Zoonoese caused by Rhabdoviruses. In: Zoonoses. Infectious diseases transmissible from animals to humans, 3rd edn. ASM Press, Washington DC, pp 112–119
72. Kunz TH, Pieson ED Bats of the world. An introduction. In: Nowak RM (ed) Walker's Bats of the World, 1994. John Hopkins University Press, Baltimore, pp 1–46
73. Hayman DTS, Bowen RA, Cryan PM et al (2013) Ecology of zoonotic infectious diseases in bats: current knowledge and future directions. Zoonses Public Health 60:2–21
74. Calisher CH, Childs JE, Field HE, Holmes KV, Schountz T (2006) Bats: important reservoir hosts of emerging viruses. Clin Microbiol Rev 19:531–545
75. Johnson PT, Townsend AR, Cleveland CC et al (2010) Linking environment and disease emergence in humans and wildlife. Ecol Appl 20:16–29
76. Daszak P, Cunningham AA, Hyatt AD (2000) Emerging infectious diseases of wildlife—threats to biodiversity and human health. Science 287:443–449
77. Wolfe ND, Daszak P, Kilpatrick AM, Burke DS (2005) Bushmeat hunting, deforestation, and prediction of zoonoses emergence. Emerg Infect Dis 11:1822–1827
78. Lo Giudice K, Ostfeld RS, Schmidt KA, Keesing F (2003) The ecology of infectious disease: effects of host diversity and community composition on Lyme disease risk. Proc Natl Acad Sci U S A 100:567–571
79. Fa JE, Juste J, Delval JP, Castroviojo J (1995) Impact of market hunting on mammal species in Equatorial-Guinea. Conserv Biol 9:1107–1115
80. Fa JE, Peres CA, Meeuwig J (2002) Bushmeat exploitation in tropical forests: an international comparison. Conserv Biol 16:232–237

Chapter 3
Swine and Avian Influenza Outbreaks in Recent Times

3.1 Introduction

Zoonotic influenza continues to pose serious threat to the welfare of the global population, and it is predicted by some experts that the next major influenza pandemic will be of avian origin from Asia. To prepare for the future, we have to learn from the past events, and this involves analysis of recent occurrences in influenza activity, including epizootic outbreaks. Rapid transportation by airplanes and a much larger global population than existed just over a century ago, during the 1918 Spanish influenza pandemic, may result in greater human suffering. Modernization of the global healthcare facilities and effective medicines and vaccines will likely prevent a repeat of the 1918 explosive mortality rate, and the brunt of a virulent influenza pandemic will be borne by the poorer countries, the debilitated elderly, pregnant women, and the growing number of people with immunosuppression.

3.2 Virology

The influenza viruses are enveloped viruses of *the Orthomyxoviridae* family, with influenza A and B causing annual seasonal excess morbidity and mortality and influenza C a milder respiratory illness [1]. Only influenza A viruses are true zoonotic agents responsible for influenza pandemics, and influenza B and C are primarily human pathogens, but influenza C occasionally infects pigs and dogs [2]. The morphology of the influenza viruses includes segmented, single-stranded, negative-sense RNA genomes, surrounded by nucleoprotein envelope and covered with surface projections or spikes [1]. These surface spikes are glycoproteins with hemagglutinin [HA] or neuraminidase [NA] activity that are the main targets of the host humoral immune response. The influenza A viruses can be subtyped according to the antigenic nature of their surface glycoproteins, with 16–17 HA and 9–10 NA identified to date.

I.W. Fong, *Emerging Zoonoses*, Emerging Infectious Diseases of the 21st Century,
DOI 10.1007/978-3-319-50890-0_3

The HA serves as the receptor-binding protein and facilitates fusion of the viral envelope with the host cell membrane [3]. NA is responsible for assisting virus entry into cell by mucus degradation [4] and release and spread of the progeny virions [4]. There are eight RNA segments within the viral genome, encoding 10–11 proteins.

Influenza A and B viruses frequently undergo antigenic variation of the surface glycoproteins [HA and NA] that allow the viruses to evade human neutralization from previous exposures and vaccinations. Variation in the viruses is caused by accumulation of point mutations in the HA and NA genes, antigenic drift [5, 6]. A variety of mutations including substitutions and deletions result in genetic variation during antigenic drift. The gradual accumulation of amino acid changes that occur with antigenic drift on the HA and NA sites allows the virus to survive, due to ineffective neutralization, and replaces existing strains as the predominant circulating virus in the population, to cause yearly outbreaks [1]. The propensity of influenza viruses for multiple mutations and rapid antigenic variation may be due to evolutionary effect of selective pressure for self-preservation. High error rate or mutations during genomic replication are typical of RNA viruses, as the RNA polymerase lacks proofreading activity, and the segmental influenza virus genome facilitates reassortment between different viral strains that infect the same cell [6–8].

A major antigenic variation, called antigenic shift, that usually precedes a pandemic occurs only in influenza A viruses and by a different mechanism. Antigenic shift results in the introduction of a new HA, with or without a new NA, to introduce a new virus to which the population is naïve, lacking any or even partial immunity [1, 6]. This pattern of replacement of previous HA and NA in new subtypes of influenza A viruses is associated with emergence of pandemic influenza outbreaks over the last century. Antigenic shift may occur by at least two different mechanisms. A zoonotic influenza virus can be transmitted from an animal reservoir without reassortment, usually by adaptation to a mammalian or human receptor site by mutation [1, 6]. The other mechanism is by reassortment, when two or more viruses coinfect the same cell and exchange one or more RNA segments to produce new progeny virus with new antigenic HA [with or without new NA] and new biological properties [7]. Swine is commonly believed to be the "mixing vessel" for genetic reassortment, as they are susceptible to infection with both avian and human strains and swine strains of influenza viruses [9]. The Asian influenza pandemic of 1957 and the Hong Kong influenza pandemic of 1968 were the result of antigenic shift with reassortment of genetic material during dual infection with circulating human influenza strains, H1N1 in 1957 and H2N2 in 1968, and avian influenza strains, probably with pigs as the mixing vessel [10, 11].

3.3 Ecology and Host Tropism

Influenza A viruses can infect several animals including humans, birds, pigs, horses, cats, and marine mammals such as seals and whales [1, 6]. Most influenza viruses are restricted to specific hosts, but some strains can circulate among several animal species, i.e., H1N1 and H3N2 viruses are endemic in humans, birds, and pigs [12]. The primary determinant of host tropism and transmission is directly related to the specificity and affinity of the viral HA for the host receptor. Sialic acid is the receptor for influenza viruses that binds to the viral HA. Avian influenza viruses preferentially bind to sialic acid molecules with specific side chains with α-2,3-linkages and mammalian viruses to α-2,6-linkages [6]. Human-adapted seasonal viruses such as H1N1 and H3N2 have high affinity for α-2,6-linked sialic acid [SA], which are expressed on the epithelium of the upper respiratory tract [URT] of humans [12]. Avian influenza viruses bind preferentially to galactose-linked α-2,3-SA which is found abundantly in the URT and gut of birds, but is also found in the lower respiratory tract epithelium of humans [12]. Hence, highly pathogenic avian influenza viruses [HPAIVs] not well adapted to human receptors can cause limited avian to human transmission with viral pneumonia by this mechanism. The tracheal epithelium of pigs contain receptors with both α-2,3-SA and α-2,6-SA linkages and be infected simultaneously with avian and human [mammalian] influenza viruses that predispose to reassortment of zoonotic and human strains [12]. See Fig. 3.1 for ecology, cycle of avian, and swine influenza A.

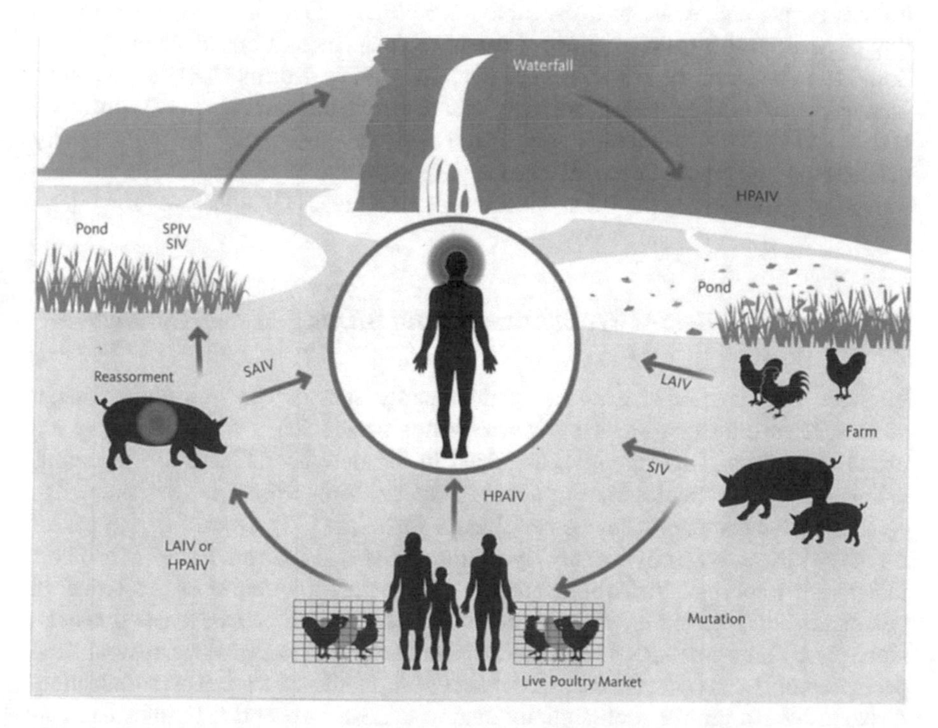

Fig. 3.1 Biological cycle of avian and swine influenza A. *LPAIV* low pathogenic avian influenza virus, *HPAIV* highly pathogenic avian influenza virus, *SIV* swine influenza virus, *SAIV* influenza virus with genetic elements from avian and swine influenza

3.4 Swine Influenza Viruses of the Twenty-First Century

The spread of influenza zoonotic viruses in animals and humans is primarily related to the interface and transmission between birds, pigs, and humans. Swine influenza viruses [SIVs] was first isolated from pigs in 1930 [13], although clinical disease resembling influenza was noted in pigs in the midwestern United States [US] in 1918, coinciding with the Spanish influenza outbreak [14]. Since then swine influenza [SI] has been recognized worldwide in the pig industry as a significant problem. The clinical signs of influenza in pigs are similar to those in humans with high herd morbidity [nearly 100%] and low mortality [<1%], but pneumonia can occur and recovery usually after a week [15]. The epizootic pattern of SI is similar to that in humans with outbreaks in late fall and early winter. The predominant SIV circulating in North America from the first isolate in 1930 to 1998 was caused by the swine H1N1 lineage [SH1N1], but there was low level of human subtype H3 virus also circulating in pigs [16]. In 1998 a severe SI outbreak occurred in pig farms in several states of the US, this was subsequently attributed to a new SIV subtype H3N2 [15]. Within a year there was widespread circulation of the SH3N2 virus in pig farms across the US, containing gene segments similar to those of human influenza subtype and the classic SI subtype [double reassortment], and another strain with the gene segments from human, swine, and avian lineages, triple reassortment [17]. The triple reassortment SIV circulated more efficiently than the double reassortment virus in the pig population. Over the years in less than a decade, multiple reassortment SIVs have been identified in North America, including H3N2 genotypes, H1N2, reassortment [r] H1N1, and H3N1 [15]. In 2006 humanlike H1 viruses genetically and antigenically distinct from the classic swine H1 lineage were identified in pigs in Canada and subsequently spread across the US swine farms as H1N1 and H1N2 viruses [15, 18].

3.4.1 Cross-Species Transmission and Mixing Vessel Concept

Zoonotic influenza A viruses are predominantly species specific, and although cross-species transmission of influenza viruses occurs fairly frequently, they are usually self-limited and rarely maintained in the new host. There are numerous examples of cross-species transmission of influenza virus from avian to mammalian species and intra-mammalian cross-species infection [15]. Specific subtypes of influenza viruses differ in their ability to cross the species barrier, and viral and host factors are important in the transmission. The segmental nature of the influenza genome is considered important in the viral evolution and cross-species transmission. Host range restriction of species transmission is governed by several viral proteins, but HA is critical to bind to host receptor to allow invasion and replication. Avian influenza viruses preferentially bind to α-2,3-SA receptors in intestinal epithelial cells, whereas human influenza viruses HA bind more favorably to respiratory epithelial receptors with α-2,3-SA [19]. Thus, avian influenza viruses usually cannot replicate effectively in humans, and birds are less susceptible to human viruses. NA also contributes to the virus-species host specificity as efficient growth of influenza A virus depends on the balance between HA receptor-binding affinity and the NA receptor-cleaving activity [9]. The viral polymerase basic protein-2 is important in virus replication and is a host range determinant [20].

The concept that swine could be a "mixing vessel" for reassortment of influenza viruses was proposed by Scholtissek et al. [21] in 1985, as pigs could be dually infected with human and avian influenza viruses. This is related to the presence of both receptor types found in the respiratory tract of pigs [22]. Documentation of primary avian influenza viruses in swine has been reported over the years in different regions of the world: in European swine in 1979, pigs from China and Asia multiple times, and in Canadian swine in 2000–2004 [9]. Wholly human influenza viruses in pigs had also been well documented a few times in Taiwan and China [23], and pig-pig transmission of human H1N1 viruses has been reproduced experimentally [24]. Prior to 2005 sustained circulation of human influenza viruses was uncommon in swine herds, but since then swine viruses containing human-origin H1 and H1N2 gene segments have become established in the US [9].

3.4.2 Reassortment of Influenza Viruses in Pigs

It has been proposed that the influenza A viruses responsible for the 1957 and 1968 human pandemics were the result of avian, human, and swine viruses reassortment, but direct evidence was lacking. The PB1 gene of influenza A virus, involved with initiation of transcription and chain elongation with other viral polymerase gene products [25], was introduced from avian species into the human pandemic strains [1957 H2N2 and 1968 H3N2], and this avian PB1was also found in pig viruses [26]. Genetic reassortment between avian and human H3N2 viruses has occurred in European pigs [27], and novel reassortment viruses were transmitted to children in the Netherlands [28]. A similar reassortment H3N2 virus was isolated from a child in Hong Kong [29].

Since 1998 double [human/swine] and triple [avian/human/swine] reassortment viruses, H3N2, H1N2, rH1N1, and H3N1, have emerged in US pigs [9]. The predominant viruses circulating in US swine are these triple reassortment H3N2, H1N2, and H1N1 viruses. With the advent of the twenty-first century, avian H9N2 and existent human H3N2 influenza viruses were co-circulating in pigs of southeastern China [30]. Subsequently, double reassortment H3N2 viruses containing human viral genes [HA and NA] and avian genes [polymerase, matrix, and nonstructural proteins] and triple reassortment H3N2 viruses carrying human, avian, and swine viral genes have emerged in pigs in China [31].

In central US a unique H2N3 influenza virus was recovered from pig farms in 2007 [32], with HA and NA sequences similar genes to avian influenza viruses [H2N3 and H4N3] and genes from US swine influenza viruses. This swine H2N3 virus with avian origin surface glycoprotein was already well adapted to the mammalian host. Of concern was that the H2 influenza viruses were absent from the human circulation since 1968, and individuals born subsequently would have had little immunity to this subtype [19], thus posing a pandemic risk to a large nonimmune human population. The HA mutation was identical to the initial human influenza virus isolates found at the beginning of the 1957 H2N2 pandemic [9].

3.4.3 Transmission of Swine Reassortment Viruses to Humans

Although some experts believe that the Spanish influenza pandemic of 1918 was caused by a reassortment swine H1N1 virus, this is controversial with no direct supporting evidence and others contend that it is of avian origin. The first swine influenza virus isolated from human was in 1974 [33], and prior to 2009 there were only sporadic cases reported. Myers et al. [34] subsequently reviewed 50 cases of zoonotic swine influenza reported up till 2005. Cases were reported from the US, Europe, Russia, Canada, and Hong Kong. Most cases [37 subjects] were civilians but there was a localized outbreak of 13 cases in the military, Fort Dix, New Jersey. Swine exposure was reported in 61% of civilians, and the case fatality rate was 14%, 7 of 50 infected. The predominant influenza A viruses were H1N1 subtype with only 4 H3N2 subtype.

In 2009 a swine-derived H1N1 influenza A reassortment virus caused a moderately mild pandemic, the "Mexican" influenza pandemic. The first confirmed cases of the pandemic virus appeared in Mexico in February 2009, followed by cases detected in California in March–April that year [35]. The speed of the pandemic was rapid by June 2009; 73 countries had reported 26,000 confirmed cases. The initial outbreak in Mexico was the most worrisome, as 6.5% of hospitalized patients became critically ill and 41% of these patients died [36]. By August 2010, nearly all countries in the world reported confirmed cases of the pandemic H1N1 influenza, but the global outbreak started to wane. Although experts considered the 2009 pandemic a mild outbreak, there was marked variation in the severity of the disease in different regions of the world [37]. Unlike seasonal influenza outbreaks, and similar to the 1918 pandemic, older adults fared relatively well, and excess mortality and adverse outcome were greater in children, young adults, and pregnant women [35]. Estimates of the influenza-related deaths worldwide, 123,000–203,000, were similar to that of mild seasonal influenza [37]. Baseline pre-existing immunity at the start of the 2009 pandemic was nonexistent in children and very low in those born after 1980, but greater in older adults [38, 39].

The influenza A H1N1 2009 pandemic virus was first detected in Canadian pigs in May 2009 and subsequently in pigs from 14 countries in the Americas, Europe, and Asia [40, 41]. Since then new reassortment events with endemic swine influenza strains were reported in pigs in Hong Kong [42], Italy [43], Germany [44], and the US [45], derived from the 2009 swine H1N1 influenza virus. In the US nine reassortment viruses representing seven genotypes were found in commercial pig farms [45]. The pandemic strain of 2009, H1N1pdmo9, was antigenically related to the 1976–1977 swine influenza virus in the Fort Dix outbreak; and the genetic makeup consisted of a number of reassortments between avian, human, and swine viruses [46].

Since the influenza pandemic of 2009, H1N1-pdm09 has continued to circulate in various regions of the world during seasonal influenza activity. Influenza A viruses recovered in the latter part of March–April, 2015, from 84 countries were reported by WHO to be H1N1-pdm09 in 48.5% of isolates [WHO, Influenza Update no. 235, April 21, 2015]. The largest outbreak of influenza A [H1N1-pdm09] in recent years has been reported in India. Since the 2009 pandemic, the H1N1-pdm09 has replaced the previous seasonal H1N1 and became established in the human population. In India H1N1-pdm09 has recently caused a localized outbreak in the northern region in 2014–2015, with at least 22,240 cases and 1194 influenza-related fatalities [47]. The influenza HA sequences from viruses isolated in India indicate that the virus has gradually evolved since 2009 and acquired mutations in the H1 antigen sites and linked to enhanced virulence and appears to be antigenically distinct from the current vaccine containing 2009 [Ca10109] H1N1 viral HA [48]. In the US 13 cases of infection with novel triple reassortment swine-origin, influenza A [H3N2], variant virus occurred between 2011 and 2012 and were mostly related to agriculture fairs [49].

3.5 Avian Influenza in the Modern Era

Although avian influenza A viruses may have caused human epidemics for centuries, this has not been well documented. In 1557 and 1580, influenza pandemics, called “chicken malady” in German because the human cough sounded like sick chickens, were not preceded or concurrent with poultry outbreaks to link the events with avian influenza [50]. However, it is possible these outbreaks could have been related to low pathogenic avian influenza viruses [LPAIVs] without symptomatic disease in poultry. The first epizootics of avian influenza in poultry were described in Northern Italy in 1789, and they were not associated with human outbreaks [51]. However, highly pathogenic avian influenza viruses [HPAIVs] became well known to veterinarians around the end of the nineteenth century, after description of the “fowl plague” by an Italian scientist [51]. Although the origin of the 1918 Spanish influenza pandemics is still not fully resolved, the virus has avian-like genome probably derived from HPAIV a decade before [52].

Avian influenza is now considered by experts to be the greatest threat to global public health to arise from animals. Before the end of the millennium, HPAIV was linked to poultry but occurred rarely with self-limiting course. Since then a marked increase in avian influenza outbreaks has occurred worldwide. It has been estimated that avian influenza outbreaks have increased 100-fold with 23 million birds affected between 1959 and 1998 and over 200 million from 1999 to 2004 [53]. Poultry outbreaks continued to emerge even in Europe and North America in the first decade of the twenty-first century, with substantial damage and cost to the poultry industry and with sporadic human infections [51].

The natural reservoir hosts of avian influenza A viruses are wild waterfowls, such as ducks, geese, swans, gulls, waders, and others, and typically are asymptomatic in the birds or cause mild disease [54]. Thus, wild waterfowls are natural hosts for LPAIV which are transmitted to domestic birds and mammals by fecal-oral route through contamination of water, soil, and the environment. Poultry also have subclinical or mild respiratory disease, but they represent the main transmitters to humans. In domestic fowls such as chickens and turkeys, LPAIV of H5 and H7 subtypes may evolve to become more virulent as HPAIVs with lethal effect and can be transmitted via fecal-oral route or through the respiratory secretions. LPAIVS and adapted variants [HPAIVs] can cause respiratory disease in mammals and humans of varying severity with respiratory transmission [51]. Cross-species transmission had resulted in human infections with LPAIV H9N2, H7N2, H7N3, and H7N7 [51]. HPAIVs have rarely been transmitted from poultry to other species, but in the past decade or more, they have caused increasing respiratory and systemic infections in humans and other animals. Cross-species transmission of HPAIVs to humans had occurred with H5N1, H7N3, and H7N7 [55–57]. In most cases transmission from poultry to humans occurred via the respiratory or ocular route, but some strains of the HPAIV H5N1 may have been transmitted by both respiratory and oral routes to mammals. To date avian influenza viruses have limited or no human-to-human transmission capability. HPAIV infections manifestations are more atypical than regular influenza and may include gastrointestinal and neurological symptoms/signs besides the usual respiratory disease with H5N1 and ocular disease [conjunctivitis] with H7 subtypes [51].

Domestic ducks are especially prone to a large diversity of LPAIV infections from consumption and contact with surface water shared with wild waterfowls [58]. Terrestrial birds [chickens] associated with dry environment may be infected via respiratory route from droplets or aerosols and fecal-oral route through contaminated fomites [59]. Humans and other mammals are infected by avian influenza viruses primarily through the respiratory tract with inhalation of droplets, fomites or aerosols from domestic fowl, or self-inoculation accidentally from the contaminated hands. Infection by digestion is unusual in mammals but has been demonstrated to occur with HPAIV H5N1 in ferrets, mice, hamsters, and cats [60, 61]. Inoculation of the conjunctiva appears to an important means of bird-to-human transmission with the H7 subtype, due to preferential tropism for ocular tissues in human [62]. Direct inoculation of the upper respiratory tract or conjunctiva while swimming in contaminated water is also possible, and this has been documented Southeast Asia infection with HPAIV H5N1 [62].

3.5.1 Tissue Tropism

LPAIV preferentially infects the epithelial cells of the distal small bowel and the cloaca of waterbirds [53]. Trypsin-like proteases in the small intestine can cleave the HA protein resulting in localized infection. In poultry LPAIV primarily infect epithelial cells of the respiratory tract of the trachea, bronchi, and the alveolar sacs. Extracellular proteases for cleavage of the HA proteins are present in the respiratory epithelium of poultry [53]. Experimentally, however, intravenous inoculation of LPAIV can result in replication of the virus in the kidney and intestinal epithelial cells of chickens [63].

HPAIV, which evolve from LPAIV in poultry by mutations, have a broad tissue tropism in domestic fowls. The nasal cavity and the respiratory epithelium are initially infected with submucosa and capillary invasion, with widespread dissemination via the circulation to infect the epithelial cells of numerous organs throughout the bird's body. This leads to severe avian illness and high mortality, as the brain, pancreas, heart, kidney, and skeletal muscle can be affected in poultry in acute infection [53]. Wild waterfowls are rarely infected with HPAIV, but since 2002 HPAIV H5N1 has spread from poultry to a wide range of wild bird species. The HA protein of HPAIVs possess multibasic cleavage sites that can be cleaved by intracellular proteases present in a diverse number of cell types in avian and mammalian species [64].

Wild birds become infected with HPAIV H5N1 via the respiratory epithelium initially, with replication and viremia and dissemination to variable number of organs depending on the species. In contrast to LPAIV the intestinal tract is not usually infected by HPAIV in wild birds. In most infected wild birds, the parenchymal cells are the main sites for H5N1 replication, and although the virus can spread to many organs, the main tissue tropism is in the brain besides the respiratory tract.

In humans and other mammals, LPAIV and HPAIV preferentially cause infection of the respiratory epithelium and especially of the lower tract. Infections of humans with LPAIV typically cause mild respiratory disease, including conjunctivitis for H7 subtypes, with resolution in 1–2 weeks [54]. Tropism of the influenza viruses is in large part determined by the receptor-binding affinity of the HA protein. Avian influenza viruses bind preferentially to receptor with α-2,3-SA to galactose, which are abundant in the intestinal and respiratory epithelium of domestic and wild birds but are also present in other tissues [heart, kidney, and brain] and endothelium in ducks and chickens [64]. Binding of avian influenza in humans occurs mainly in the lower respiratory tract where α-2,3-SA linkages are present focally in bronchiolar epithelial cells, type 11 pneumocytes, alveolar macrophages, acinar cells of the submucosal glands of the trachea and bronchi, and epithelial cells of the eye [65–67].

3.5.2 *Highly Pathogenic Avian Influenza H5N1*

The HPAIV subtype H5N1 was first described in Southeast Asia in 1996, and since then it has spread to at least 63 countries in Asia, Europe, Africa, and the Middle East [68]. Although the virus likely originated from poultry as a result of mutations of a LPAIV, it appears to have infected wild birds, which subsequently caused spill-back infection to poultry in distant regions. Migrating aquatic wild birds were considered responsible for long-distance dispersal of the HPAIV H5N1 from Qinghai Lake [China] to Europe, Russia, and Africa [69]. The ancestor virus was initially isolated from domestic geese in China, but the prime long-distance vector appeared to be the wild mallard ducks, as they showed abundant viruses excretion without clinical or debilitating disease [70]. Over 100 million birds have died from the infection either naturally or from culling to limit the spread of the disease. In Thailand which had seven waves of HPAIV H5N1 outbreaks, >62 million poultry have died and outbreaks in poultry were associated with increased infection in wild birds in the preceding months [71].

Genomic analysis of H5N1 isolates form birds and humans in 2005 showed two distinct clades from separate noncontiguous regions [72]. All the genes were of avian origin with no evidence of reassortment with human influenza virus. The human isolates were resistant to amantadine but were susceptible to the neuraminidase inhibitors [72]. Isolation of the first human HPAIV H5N1 occurred in a child in Hong Kong in 1997, and this raised fears of an impending H5N1 pandemic [73]. Subsequently only 608 human cases had occurred up to August 2012 from 15 countries [74]. Most cases were related to contact with poultry or poultry products, and occasionally from contaminated water and human transmission appeared to be rare [74]. No human cases were reported from Western Europe or the Americas. The high case fatality rate reported to the WHO of about 59% was most likely an over estimate from unrecognized mild infections or subclinical cases. A recent prospective serological epidemiology study from Egypt, where most cases of H5N1 were reported since 2009 from backyard poultry producers, found that most seroconverters were asymptomatic or had mild disease [56]. Thus the true case fatality rate is likely very low. Although HPAIV H5N1 continue to circulate in poultry, spill over infection in humans and other mammals have remained rare. Hence so far, the HPAIV H5N1 has not mutated to allow facile transmission from poultry to humans or human to human.

3.5.3 Emergence of Avian Influenza A H7N9

Over the past several years, other avian influenza A viruses of subtypes H6, H7, H9, and H10 have crossed the species barrier and caused mainly sporadic, nonfatal cases of human infections [55, 75–79]. Some of these cases were secondary to low pathogenic strains that caused disease in persons with immunosuppression. However, a H7N7 strain caused human conjunctivitis and a case of fatal respiratory distress syndrome in the Netherlands [79]. Of major global public health concern is the emergence of a novel avian influenza A H7N9 virus in China in March 2013. Within two months of its appearance in the human population, the cumulative number of human cases in China was almost three times as high as the number caused by the H5N1 outbreak during a similar period of time [80]. By June 2013, there were 132 symptomatic cases and one asymptomatic case with 40 attributable deaths [81]. Infection with the H7N9 virus has been associated with a high incidence of severe disease, with rapidly progressive pneumonia and multiorgan failure associated with cytokine "storm" or severe dysfunction [81, 82]. In a review of 139 confirmed cases in China that occurred in the first 9 months of the outbreak, 99% were hospitalized,

90% had severe pneumonia or respiratory failure, and 63% required intensive care and 34% of the total cases died [83]. The first cases of H7N9 influenza infection appeared in eastern China around the Yangtze River delta and subsequently spread to 12 regions in China along an avian migratory pathway [81, 84]. Most cases were associated with exposure to poultry [82%], but in four family clusters, non-sustained human-to-human transmission could not be excluded [83]. At the onset of the outbreak in eastern China, there was no apparent outbreak in poultry or in wild birds. Analysis of the virus showed all gene segments were of avian origin, and the H7 isolated virus was closest to that of H7N3 virus from domestic ducks in Zhejiang, but the N9 was closest to that of the wild bird H7N9 virus in South Korea [85]. The H7N9 virus has been isolated from live poultry and the environment of poultry markets; and case-control studies confirmed the association of human infection with visits to these markets [83]. Moreover, closure of live poultry markets have resulted in the reduction of confirmed H7N9 influenza cases. The H7N9 avian influenza virus potentially poses a high risk to human populations, which are naïve to the virus, as the virus has biological properties conducive to aggressive disease in mammals. Studies have confirmed that the H7N9 virus can bind to both avian and human receptor and it can replicate efficiently in several mammalian cell lines, including human lower respiratory tract epithelial cells and type 11 pneumocytes of alveoli [86]. H7N9 avian virus replicated to higher titer in human respiratory epithelial cells and respiratory tract of ferrets compared to seasonal influenza H3N2 virus and produced greater infectivity and lethality in mice compared to other genetically related virus [87].

3.5.4 Current Status of the Emerging Pathogenic Avian Influenza Viruses

Epizootic HPAIV H5N1 continues to circulate in poultry in several countries of the world, and the virus has become endemic in Indonesia and Egypt since 2006–2008 [88]. The largest number of poultry afflicted from 2003 to May 2015 were in Vietnam, Thailand, and Egypt [88]. Fifteen countries continued to report H5N1 infection in poultry for the first 5 months of 2015.From February 2003 to March 2015, there have been 826 symptomatic [severe] cases of human influenza with HPAIV H5N1 recognized from 16 countries with 440 fatal, resulting in a case fatality rate of 53% in these clinically recognized cases. However, these diagnosed cases likely represent only a fraction of the total infected human subjects. In the first 4 months of 2015, only five confirmed cases of H5N1 influenza infection have been reported from the Western Pacific Region [89].

A unique feature of avian influenza A H7N9 human outbreak has been the absence of preceding bird epidemics with die off of poultry or wild birds. Thus, H7N9 appears to be a LPAIV with asymptomatic or mild infection in domestic and wild birds. This is attributable to the absence of a multibasic cleavage site of the HA, which is a virulence marker in birds but not in humans [85]. Also previous infection in poultry with a closely related LPAIV [H7N3] may have elicited cross-protection in birds [81]. Hence, the distribution of H7N9 in poultry and wild birds is poorly characterized and is only retrospectively recognized after human outbreaks. Thus the H7N9 avian virus potentially is a greater risk than the H5N1 to produce widespread epidemics in human populations because of its stealth. So far human infection occurrence has been limited to mainland China, including 19 cases diagnosed in Taipei [Taiwan], Hong Kong, and Kuala Lumpur in Malaysia [90]. As of February 2015, a total 571 confirmed human cases of H7N9 infection has been reported to WHO with 212 [37.1%] deaths [91]. However, the true case fatality rate appears to be much lower as only the severe cases are recognized clinically. A recent seroprevalence study from southern China in poultry workers found that 7.2% in spring and 14.9% in winter have been infected with the H7N9 influenza virus with subclinical or mild infection [92].

Overt clinical influenza infection with the avian H7N9 virus appeared in three waves starting in the winter to the spring and with the first wave in February–May 2013 [see Fig. 3.2]. Since February 2015, 20 additional confirmed human cases have been reported from China and with four deaths [http://www.who.int/csr/don/14-April-2015-avian-influenza-china/en]. The majority of reported cases have had exposure to poultry and overall the public health risk from the avian H7N9 virus has not changed. Thus, the virus has not mutated to cause efficient human-to-human transmission. Although the extent of transmission in poultry is unclear, the H7N9 virus has persisted in domestic fowls with a seasonal pattern similar to that of other avian influenza viruses, circulating at higher levels in cold weather compared to warm seasons.

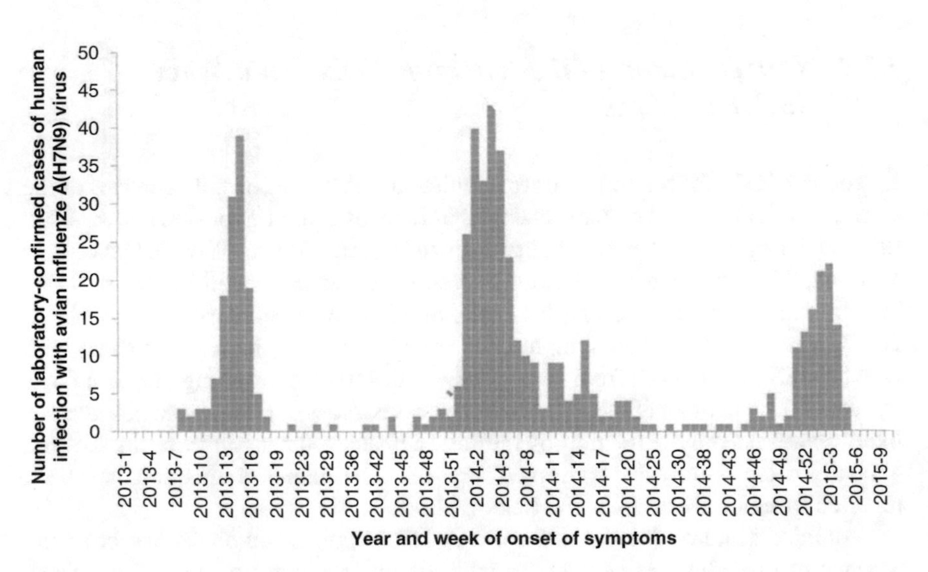

Fig. 3.2 Laboratory confirmed cases of human infection with influenza A [H7N9] virus by week of onset. World Health Organization, 23 February 2015

3.6 Strategies to Contain Zoonotic Influenza A

Measures to contain zoonotic influenza are already in place in many developed countries but are lacking or incomplete in many developing and middle-income countries. The exact sequence of events and break in control measures that led to the recent swine H1N1 pandemic is still unclear. The outbreak appeared to have originated in Mexico with the virus jumping the species barrier in local pig farms. Most developed countries have highly regulated, hygienic swine and poultry farms and backyard animal farms are not allowed. Culling of sick animals is regularly used to control pathogenic zoonotic infections. LPAIV would, however, avoid detection but hygienic infection control measures theoretically should limit cross-species transmission.

Live poultry or animal markets, which are common in Asia and other developing regions of the world, have been shown to be a primary source of avian influenza outbreaks in humans, and this cultural practice represents a major obstacle to the prevention of avian influenza and other zoonoses cross-species transmission. Disbanding these markets would be the only permanent solution, but so far countries such as China and others have not implemented any such measure. However, closure of these animal markets would not eliminate the risk for zoonotic influenza outbreaks in communities. These zoonotic viruses likely arise in farms, and the entire housing and transportation chain would be contaminated [90]. Moreover, LPAIVs would avoid detection in farms to alert farm workers and transportation staff of the risk of cross-species transmission. An example is the H7N9 outbreak, where the first positive farm detected with the virus was in Guangdong Province reported in March 2014, a year after recognition of human cases [90].

Local measures have been implemented in Hong Kong that reduced the epizootic spread of avian influenza and human outbreaks. In 1997 the H5N1 human outbreak was halted in Hong Kong after culling all poultry and restriction of importing chickens from mainland China only to farms with stringent biosecurity measures [93]. Where HPAIV outbreaks had occurred, all live poultry and poultry products from the affected province would be suspended for 21 days; and unaffected farms within 3 km of index farms would also have suspension for live poultry/products for 90 days. In live poultry markets, several control measures were instituted: segregation of poultry species to reduce the risk of genetic reassortment, regular cleaning of transport cages to limit trafficking of viruses from farms to markets and interrupt amplification of the viruses, and banning of overnight poultry storage in markets [93, 94]. Public education to avoid contact with poultry, proper hand hygiene after contact with poultry products, encouragement to purchase frozen chicken instead of live ones, and banning the possession of live poultry in the household were also implemented. These measures resulted in dramatic reduction in the isolation rate of HPAIV H9N2, and no local cases of avian H5N1virus infection has been identified in Hong Kong since 2007 [81].

3.6.1 Vaccines for Zoonotic Influenza A

It is generally considered that influenza vaccines are the most effective means of preventing or limiting the spread of influenza outbreak. However, there are several limitations to this approach, including our inability to predict the next pandemic strain of virus in order to produce sufficient vaccines in time to supply the global needs. This operational shortcoming was evident in the 2009 H1N1 influenza pandemic, as by the time 77 countries received adequate supply of vaccines [78 million doses] the outbreak was already waning [35].

Vaccines have also been used in animals since the 1990s to control highly pathogenic influenza epizootics, along with other methods. HPAIVs were first recognized as a cause of fowl plague in 1955, and since then 30 epizootics have occurred globally [95]. It is estimated that 58 billion poultry are raised each year in the world, and fowl plague may have affected only a fraction of about 250 million birds per year. Most HPAIV epizootics involved single countries, and only two have embroiled multiple countries, H5N1 since 1996 to present in 63 countries of Asia, Europe, Middle East, and Africa and H7N7 in the Netherlands, Belgium, and Germany [95]. Vaccination of poultry had been used in four epizootics to control the outbreaks because of inadequate control with traditional means. These include: Mexico in1994–1995 for H5N1 epizootic, Pakistan in 1995–2004 for H7N3 epizootics, Asia/Africa/Europe from 1996 to present to control the ongoing H5N1 epizootics, and North Korea in 2005 for a H7N7 epizootic [95]. From 2002 to 2010, over 13 billion doses of avian influenza vaccines had been used in poultry, inactivated whole virus vaccines in 95.5% and live vectored vaccines in 4.5% [96]. Most vaccines, 91.9%, were used to control HPAIVs, and only 8.1% were used for LPAIVs, H5 and H7 strains. Over 99% of vaccines used for HPAIV H5N1 were for the four enzootic countries: China, including Hong Kong [91%], Egypt [4.7%], Indonesia [2.3%], and Vietnam [1.4%] where vaccination programs have been routine and nationwide to all poultry [96]. Bangladesh and Eastern India have enzootic H5N1 HPAIV but have not used vaccination in their control programs.

Overall poultry vaccination has been found to be beneficial in reducing clinical disease and mortality in chickens and ducks and lessens the risk of human infection and economically appears to be cost-effective [95]. A recent meta-analysis on the efficacy of avian influenza vaccines [against H5N1 or H5N2 viruses] reported efficacy on four outcomes for homologous inactivated vaccines: protection against mortality 92%, morbidity 94%, reduction in respiratory virus 54%, and reduction in virus excretion from the cloaca 88% and somewhat less for inactivated heterologous vaccines [97]. Field outbreaks have occurred in vaccinating countries mainly because of inadequate vaccine coverage of susceptible fowls or only after a single dose, but vaccine failures have occurred following antigenic drift in the four main vaccinating countries [95]. Influenza vaccines for swine and poultry are primarily conventional inactivated preparations, but there are novel vaccines in the field and under development ranging from nuclei acid-based vaccines, replicon particles, subunits and virus-like particles, vectored vaccines, and live attenuated vaccines [98].

Development of human vaccines for avian influenza H5N1 and H7N9 has gained interest in recent years because of their perceived potential to cause pandemic outbreaks. However, vaccines in development against these viruses have been weakly immunogenic. Two recent studies using adjuvanted vaccines showed promise for the development of effective avian influenza vaccines in humans, but likely with multiple doses. Various doses of a monovalent inactivated surface antigen H5N1 influenza vaccine, with or without the MF59 adjuvant, were tested in 565 vaccine-naïve adults and 72 subjects who were vaccinated a year before with an older Vietnam vaccine [99]. Low-dose adjuvanted vaccine was more effective than high-dose unadjuvanted vaccine after two doses given 28 days apart in generating effective hemagglutinin inhibition titers of >1:40, but local and systemic reactions were higher. A single dose of vaccine with or without adjuvant had a boosting effect on subjects vaccinated a year before in 21–50% [99]. Another recent study assessed the immune response to a split virus inactivated monovalent avian influenza H7N9 vaccine with or without the same adjuvant [100]. The vaccine was given twice 28 days apart in various doses in 700 participants in seven groups. At the lowest dose with adjuvant seroconversion occurred in 59% of subjects, but there was no data on antibody titer after 42 days. There was no serious reaction to the vaccine, but subjects given adjuvants had greater local reaction than those without. An unexpected finding was the attenuated response in participants who received recent seasonal influenza vaccine, similar to the response in older subjects. Another approach is the administration of priming, live attenuated influenza vaccine [pLAIV] against influenza A [H7N9], followed 12 weeks after with the candidate pandemic inactivated unadjuvanted influenza vaccine [pIIV]. A study in healthy young and older [18–49 years and 50–70 years] volunteers demonstrated strong immune memory with subsequent antigenic challenge [101].

Further research is needed to develop more effective single dose avian influenza vaccines, as any multiple dose vaccination program for large-scale use in an emergency setting, such as an impending pandemic, would be an obstacle to achieve adequate protection of the global population. Furthermore, it is unknown whether or not a hemagglutinin inhibition or neutralizing antibody titer of >1:40 would be effective to prevent human infection with the avian influenza viruses [102]. A model on the efficacy of vaccination program in the setting of an emerging influenza pandemic found that timeliness of vaccine production and administration would have the greatest impact even for an effective vaccine [103]. Starting a vaccination program in the US 16 weeks before the onset of a major epidemic, with an estimated 30% clinical attack rate and production of 30 million doses per week, would result in 38% reduction in hospitalizations and deaths. Delaying the start of the vaccination program to the same week of an outbreak decreases the reductions in severe morbidity and mortality to only 18% [103]. In addition, administering only 10 million doses per week would result in lower benefit to 21% with an early program and to 6% when delayed.

3.6.2 Treatment of Zoonotic Influenza

Neuraminidase inhibitors [NAIs], oseltamivir, peramivir, and zanamivir, are the only licensed agents for treatment of circulating influenza and for zoonotic strains. There is debate about their effectiveness in uncomplicated influenza in healthy people, and randomized studies only showed a reduction of symptoms by 1 day [104–106]. However, observational case-control studies indicate that early administration [within 48 h of symptoms] of oseltamivir can reduce the morbidity and mortality of severe infections in hospitalized patients with influenza [107, 108]. A recent meta-analysis of observational studies on the benefit of NAIs during the swine H1N1 influenza pandemic concluded that early treatment was effective in reducing the mortality and severe outcomes compared to late or no treatment, odds ratios 0.35 and 0.41, respectively [108].

In a study from CDC, a spread-sheet model was used to calculate the potential benefit of NAIs in a severe H7N9 avian influenza outbreak in the US. It was estimated that the demands could be met with the current supply available. Early treatment with these antivirals could prevent 5200–248,000 deaths and 4800–504,000 hospitalizations, but there still would be large number of deaths [25,000–425,000] and hospitalization [500,000–3,700,000] [109]. Although there is no good clinical evidence on the efficacy of NAIs against the avian influenza strains H5N1 and H7N9, these viruses have been found to be susceptible in vitro. Data collected on patients with severe H5N1 avian influenza infection treated with oseltamivir showed limited efficacy with mortality rate still around 50% [110, 111]. Furthermore, during the first wave of human infection with H7N9, six patients treated with NAI had developed resistant variants, and three died [112]. A NA-R292 K mutation that confers broad-spectrum NAIs resistance after treatment was detected. Prospective data on all patients infected with H7N9 and treated with a NAI should be collected and screened for mutations to determine the frequency of antiviral resistance.

3.7 Future Directions

There are several issues and exigencies that should be addressed in order to prepare adequately for a severe avian influenza pandemic. These include: (1) measures to prevent and reduce cross-species transmission, (2) means of improving vaccine efficacy and rapid production of large supplies, (3) development of more effective novel antiviral agents, and (4) improvement in the coordinated global emergency response to an emerging influenza pandemic. In order to prevent cross-species transmission of zoonotic influenza, further research is needed to elucidate the mechanisms and mutations necessary for this to occur. The pragmatic approach taken in Hong Kong when faced by an outbreak of zoonotic influenza does not provide a permanent solution. Modern hygienic facilities for animal husbandry are needed in developing countries, and plans to fade out live animal markets and local backyard farms need to be gradually implemented.

Use of animal vaccines for epizootic and enzootic influenza viruses should be expanded to more countries and greater proportion of animals for both HPAIV and LPAIV. However, the constant antigenic shift and drift of the influenza viruses poses significant challenges to develop and mass produce new vaccines every year for both humans and animals. The ideal solution would be the development of a universal vaccine that covers all strains with development of antibodies to non-variable and highly immunogenic epitopes. Current vaccines are based on administration of purified immunodominant envelope glycoproteins [HA and NA], which are highly variable and drift under immune pressure. In addition most neutralizing antibodies are directed against the strain-specific globular head of HA [98]. Development of more broadly neutralizing and cross-reactive antibodies with novel vaccines should be feasible, against receptor-binding site on HA1 subunit and the fusion machinery of the HA2 subunit [113]. The most pressing need, however, as demonstrated by the 2009 pandemic, is to produce sufficient influenza vaccines for a novel strain in a short period of time for an impending epidemic. The traditional influenza vaccines are produced in embryonated chicken eggs cultures and require a lead time of 6 months for production of a new vaccine in sufficient quantities. The cell culture systems can produce the vaccines more rapidly, but the amount of antigens produced is less than with eggs. Several approaches will be needed to overcome these limitations such as licensing more companies to make influenza vaccines and encourage rapidly developing countries [China, India, and Brazil] to produce influenza vaccines for their own and the regional needs and more reliance on cell culture methods. Also lower doses of vaccines can be given by intradermal injection and may be just as or more effective than intramuscular administration and hence provide a greater supply of vaccines. The safety and efficacy of live attenuated avian influenza vaccines in humans should be assessed. Currently live attenuated intranasal seasonal influenza vaccines are available for children and young adults. The present evidence indicates that these vaccines are safe and more immunogenic than conventional inactivated vaccines and can produce longer-lasting antibodies. The fear that the live virus in these vaccines could mutate to become more virulent and cause influenza disease has not been found to date.

Development of new classes of antivirals is important for preparing for future influenza pandemics, especially for treatment of severe cases and in high-risk subjects. Peramivir is now available in the US for intravenous therapy, but this agent has the disadvantages of the NAI-class, questionable efficacy for avian influenza and increasing reports of drug resistance. A promising drug is a new broad-spectrum antiviral agent favipiravir [Toyama Chemical], which is approved for emergency treatment of influenza virus in Japan [114]. Potential targets for development of new agents include polymerase inhibitors [i.e., T-705] and attachment inhibitors [i.e., DAS-181] [115]. Patients with acute lung injury [ARDS] secondary to avian influenza have a high mortality, and antihuman anticomplement C5a antibody may be an effective novel treatment that should be tested in humans. A recent study in monkeys with avian H7N9 influenza virus infection demonstrated that anti-C5a significantly reduced the systemic inflammatory response and the viral load in the lungs [116]. Intravenous immunoglobulin [IVIG] contains broadly cross-reactive antibody-dependent cellular cytotoxicity against heterologous influenza strains [including the swine H1N1 strain] and has been proposed for use in critically ill patients with influenza [117]. However, for patients infected with an emerging avian influenza virus, the source of the IVIG would preferably be derived from previously exposed subjects with evidence of existing antibodies to the avian virus.

3.8 Conclusion

Zoonotic influenza A viruses, particularly pathogenic avian viruses, will continue to pose a serious global public health risk for the foreseeable future. Better understanding of mechanisms of cross-species transmission is direly needed in order to prevent these happenings. Although current data shows no impending risk of a major avian or zoonotic influenza major outbreak, continued surveillance and vigilance should be maintained. Presently the two avian viruses of public health interest [H5N1 and H7N9] have not mutated to become more easily transmissible from birds to humans or humans to humans. However, of some concern is a recent report from China of a probable nosocomial transmission of avian influenza A [H7N9]. A 57-year-old male, with a history of chronic obstructive lung disease, developed acute influenza A after sharing the same hospital ward with an index patient for 5 days. The index patient became ill 7 days after visiting a poultry market, but the secondary case had no bird contact. Both patients died and the influenza A [H7N9] isolated from both patients genome sequences were nearly identical and genetically similar to the virus isolated from the live poultry market [118]. Thus, continuous monitoring for further cases of human-to-human transmission for this avian influenza virus is critical.

References

1. Treanor JJ (2000) Influenza virus. In: Mandell GL, Bennett JC, Dolin R (eds) Principles and practice of infectious diseases, 5th edn. Churchill Livingstone, Philadelphia, pp 1823–1849
2. Ohwada K, Kitame F, Sugawara K, Nishimura H, Homma M, Nakamura K (1987) Distribution of the antibody to influenza C virus in dogs and pigs in Yamagata Prefecture, Japan. Microbiol Immunol 31:1173–1180
3. Skehel JJ, Wiley DC (2000) Receptor binding and membrane fusion in virus entry: the influenza hemagglutinin. Annu Rev Biochem 69:531–569
4. Matrosovich MN, Matrosovich TY, Gray T, Roberts NA, Klenk HD (2004) Neuramidase is important for the initiation of influenza virus infection in human airway epithelium. J Virol 78:12665–12667
5. Ina Y, Gobojori T (1994) Statistical analysis of nucleotide sequences of the hemagglutinin gene of human influenza A viruses. Proc Natl Acad Sci U S A 91:8388–8392
6. Cox NJ, Subbarao K (2000) Global epidemiology of influenza: past and present. Annu Rev Med 51:407–421
7. Stein RA (2009) Lessons from outbreaks of H1N1 influenza. Ann Intern Med 151:407–408
8. Fitch WM, Bush RM, Bender CA et al (1997) Long term trends in the evolution of H[3] HA1 human influenza type A. Proc Natl Acad Sci U S A 94:7712–7718
9. Ma W, Lager KM, Vincent AL, Jarke BH, Gramer MR, Richt RA (2009) The role of swine in the generation of novel influenza viruses. Zoonoses Publ Hlth 56:326–337
10. Belshe RB (2005) The origins of pandemic influenza lessons from the 1918 virus. N Engl J Med 353:2209–2111
11. Fang R, Jou WM, Huyebroeck D, Devos R, Fiers W (1981) Complete structure of A/duck/Ukraine/63 influenza hemagglutinin gene: animal virus as progenitor of human H3 Hong Kong 1968 influenza hemagglutinin. Cell 25:315–323

12. Medina RA, Garcia-Sastre A (2011) Influenza A viruses: new research developments. Nat Rev Microbiol 9:590–603
13. Shope RE (1931) Swine influenza. Filtration experiments and etiology. J Exp Med 54:375–385
14. Webster RG (2002) The importance of animal influenza for human disease. Vaccine 20(Suppl. 2):S16–S20
15. Vincent AL, Ma W, Larger KM, Jarke BH, Richt JA (2008) Swine influenza viruses: a North American perspective. Adv Virus Res 72:127–154
16. Chamber TM, Hinshaw VS, Kawaoka Y, Easterday BC, Webster RG (1991) Influenza viral infection of swine in the United States 1988–1989. Arch Virol 116:261–265
17. Webby RJ, Rossow K, Erickson G, Sims Y, Webster R (2004) Multiple lineages of antigenically and genetically diverse influenza A virus co-circulate in the United States swine population. Virus Res 103:67–73
18. Karasin AI, Carman SA, Olsen CW (2006) Identification of human H1N2 and human-swine reassortment H1N2 and H1N1 influenza A viruses among pigs in Ontario, Canada [2003–2005]. J Clin Microbiol 44:1123–1126
19. Rogers GN, Paulson RS, Daniels RS, Skehel JJ, Wilson IA, Wiley DC (1983) Single amino acid substitution in influenza hemagglutinin change receptor binding specificity. Nature 304:76–78
20. Subbarao EK, London W, Murphy BR (1993) A single amino acid in the PB2 gene of influenza virus is a determinant of host range. J Virol 67:1761–1764
21. Scholtissek C, Burger H, Kistner O, Shortridge KF (1985) The nucleoprotein as a possible major factor in determining host specificity of influenza H3N2 viruses. Virology 147:287–294
22. Ito T, Couceiro JN, Kelm S et al (1998) Molecular basis for the generation in pigs of influenza A viruses with pandemic potential. J Virol 77:7367–7373
23. Yu G, Zhang GH, Hua RH, Zhang Q, Liu TQ, Liao M, Tong GZ (2007) Isolation and genetic analysis of human origin H1N1 and H3N2 influenza viruses from pigs in China. Biochem Biophys Res Commun 356:91–96
24. Kundin WD, Easterday BC (1972) Hong Kong influenza infection in swine: experimental and field observations. Bull World Health Organ 47:489–491
25. Detjen BM, St. Angelo C, Katze MG, Krucy KM (1987) The three influenza virus polymerase [P] proteins not associated with viral nucleocapsids in the infected cells are in the form of a complex. J Virol 61:16–22
26. Kawaoka Y, Krauss S, Webster RG (1989) Avian-to-human transmission of the PB1 gene of influenza A viruses in the 1957 and 1968 pandemics. J Virol 63:4603–4608
27. Brown IH, Harris PA, Mc Cauley JW, Alexander DJ (1998) Multiple genetic reassortment of avian and human influenza A viruses in European pigs, resulting in emergence of an H1N2 novel genotype. J Gen Virol 79:2947–2955
28. Claas EC, Kawaoka Y, de Jong JC, Mausrel N, Webster RG (1994) Infection of children with avian-human reassortment influenza virus from pigs in Europe. Virology 204:453–457
29. Gregory V, Lim W, Cameron K et al (2001) Infection of a child in Hong Kong by an influenza A H3N2 virus closely related to viruses circulating in European pigs. J Gen Virol 82:1397–1406
30. Peiris JS, Guan Y, Markwell D, Ghose P, Webster RG, Shortridge KF (2001) Cocirculation of avian H9N2 and contemporary 'human' H3N2 influenza A viruses in pigs in southeastern China: potential for genetic reassortment? J Virol 75:9679–9686
31. Yu H, Hua RH, Zhang Q, Liu TQ, Liu HL, Li GX, Tong GZ (2008) Genetic evolution of swine influenza A [H3N2] viruses in China from 1970 to 2006. J Clin Microbiol 46:1067–1075
32. Ma W, Vincent AL, Gramer MR et al (2007) Identification of H2N3 influenza A viruses from swine in the United States. Proc Natl Acad Sci U S A 104:20949–20954
33. Smith TF, Burgert EO Jr, Dowdle WR, Noble GR, Campbell RJ, Van Scoy RE (1976) Isolation of swine influenza virus from autopsy lung tissue of man. N Engl J Med 294:708–710

34. Myers KP, Olsen CW, Gray GC (2007) Cases of swine influenza in humans: a review of the literature. Clin Infect Dis 44:1084–1088
35. Fineberg HV (2014) Pandemic preparedness and response lessons from the H1N1 influenza of 2009. N Engl J Med 370:1335–1342
36. Dominguez-Cheriot G, Lapinsky SE, Marcias AE et al (2009) Critically ill patients with 2009 influenza A [H1N1] in Mexico. JAMA 302:1880–1887
37. Simonsen L, Spreeuwenberg P, Lustig R et al (2013) Global mortality estimates for the 2009 influenza pandemic from the GLaMOR project: a modeling study. PLoS Med 10:e1001558
38. Hancock K, Veguilla V, Lu X et al (2009) Cross reactive antibody responses to the 2009 pandemic H1N1 influenza virus. N Engl J Med 361:1945–1952
39. Trauer JM, Laurie KL, Mc Donnell J, Kelso A, Markey PG (2011) Differential effects of pandemic [H1N1] 2009 on remote and indigenous groups, Northern Territory, Australia, 2009. Emerg Infect Dis 17:1615–1623
40. Howclen KJ, Brockhoff EJ, Caya FD et al (2009) An investigation into human pandemic influenza virus [H1N1] 2009 on an Alberta swine farm. Can Vet J 50:1153–1161
41. World Organization for Animal Health. World Animal Health Information Database [WAHID] Interface. Weekly disease information. [cited2010July14]. http://web.oie.int/wahis/public.php?page=weekly-reort-index&admin=0
42. Vijaykrishna D, Poon LL, Zhu HC et al (2010) Reaasortment of pandemic H1N1/2009 influenza A virus in swine. Science 328:1529
43. Moreno A, Di Trani L, Faccini S et al (2011) Novel H1N2 swine influenza reassortment strain in pigs derived from the pandemic H1N1/2009 virus. Vet Microbiol 149:472–477
44. Starick E, Lange E, Feneidouni S et al (2011) Reassortment pandemic [H1N1] 2009 influenza A virus discovered from pigs in Germany. J Gen Virol 92:1184–1188
45. Ducatez MF, Hause B, Stigger-Rosser E et al (2011) Multiple reassortment between pandemic [H1N1] 2009 and endemic influenza viruses in pigs, United States. Emerg Infect Dis 17:1624–1629
46. Garten RJ, Davis CT, Russell CA et al (2009) Antigenic and genetic characteristics of swine origin A [H1N1] influenza viruses circulating in humans. Science 325:197–201
47. Bagchi S (2015) India tackles H1N1 influenza outbreak. Lancet 387:e21
48. Epperson S, Jhung M, Richards S et al (2013) Human infections with influenza A [H3N2] variant virus in the United States, 2011–2012. Clin Infect Dis 57(S1):S4–S11
49. Morens DM, Taubenberger JK (2011) Pandemic influenza: certain uncertainties. Rev Med Virol 21:262–284
50. Capua I, Alexander DJ (2008) Ecology, epidemiology and human health implications of avian influenza viruses: why do we need to share genetic data? Zoonoses Publ Hlth 55:2–15
51. Taubenberger JK, Reid AH, Lourens RM, Wang R, Jin G, Fanning TG (2005) Characterization of the 1918 influenza virus polymerase genes. Nature 437:889–893
52. Capua I, Alexander DJ (2004) Avian influenza: recent developments. Avian Pathol 33:393–404
53. Resperant LA, Kuiken T, Osterhaus AD (2012) Influenza vaccines. From birds to humans. Hum Vaccines Immunother 8:7–16
54. Webster RG, Bean WJ, Gorman OT, Chambers TM, Kawoaka Y (1992) Evolution and ecology of influenza A viruses. Microbiol Rev 56:152–179
55. Fouchier RA, Schneeberger PM, Rozendaal FW et al (2004) Avian influenza A virus [H7N7] associated with human conjunctivitis and a fatal case of respiratory distress syndrome. Proc Natl Acad Sci U S A 101:1356–1361
56. Tweed SA, Skowronski DM, David ST et al (2004) Human illness from avian influenza H7N3, British Columbia. Emerg Infect Dis 10:2196–2199
57. Alexander DJ (2000) A review of avian influenza in different bird species. Vet Microbiol 74:3–13
58. Yee KS, Carpenter TE, Farver TB, Cardona CJ (2009) An evaluation of transmission routes for low pathogenicity avian influenza virus among chickens sold in live bird markets. Virology 394:19–27

59. Lipatov AS, Kwon YK, Pantin-Jackwood MJ, Swayne DE (2009) Pathogenesis of H5N1 influenza virus infection in mice and ferret models differs according to respiratory tract or digestive system exposure. J Infect Dis 199:717–725
60. Resperant LA, van de Bildt MW, van Amerongen G et al (2012) Marked endotheliotropism of highly pathogenic avian influenza virus H5N1 following intestinal inoculation in cats. J Virol 86:1158–1165
61. Hayden F, Croisier A (2005) Transmission of avian influenza viruses to and between humans. J Infect Dis 192:1311–1314
62. Skemons RD, Swayne DE (1990) Replication of a waterfowl origin influenza virus in the kidney and intestine of chickens. Avian Dis 34:277–284
63. Steinbauer DA (1999) Role of hemagglutinin cleavage for the pathogenicity of influenza virus. Virology 258:1–20
64. Baigent SJ, Mc Cauley JW (2003) Influenza type A in humans, mammals and birds: determinants of virus virulence, host range and interspecies transmission. BioEssays 25:657–671
65. Shinya K, Ebina M, Yamada S, Ono M, Kasai N, Kawaoka Y (2006) Avian flu: influenza virus receptors in the human airway. Nature 440:435–436
66. Van Riel D, Munster VJ, de Wit E et al (2006) H5N1 virus attachment to lower respiratory tract. Science 312:399
67. World Organization for Animal Health. 63 countries report H5N1 avian influenza in domestic poultry/wildlife 2003–2010. [cited2010Mar11]. http://www.oie.int/eng/info-ev/en-AI-factoids-2.htm
68. Liu J, Xiao H, Lei F et al (2005) Highly pathogenic H5N1 influenza virus infection in migratory birds. Science 309:1206
69. Keawcharoen J, van Riel D, van Amerongen G et al (2008) Wild ducks as long–distance vectors of highly pathogenic avian influenza [H5N1]. Emerg Infect Dis 14:600–607
70. Keawcharoen J, van den Broek J, Bouma A, Trensin T, Osterhaus AD, Heersterbeck H (2011) Wild birds and increased transmission of highly pathogenic avian influenza [H5N1] among poultry, Thailand. Emerg Infect Dis 17:1016–1022
71. World Health Organization Global Influenza Program Surveillance Network (2005) Evolution of H5N1 avian influenza virus in Asia. Emerg Infect Dis 11:1515–1521
72. Claas ECJ, Osterhaus AD, van Beck R et al (1998) Human influenza A H5N1 virus related to a highly pathogenic avian influenza virus. Lancet 351:472–477
73. World Health Organization. Cumulative number of confirmed cases of avian influenza A [H5N1] reported to WHO, 2003–2013. http://www.who.int/influenza/human_animal_interface/EN_GIP_20130604CumulativeNumberH5N1cases.pdf
74. Gomer MR, Kayed AS, Elabd MA et al (2015) Avian influenza A [H5Ni] and A [H9N2] seroprevalence and risk factors for infection among Egyptians: a prospective, controlled seroepidemiological study. J Infect Dis 211:1399–1407
75. Chang VC, Chan TF, Wen X et al (2011) Infection of immunocompromised patients by avian H9N2 influenza A virus. J Infect Dis 62:394–399
76. Ostrowsky B, Huang A, Terry W et al (2012) Low pathogenic avian influenza A [H7N2] virus infection in immunocompromised adults, New York, USA, 2003. Emerg Infect Dis 18:1128–1131
77. Arzey GG, Kirkland PD, Arzey KE et al (2012) Influenza virus A [H10N7] in chickens and poultry abattoir workers, Australia. Emerg Infect Dis 18:814–816
78. Centers for Disease Control, Republic of China [Taiwan]. Laboratory-confirmed case of human infection with avian influenza A [H6N1] virus in Taiwan recovered: Taiwan CDC urges public to take precautions to say healthy. http://www.cddc.gov.tw/english/index.asp [June 2013]
79. Wang Y (2013) The H7N9 influenza virus in China changes since SARS. N Engl J Med 368:2348–2349
80. To KKW, Chan JFW, Chen H, Li L, Yuen KY (2013) The emergence of influenza A H7N9 in human beings 16 years after influenza A H5N1: a tale of two cities. Lancet 13:809–821

81. Gao R, Cao B, Hu Y et al (2013) Human infection with a novel avian-origin influenza A [H7N9] virus. N Engl J Med 368:1888–1897
82. Li Q, Zhou L, Zhou M et al (2014) Epidemiology of human infection with avian influenza [H7N9] virus in China. N Engl J Med 370:520–532
83. Butler D (2013) Mapping the H7N9 avian influenza outbreaks. Nature 496:145–146
84. Chen Y, Liang W, Yang S et al (2013) Human infections with the emerging avian influenza A H7N9 virus from wet market poultry: clinical analysis and characterization of viral genome. Lancet 381:1916–1925
85. Zhou H, Gao R, Zhao B et al (2013) Biological features of novel avian influenza A [H7N9] virus. Nature 499:500–505
86. Belser JA, Gustin KM, Pearce MB et al (2013) Pathogenesis and transmission of avian influenza A [H7N9] virus in ferrets and mice. Nature 501:556–560
87. World Organization for Animal Health. Update on highly pathogenic avian influenza in animals [type H5 and H7]. http://www.oie.int./animal-health-in-the-world/update-on-avian-influenza/2015. Accessed 25 May 2015
88. World Health Organization. (2015) Avian influenza weekly update, No. 480. http://www.who.int/influenza/human animal interface/en/
89. Vong S (2014) Some perspectives regarding risk factors for A [H7N9] influenza virus infection in humans. Clin Infect Dis 59:795–797
90. World Health Organization (2015) WHO risk assessment of human infections with avian influenza A [H7N9] virus. http://www.who.int./influenza/human animal interface/influenzah7n9/en/
91. Wang X, Fang S, Lu X et al (2014) Seroprevalence to avian influenza A [H7N9] virus among poultry workers and the general population in Southern China: a longitudinal study. Clin Infect Dis 59:e76–e83
92. Guan Y, Chen H, Li K et al (2007) A model to control the epidemic of H5N1 at the source. BMC Infect Dis 7:132
93. Leung YH, Lau EH, Zhang LJ, Guan Y, Cowling BJ, Peiris JS (2012) Avian influenza and ban on overnight poultry storage in live poultry markets, Hong Kong. Emerg Infect Dis 18:1339–1341
94. Swayne DE (2012) Impact of vaccines and vaccination on global control of avian influenza. Avian Dis 56:818–828
95. Swayne DE, Spackman E, Pantin-Jackwood M (2014) Success factors for avian influenza vaccine use in poultry and potential impact at the wild bird-agriculture interface. EcoHealth 11:94–108
96. Hsu SM, Chen TH, Wang CH (2010) Efficacy of avian influenza vaccine in poultry: a meta-analysis. Avian Dis 54:1197–1209
97. Rahni J, Hoffman D, Harder TC, Beer M (2015) Vaccines against influenza A viruses in poultry and swine: status and future developments. Vaccine 33:2414–2424
98. Belshe RB, Frey SE, Graham IL et al (2014) Immunogenicity of avian influenza A/Anhui/01/2005 [H5N1] vaccine with MF59 adjuvant. A randomized clinical trial. JAMA 312:1420–1428
99. Mulligan MJ, Bernstein DI, Winokur P et al (2014) Serological responses to an avian influenza A/H7N9 vaccine mixed at the point of use with MF59 adjuvant. A randomized clinical trial. JAMA 312:1409–1419
100. Sobhanie M, Matsuoka Y, Jegaskanda S et al (2016) Evaluation of the safety and immunogenicity of a live attenuated influenza vaccine [pLAIV] against influenza [H7N9]. J Infect Dis 213:922–929
101. Treanor JJ (2014) Expanding options for confronting pandemic influenza. JAMA 312:1401–1402
102. Biggerstaff M, Reed C, Swerdlow DL et al (2015) Estimating the potential effects of a vaccine program against an impending influenza pandemic United States. Clin Infect Dis 60(S1):S20–S29

103. Lalezani J, Campion K, Keene O, Silagg C (2001) Zanamavir for the treatment of influenza A and B infection in high-risk patients: a pooled analysis of randomized controlled trials. Arch Intern Med 161:212–217
104. Nicholson KG, Aoki FY, Osterhaus AD, Neuramidase Inhibitor Flu treatment Investigator Group et al (2000) Efficacy and safety of oseltamivir in treatment of acute influenza: a randomized controlled trial. Lancet 283:1016–1024
105. Ag D, Arachi D, Hudgins J, Tsafnat G, Coiera E, Bourgeois FT (2014) Financial conflicts of interest and conclusion about neuramidase inhibitors for influenza. An analysis of systematic reviews. Ann Intern Med 161:513–518
106. Hsu J, Santesso N, Mustafa R et al (2012) Antivirals for treatment of influenza: a systematic review and meta-analysis of observational studies. Ann Intern Med 156:512–524
107. Muthuri SG, Myles PR, Venkutesan S, Leonardi-Bee J, Nguyen-Van-Tam JS (2013) Impact of neuramidase inhibitor treatment on outcomes of public health importance during the 2009–2010 influenza A [H1N1] pandemic: a systematic review and meta-analysis in hospitalized patients. J Infect Dis 207:553–563
108. O'Hagan JJ, Wong KK, Campbell AP et al (2015) Estimating the United States demand for influenza antivirals and the effect on severe influenza disease during a potential pandemic. Clin Infect Dis 60(S1):S30–S41
109. Update (2007) WHO-confirmed human cases of avian influenza A [H5N1] infection, 25 November 2003–24 November 2006. Wkly Epidemiol Rec 82:41–47
110. Writing Committee of the Second World Health Organization Consultation on Clinical Aspects of Human Infection with Avian Influenza A [H5N1] Virus (2008) Update on avian influenza A [H5N1] virus infection in humans. N Engl J Med 358:261–273
111. Yen H-L, Zhou J, Choy K-T et al (2014) The R292 K mutation that confers resistance to neuramidase inhibitors lead to competitive fitness loss of A/Shanghai/1/2013 [H7N9] influenza virus in ferrets. J Infect Dis 210:1900–1908
112. Ekiert DC, Wilson IA (2012) Broadly neutralizing antibodies against influenza virus and prospects for universal therapies. Curr Opin Virol 2:134–141
113. Furuta Y, Gowen BB, Takahashi K, Shiaraki K, Smee DF, Barnard DL (2013) Favipiravir [T-705], a novel viral RNA polymerase inhibitor. Antivir Res 100:446–454
114. Hayden F (2009) Developing new antiviral agents for influenza treatment: what does the future hold? Clin Infect Dis 48(Suppl. 1):S3–S13
115. Sun S, Zhao G, Liu C et al (2015) Treatment with anti-C5a antibody improves the outcome of H7N9 virus infection in African green monkeys. Clin Infect Dis 60:486–495
116. Jegaskanda S, Vandenberg K, Laurie KL et al (2014) Cross-reactive influenza-specific antibody dependent cellular cytotoxicity in intravenous immunoglobulin as a potential therapeutic against emerging influenza viruses. J Infect Dis 210:1811–1822
117. Fang CF, Ma MJ, Zhan BD et al (2015) Nosocomial transmission of avian influenza A [H7N9] virus in China: epidemiological investigation. BMJ 351:h5765. doi:10.1136/bmj.h5765
118. Tharakaraman K, Sasisekharan R (2015) Influenza surveillance: 2014–2015 H1N1 'swine'-derived influenza viruses from India. Cell Host Microbe 17:279–282

Chapter 4
Emerging Animal Coronaviruses: First SARS and Now MERS

4.1 Introduction

The severe acute respiratory syndrome [SARS] first appeared in southern China [Guangdong Province] in November 2002, as an atypical community pneumonia [1]. Within a year, the World Health Organization [WHO] reported 8096 cases with 774 deaths [9.6% fatality] in >30 countries from five continents, and the outbreak was declared a pandemic infection [1]. In the elderly and subjects with significant comorbid illness, the mortality rate was up to 50%. Investigations revealed that SARS was due to a novel coronavirus [SARS-CoV] which was circulating among wild game animals in wet markets of southern China. The palm civet, a wild feline, was considered the amplification host that transmitted the virus to humans from occupational contact and handling by consumers during the preparation for consumption as a delicacy [2, 3]. SARS-CoV mainly spread from human to human by respiratory droplets or by contact of mucosae with contaminated fomites. Spread of SARS-CoV from mainland China to Hong Kong and other countries was from interpersonal transmission in healthcare facilities, homes, workplaces, and public transports [1].

The Middle East respiratory syndrome [MERS] coronavirus [MERS-CoV] was first isolated from a patient with fatal pneumonia in September 2012, in Jeddah, Saudi Arabia [4]. From September 2012 through July 2014, WHO reported at least 834 laboratory-confirmed cases of MERS with 288 deaths [34.5% fatality]; and known cases were directly or indirectly linked to countries in the Arabian Peninsula [5]. However, an outbreak of MERS was subsequently reported from South Korea in June 2015. The illness occurred in an older man who returned from traveling to Saudi Arabia, and within a month 17 secondary cases occurred in South Korea which was connected to the index case [6]. Within 2 months the local outbreak resulted in a total of 186 confirmed cases of MERS in South Korea, except for one case exported to China. There were 36 deaths attributed to MERS-CoV infection with a mortality rate of 19.4% [6].

I.W. Fong, *Emerging Zoonoses*, Emerging Infectious Diseases of the 21st Century,
DOI 10.1007/978-3-319-50890-0_4

4.2 Virology

Coronaviruses are the largest RNA viruses and they are enveloped with positive-strand genomes of 26–32 kb, and they are distributed globally in a wide range of animals and humans [7]. The coronaviruses are classified in four genera based on phylogenetic analysis: alpha-coronaviruses, beta-coronaviruses, gamma-coronaviruses, and delta-coronaviruses [7]. Some strains of coronaviruses, HCoV-OC43 [a beta-coronavirus] and HCoV-229E [an alpha-coronavirus], are causative agents of the common cold and rarely severe respiratory disease [8, 9]. Other newly discovered human coronaviruses, HCoV-NL634 and HCoV-HKU1, are occasionally associated with severe lower respiratory tract infection in infants and immunocompromised patients [10, 11]. SARS-CoV is a lineage B beta-coronavirus and MERS-CoV is a novel beta-coronavirus of lineage C, and both these viruses appear to have crossed the species barrier from bats to humans [12, 13]. Bats are reservoirs for many mammalian coronaviruses [13–15]. Various SARS-like coronaviruses have been found in bats from China, Asia, and Europe, but none were considered as direct progenitor of SARS-CoV. Recently, however, investigators from China reported whole genome sequences of two novel bat coronaviruses from Chinese horseshoe bats closely related to SARS-CoV [16]. The receptor-binding domain of the spike protein was very similar to that of SARS-CoV, and one of the isolates [bat SL-CoV-w1V1] uses the angiotensin-converting enzyme 2 [ACE2] from humans, civets, and Chinese horseshoe bats for cell entry [16]. SARS-CoV was previously shown to use human ACE2 molecule as its entry receptor, an outstanding feature of its cross-species transmissibility [17]. Thus, the bat SL-CoV-WIVI may be the ancestor virus that precedes the evolution of SARS-CoV in humans.

The respiratory epithelial cells of humans are the main targets of SARS-CoV, but the virus can also be found in immune cells in the circulation [lymphocytes and macrophages] and in various organs [18]. MERS-CoV, on the other hand, uses human and bat dipeptidyl peptidase-4 [DPP4] as receptor for cell entry [19]. Replication and RNA protein synthesis by MERS-CoV can occur in human airway epithelial cells, lung fibroblasts, microvascular endothelium, and alveolar type II pneumocytes [20]. Hence MERS-CoV has a broader tissue tropism than the SARS-CoV. The mechanism of the SARS-CoV and MERS-CoV interspecies transmission appears to be mediated by the S protein, by mediating receptor recognition and membrane fusion, a key factor in host specificity [21].

4.3 Pathogenesis

It has been postulated that the primary mechanism of SARS is immune suppression resulting from damage to the immune cells of the spleen, lymph nodes, and lymphoid tissue with severe lymphopenia [18]. Furthermore, Gu et al. [18] estimated that the extent of immune cell damage is a better predictor of outcome than the damage to the lungs. Other investigators judge the lung injury and subsequent acute respiratory distress syndrome [ARDS] and pulmonary fibrosis as the primary events leading to adverse outcome and death. Using multiple modes of investigation, modeling with gene sets, proteomic analysis, and histopathology correlation, these investigators conclude that the urokinase and the extracellular fibrinolytic pathways are the primary mechanisms involved in lung damage and overall SARS-CoV infection pathogenesis [22]. Events driven by these pathways result in imbalance between the host coagulation and fibrinolysin pathways, ultimately leading to diffuse alveolar and acute lung damage. It is likely, however, that both mechanisms are involved in the pathogenesis of SARS.

MERS-CoV infection has been associated with severe pneumonia and multiorgan dysfunction with higher mortality rate than SARS. A recent study compared the viral replication, cytokine/chemokine response, and antigen presentation in MERS-CoV-infected human monocyte-derived macrophages [MDMs] and SARS-CoV-infected MDMs [23]. Only MERS-CoV and not SARS-CoV could replicate in human macrophages, but both viruses could not stimulate the expression of antiviral cytokines, interferon-alpha [IFN-α], and IFN-β. Comparable levels of tumor necrosis factor [TNF]-α and interleukin [IL]-6 could be induced by both viruses. However, MERS-CoV induced significantly higher levels of IL-12, IFN-¥, and several chemokines than SARS-CoV [23]. In addition, the expression of major histocompatibility complex [HLA] class 1 and costimulatory molecules were greater in MERS-CoV-infected MDMs than SARS-CoV-infected cells. The establishment of productive infection in macrophages and dendritic cells results in impaired antigen-presenting pathway, and greater aberrant induction of cytokines/chemokines by MERS-CoV could explain the higher severity of infection and greater mortality than SARS-CoV infection [23]. Furthermore, it has recently been shown that MERS-CoV has the capacity to infect T lymphocytes and induce apoptosis of these cells by activation of the extrinsic and intrinsic apoptosis pathways [24]. MERS-CoV-derived proteins inhibit IFN-α/IFN-β expression [25], resulting in lower IFN levels in the respiratory tract [26], and lower expression of type 1 IFN in fatal cases [27]. Persistent expression of proinflammatory cytokines, neutrophil activation, and chemotactic response can result in damage to surrounding uninfected lung tissues [27]. Furthermore, MERS-CoV can impair activation of the adaptive immunity through multiple mechanisms: downregulation of antigen-presenting pathways to inhibit activation of T-cells [28], infection of CD4+ and CD8+ T-cells of peripheral blood lymphoid organs [tonsils and spleen], and extensive apoptosis of T-cells can result in impaired T- and B-cell function as the helper T-cells are also affected.

The spike [S] and nucleocapsid [N] proteins are the major immunogenic components of the coronaviruses and are produced in large quantities during infection. Antibodies against the S and N proteins have diagnostic and therapeutic potentials [28–30]. The S protein appears to be the main determinant of protective immunity and cross-species transmission in SARS-CoV and other emerging animal coronaviruses [31]. In mice antibodies against S protein protect against SARS-CoV challenge but antibodies against N protein produced limited protection [32].

4.4 Transmission

The initial cases of SARS in 2003 and the Guangdong outbreak of 2004 were related to game animals contact in wet markets, or handling for consumption [3]. However, secondary cases [the majority] were from exposure to infected droplets or contaminated fomites from other infected patients. Airborne transmission of SARS-CoV was considered rare or unlikely, but may have occurred in one community outbreak from generation of negative pressure by exhaust fans with dissemination of contaminated aerosols from sewage drains [33]. Nosocomial transmission of SARS-CoV, which was a major source of the outbreaks in Hong Kong and Toronto, was enhanced with the use of nebulizers, suction, intubation, bronchoscopy, or cardiopulmonary resuscitation, from generation of high amount of infectious droplets [1]. It was estimated during the height of the SARS pandemic that each single case resulted in an average of 2–4 secondary cases, but some patients were superspreaders of the virus and could infect larger number of people [1].

MERS-CoV infection outbreak epidemiology strongly suggested zoonotic transmission through an intermediate animal host. Genomic analysis of the virus indicated that the MERS-CoV arose from a bat coronavirus [34, 35]. This is supported by the presence of a small fragment of genomic sequences identical to the MERS-CoV Essen isolate [KC875821] in an Egyptian tomb bat [*Taphozous perforatus*] found in Saudi Arabia [36]. There is increasing evidence that camels are the intermediate hosts responsible for cross-species transmission to humans. Cross-reactive antibodies to MERS-CoV have been found in dromedary camels in Oman, Canary Islands, and Egypt [37, 38]. Further studies have also confirmed the presence of MERS-CoV RNA by real-time reverse transcriptase polymerase chain reactor [RT-PCR] assay and partial genome sequencing of viral RNA in 3 of 14 nasal samples collected from 14 camels in Qatar and two subjects from the same farm [39]. In addition, a fatal case of human MERS-CoV infection was transmitted through contact with an infected camel with rhinorrhea, and the full genome sequence of the isolates from the patient and the camel was identical [40]. It has been recently reported that a high proportion of dromedaries at a slaughterhouse shed nasal MERS-CoV, with a high-risk of human exposure and potential of driving the epidemic [41].

Similar to the epidemiology of SARS, most cases of MERS occurred from human-to-human transmission from respiratory droplets, contamination of mucosae with infected fomites, and direct contact in various settings. There is no direct evidence of airborne transmission of infectious aerosols. Household contact clusters were associated with 26 index patients infected with MERS-CoV in 2013 in Saudi Arabia. Investigations for secondary transmission to 280 household contacts, using serology and RT-PCR from throat swabs, identified secondary transmission in 6 of 26 clusters [23%], but only 12 secondary infected persons for a transmission rate of only 5% [42]. However, most cases of MERS in the Arabian Peninsula resulted from transmission were associated with direct or indirect contact with healthcare facilities, from patients with unrecognized infection to other patients, visitors, and healthcare personnel [39]. In the largest single outbreak in Jeddah,

Saudi Arabia, in 2014, the majority of patients with MERS-CoV infection had contact with healthcare facility, other patients, or both [44]. There were 255 laboratory-confirmed MERS-CoV infection of which 64 subjects [25.1%] were asymptomatic and 93 of the ill patients died, with an overall mortality of 36.5%. However, many of "asymptomatic" subjects could recall symptoms consistent with a mild respiratory infection or illness. Over 90% of the symptomatic patients [excluding healthcare personnel] had contact with healthcare facility, persons with confirmed MERS, or someone with severe respiratory illness in the preceding 14 days before onset of their illness [44].

The largest outbreak of MERS outside of the Middle East occurred in South Korea in 2015, with 186 confirmed cases and 36 [19%] deaths. Healthcare facilities were the major sources of the outbreak, with four hospital clusters accounting for 82% of all the cases. Investigation in isolation wards found extensive viable MERS-CoV contamination of the air and surrounding materials in MERS units, raising concern of the adequacy of current infection control procedures [45].

The interhuman transmissibility of MERS-CoV has been estimated by using Bayesian analysis to calculate the basic reproduction number [Ro]. The estimated Ro for MERS-CoV was 0.60–0.69 [95% confidence interval (CI) 0.42–0.92], whereas prepandemic SARS-CoV Ro was 0.80, CI 0.54–1.13 [46]. When Ro is above 1.0, a pandemic potential exists; hence MERS-CoV was not considered of pandemic potential. The cycles of transmission of SARS-CoV and MERS-CoV are demonstrated in Fig. 4.1a, b.

4.5 Clinical Features

The clinical features of SARS and MERS are very similar and usually mimic influenza infection or nonspecific viral illness initially. The incubation period of SARS was estimated to be 2–14 days with most cases occurring within 10 days after exposure, and transmission from symptomatic patients usually occurred after 5 days of illness [1]. This was related to the rising viral load in nasopharyngeal secretions, which peaked around day 10 of illness. In SARS the initial symptoms were fever, chills, myalgia, malaise, and nonproductive cough, and sore throat and rhinorrhea were less frequent [1]. Clinical deterioration in those with severe illness usually occurred after several days of infection, often heralded by development of diarrhea, evidence of pneumonia, and then respiratory distress. The most common extrapulmonary manifestations of SARS were diarrhea, hepatic dysfunction, cardiac impairment, myositis, and seizures [1]. Milder disease occurred in children in both SARS and MERS, and increased morbidity and mortality were seen in the elderly, those with significant comorbid illnesses [diabetes, heart disease, etc.], and pregnancy. In severe cases with ARDS requiring mechanical ventilation, renal failure was a complication. Survivors of severe SARS could develop residual pulmonary fibrosis, muscle weakness, and depression even 6 months after the acute illness [1].

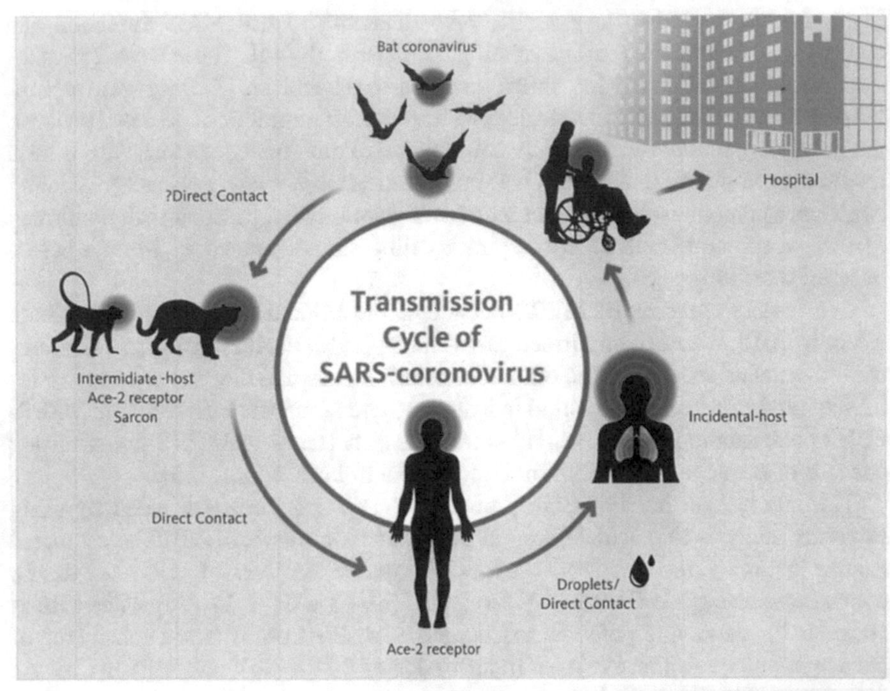

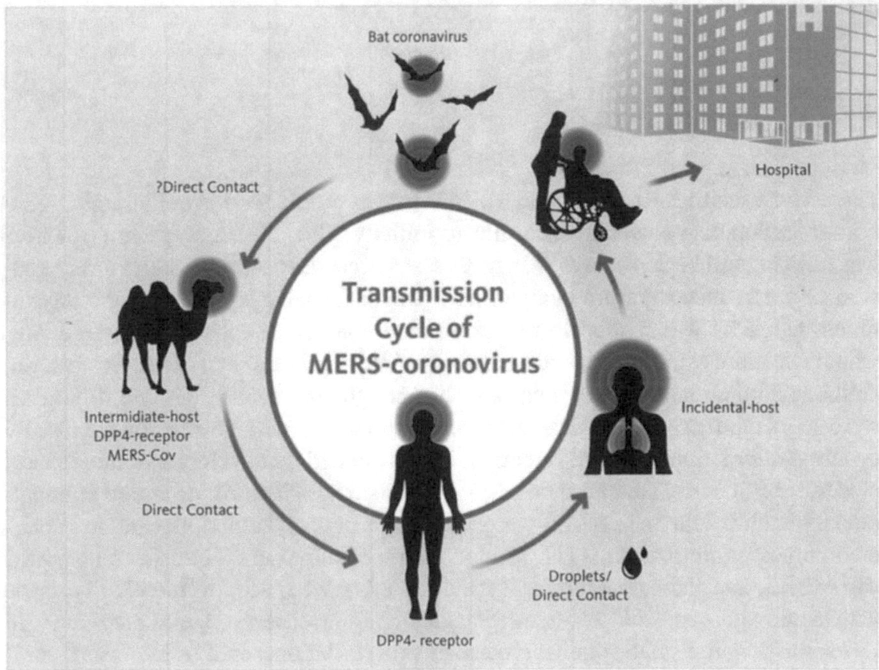

Fig. 4.1 (**a**) Transmission cycle of SARS coronavirus. (**b**) Transmission cycle of MERS-coronavirus

Infection with MERS-CoV results in disease clinically indistinguishable from SARS but a greater risk for severe pneumonia. The incubation period is also very similar, ranging from 1.9 to 14.7 days, with a median time of 5 days [43]. Both MERS-CoV and SARS-CoV were more likely to be transmitted from symptomatic patients, as the viral concentration in pharyngeal secretion for those with asymptomatic or mild infection was very low. Current data indicate that most people with MERS-CoV infection develop clinical illness. In symptomatic patients the most common presentations are fever [62–89%], cough [50–89%], shortness of breath [42–56%], chest pain, fatigue [35%], nausea and vomiting [23%], rhinorrhea and sore throat [19% each], diarrhea [15%], and muscle pain or headaches [12% each] [44, 47, 48]. Laboratory findings in MERS may include leucopenia or lymphopenia, thrombocytopenia, elevated liver enzymes, and elevated serum creatinine [47–49]. A combination of more than one system involvement was present in 88%. Pneumonia developed in about two-thirds of the patients with MERS-CoV infection [41], which could lead to respiratory failure, ARDS, acute renal failure, and death.

4.6 Diagnosis

During the SARS outbreak, the diagnosis was based on the potential exposure of the individual to the SARS-CoV at the time and presentation with acute febrile flu-like illness. The diagnosis was confirmed by real-time RT-PCR from a nasopharyngeal aspirate, with a sensitivity of 80% in the first 3 days of illness but very high specificity [50]. Antibody testing was done by various methods, and the virus could be recovered from respiratory secretions, fecal, and occasionally urine specimens by viral culture [1]. Neutralizing antibodies from acute and convalescent sera 3–4 weeks later could confirm the diagnosis, and indirect immunofluorescent antibody test was more commonly used [1]. Enzyme immunoassay [EIA] using recombinant nucleocapsid was a rapid screening test with high sensitivity after 5 days of illness, but could cross-react with other human coronaviruses and needed Western blot test for confirmation [51].

MERS diagnosis was based on clinical presentation with flu-like illness in the appropriate epidemiological setting or recent travel to the Arabian Peninsula or countries with local outbreaks. Real-time RT-PCR from a throat swab has been the primary means of confirming the diagnosis, and serological tests were used primarily for epidemiological investigations [42, 43]. Serological tests include recombinant ELISA with the use of the S1 domain of the MERS-CoV spike protein, recombinant immunofluorescence assay with the full spike protein, and plaque-reduction neutralizing assay. RT-PCR is the preferred diagnostic test for acute cases and the virus can also be cultured in Vero cells [40]. Testing for MERS-CoV should be done not only from nasopharyngeal secretions but also from lower respiratory secretions and serum, as detectable virus had been found on occasion from these sites with negative test from the upper respiratory secretions [52]. This is related to the higher viral load of MERS-CoV in lower respiratory tract than the upper respiratory tract. The WHO recommends RT-PCR targets of upE, ORF1a, or ORFb, but recent evaluation indicates targeting ORF1b is less sensitive and should not be used for diagnosis [53].

4.7 Pathology and Immunology

There is extensive data on the lung pathology and immunology in patients with SARS, but very little so far on patients with MERS. However, it is expected that the lung pathology of severe MERS cases will be similar to that of severe SARS cases. In patients who died within 10 days of onset of SARS, diffuse alveolar damage with edema was the prominent findings [1]. These findings were accompanied by other changes that can be found in ARDS, hyaline membranes, interstitial infiltrates with inflammatory cells, bronchiolar epithelial cells injury with denudation and loss of cilia, fibrin deposition, and exposure of basement membrane. After 10 days of illness, the pathological changes consist of a mixture of acute changes and reactive process: interstitial and airspace fibroblast proliferation, type II pneumocytes hyperplasia, squamous metaplasia of the bronchial epithelium, alveolar infiltration with macrophages, desquamated pneumocytes, and multinucleated cells [1]. Some cases reveal hemophagocytosis in the alveolar exudate and thrombosis of venules. Rarely [one report], histology of the lungs had revealed vasculitis of the walls of the small veins with edema, fibrinoid necrosis, and infiltration with lymphocytes, monocytes, and plasma cells [54].

Pathological changes outside the lungs in SARS consist predominantly of necrosis and atrophy of lymph nodes and the white pulp of the spleen. Even though the virus can be detected in the enterocytes of the intestines, there was no cellular damage or inflammation. Studies on patients with severe SARS soon after hospitalization had shown decreased natural killer cells, CD4+ and CD8+ T lymphocytes, and B lymphocytes [55, 56]. During the first 2 weeks of SARS, there is intense inflammatory response with elevated proinflammatory cytokines and high viral load [1]. Specific serum antibodies, detected by indirect immunofluorescence or neutralization, appeared around day 10, peak and plateau at about the second month, and persisted for more than 12 months [1].

There is limited data on the pathology and the immune response to MERS-CoV infection in humans. In one study of two patients with MERS, one died and the other recovered; there was evidence that IFN-α generation was critical to initiate a robust immune response [57]. IFN-α usually promotes antigen presentation to drive the antiviral Th1 immune response, mediated by IL-12 and IFN-¥ to clear the virus. MERS-CoV could also upregulate IL-17 expression in humans. In the patient who died, there were low IFN-α and regulatory factors that are involved in the recognition of the virus, whereas these molecules were elevated in the survivor. In addition there were elevated chemokine ligand levels, CXCL10, and IL-10, associated with low IFN-¥ expression in the non-survivor [57]. It is unclear from this report whether the difference in immune response described in the two patients with different outcome was related to differences in innate immunity or due to viral factors with overwhelming infection suppressing the immunity in the non-survivor. In vitro studies suggest that MERS-CoV induces greater dysfunction of the immune response than SARS-CoV, with downregulation of genes involved in the antigen presentation pathway [28]. To date there has not been any detailed pathological findings of severe MERS cases, probably because of religious customs in the Middle East.

4.8 Management

Clinical management of SARS and MERS were largely supportive care, depending on the severity of the illness. In healthcare settings prompt diagnosis, single room accommodation, and droplet and contact precautions were necessary to prevent nosocomial transmission. Special precautions to prevent airborne transmission were recognized to be important during the SARS outbreak for certain settings in the hospital, during tracheal suctioning, use of nebulizer, bronchoscopy, etc. [58]. Eye protection and airborne precautions should also be applied when caring for proven or suspected MERS-CoV-infected patients when performing aerosol-generating procedures [59, 60]. Antibiotics for treatment of possible community-acquired bacterial pneumonia were usually implemented until the diagnosis of coronavirus infection was confirmed. Severe cases of SARS or MERS usually require intensive care and management of fluid and electrolyte disturbances, mechanical ventilation for respiratory failure, and hemodialysis for renal failure in some cases.

No specific antiviral agents or immune modulators, such as corticosteroids, were of any significant value during the SARS outbreak. Although there was evidence of in vitro activity of IFN-α and IFN-β against SARS-CoV, the results of studies were inconsistent, and in vitro activity of the antiviral agent, ribavirin, used in combination with IFN-α was actually low [1]. However, there was a report of synergistic activity with the combination of IFN and ribavirin against SARS-CoV [61]. Pegylated IFN-α-2a was shown to be effective in reducing viral load and lung pathology in early treatment of SARS-CoV infection in a nonhuman primate model [62]. Treatment of severe cases of SARS with convalescent plasma with high neutralizing antibodies had been used with questionable value [63].

Similar to the experience during the SARS outbreak, no specific therapy had been shown to be of any definite value in severe MERS cases. In a retrospective cohort study of severely MERS-CoV-infected patients, 20 subjects were treated with ribavirin combined with IFN-α-2a compared with 24 patients treated only with supportive care. There was improved survival at 14 days but not at 28 days [64]. In another more recent report, 32 cases of MERS were treated with the combination of ribavirin and IFN-α-2a or IFN-β, with no promising results as the overall mortality rate was 69% [65]. Factors that were associated with increased mortality included age >50 years [odds ratio (OR) = 26.1], diabetes mellitus [OR = 15.74], renal failure requiring dialysis [100% mortality], and a positive plasma PCR for MERS-CoV, 90% mortality compared to those with negative plasma virus with a mortality of 44% [65].

4.8.1 Animal Experiments

Nonhuman primate models have shown varying response to SARS-CoV challenge. Cynomolgus macaques [*Macaca fascicularis*] demonstrated clinical and pathological features similar to humans infected with SARS-CoV [66]. However, other studies have not reported any overt disease in SARS-CoV-infected cynomolgus, rhesus, and African green monkeys [67, 68]. In the African green monkeys, pathology demonstrated a mild interstitial pneumonitis which resolved by 4 days [68]. A diverse range of animals had been shown to be susceptible to SARS-CoV infection including palm civets, pigs, raccoons, dogs, ferrets, and golden Syrian hamsters; but while viral replication could occur in domestic cats and BALB/c mice, they remain asymptomatic [1].

Several animal species have been experimentally infected with MERS-CoV, rhesus macaques, cynomolgus macaques, marmosets, ferrets, mice, Syrian hamsters, rabbits, and dromedary camels [69]. The outcome and development of lower respiratory tract disease were quite variable in these models. Infection of rhesus macaques resulted in transient clinical signs such as increased body temperature, increased respiratory rate, and cough [70]. Localized pulmonary infiltration and interstitial markings were visible on radiographic imaging. Histopathology after 3 days post-inoculation revealed mild-moderate interstitial pneumonia, with little inflammation in the septa but thickening with edema and fibrin; intra-alveolar infiltration with macrophages, neutrophils, multinucleated giant cells, fibrin, and sloughed epithelial cells; and perivascular inflammatory infiltrate in the interstitium [70]. At day 3 the MERS-CoV could be detected in the lungs by RT-PCR but not in extrapulmonary organs, oropharyngeal and rectal swabs. At day 6 post-infection, there was type II pneumocyte hyperplasia with alveolar edema, fibrin deposition, and hyaline membrane [71]. The viral RNA and antigen could be detected in type I and II pneumocytes and alveolar macrophages. Increased levels of proinflammatory cytokines and chemokines, such as IL-6, CXCL1, and matrix metalloproteinase, were found in serum [71].

Marmosets infected with MERS-CoV developed more severe disease, with clinical signs of respiratory distress, progressive interstitial pulmonary infiltrate visible on imaging but with resolution by day 13 [72]. In other animal species including camels, the virus caused mild or no clinical disease. In dromedary camels nasal discharge with nasal excretion of the virus can be present for 2–14 days [69, 73]. Commonly used laboratory animals such as mice, Syrian hamsters, and ferrets are not susceptible to MERS-CoV infection because of differences in the receptor dipeptidyl peptidase 4 [74]. However, transgenic mice with expression of human DPP4 had been developed that demonstrated severe and lethal respiratory disease with MERS-CoV infection [74]. Thus transgenic mice with humanized DPP4 receptor are a suitable animal model to study new therapeutics and vaccines for MERS-CoV infection.

4.9 Experimental Antivirals and Vaccines

Several agents were found to have in vitro antiviral activity against SARS-CoV including glycyrrhizin, baicalin, reserpine, niclosamide, chloroquine, and nelfinavir [1], but were never tested in a suitable animal model or developed further probably because of cessation of the SARS outbreak and no further cases. Investigation in animal model also demonstrated that it was feasible to develop a beneficial vaccine for SARS. A protective antibody response could be generated by targeting the viral spike [S] antigen. Mucosal immunization of the African green monkey with a recombinant attenuated parainfluenza-SARS-CoV spike protein chimeric virus resulted in significant neutralizing antibodies to protect against virus replication in the upper and lower respiratory tract after SARS-CoV challenge [67]. Several other methods to deliver the S protein or nucleoprotein were investigated: adenoviral vector in rhesus macaques, inactivated whole virus vaccine in mice, S protein fragments in mice and rabbits, DNA vaccination with nucleoprotein in mice, and plasmid DNA vaccine carrying S protein encoded by human codons in a mouse model [1]. None of these studies used animal models with clinical pneumonia or showed protection against clinical disease. Further development of these vaccines was not pursued, but the studies provided some evidence of proof of concept.

Intense investigations have been implemented in various research centers to identify new therapeutic agents to combat MERS-CoV since the recognition of the outbreak in the Middle East. However, development of new drugs and vaccines take many years to become readily available, usually 10–12 years. A pragmatic approach to meet current needs or demand for the near future is to assess drugs or compounds already developed. In one such study, the in vitro activity of IFN products, ribavirin, and mycophenolic acid against MERS-CoV were assessed [75]. Of all the IFNs tested, IFN-β showed the greatest activity, 41-fold more potent than IFN-α-2b. Ribavirin did not inhibit the virus at concentrations achieved by doses used in humans. Mycophenolic acid showed marked inhibition of MERS-CoV [75]. However, mycophenolate mofetil has immunosuppressive properties and is used in organ transplant patients and could lead to superinfection and adverse outcome. In another study by the same group of investigators, 290 developed pharmaceutical compounds were screened for in vitro activity against the MERS-CoV [76]. A total of 27 agents showed antiviral activity against both SARS-CoV and MERS-CoV, from 13 different classes of pharmaceuticals, including inhibitors of dopamine receptors used as antipsychotics and inhibitors of estrogen receptors used for cancer treatment [76].

Probably the most clinically relevant study on repurposed drugs for therapeutics in severe MERS-CoV infection was just recently published. Three commercially available drugs with potent in vitro activity against MERS-CoV were assessed in the common marmosets with severe disease resembling MERS in humans. The lopinavir/ritonavir [a protease inhibitor combination used for treating human immunodeficiency virus (HIV)] and IFN-β-1b-treated animals demonstrated significantly better outcome than untreated animals, with improved clinical, radiological, and pathological findings, and lower mean viral load in lungs and other tissues [77]. Animals treated with mycophenolate mofetil, in contrast, developed severe and fatal disease with higher mean viral loads than the untreated animals. Hence, clinical trials or pilot assessment in patients with severe MERS warrant trial of lopinavir/ritonavir and IFN-β-1b in combination or alone.

A novel approach for treatment and prevention of severe MERS is the use of neutralizing monoclonal antibodies [MAbs]. One such MAb, designated MERSMab1, potently blocks MERS-CoV entry into human cells [78]. MERSMab1 specifically binds to the receptor-binding domain of the MERS-CoV S protein, to block the binding to the cellular receptor DPP4. Thus, development of a humanized monoclonal antibody could be used therapeutically and prophylactically in healthcare workers and family members exposed to a patients with MERS-CoV infection [78]. Further development in this area included the isolation of a potent MERS-CoV neutralizing antibody from memory B lymphocytes of an infected subject [79]. The antibody, labeled LCA60, interfered with the binding to the cellular receptor CD26 [DPP4] and also could be used for treatment or prophylaxis. This is particularly relevant as, during the most recent MERS outbreak in South Korea, secondary and tertiary cases were largely from transmission to non-healthcare workers [80]. For more urgent need in future severe cases of MERS, it is reasonable to administer convalescent sera from previously infected and recovered subjects with MERS. It has been shown in experimental animals that MERS-immune sera from infected camel augment MERS-CoV clearance and reduced the pathological changes in the infected lungs [81].

Although it is feasible to develop an effective vaccine for the MERS-CoV, as supported by recent experiments with subunit or full-length MERS-CoV protein/antigen [82, 83], it appears that this is unlikely to occur. There are too few cases of MERS and the virus has so far not mutated to become more easily transmissible; thus development of a vaccine would not be a commercially viable enterprise. Table 4.1 summarizes the comparative features of SARS and MERS.

4.9.1 Future Direction

Although the future is unpredictable, it seems more likely that the MERS-CoV, unlike SARS-CoV, may continue to cause sporadic human infections or local outbreaks, as the virus appears to be entrenched or endemic in dromedary camels of the Middle East. It also has the potential to spread and maintain reservoirs in other animal in the region that carry similar DPP4 receptors such as horses, goats, sheep, and cows [84, 85].

Table 4.1 Comparative features of SARS and MERS

Features	SARS	MERS
Etiology	Zoonotic coronavirus [SARS-CoV]	Zoonotic coronavirus [MERS-CoV]
Source	Bats	Bats
Transmitting host	Palm civet feline	Dromedary camels
Country of origin	Southern China	Saudi Arabia
Human to human	Droplets/direct contact	Droplets/direct contact
Transmissibility	High/pandemic potential	Medium/non-pandemic
Incubation	2–14 days/median 5 days	2–14 days/median 5–7 days
Clinical aspects	Flu-like illness/severe pneumonia	Flu-like illness/severe pneumonia
Diagnosis	Real-time RT-PCR/viral culture	Real-time RT-PCR/viral culture
Management	Supportive care/ventilation	Supportive care/ventilation
Prevention	Contact/droplet isolation	Contact/droplet isolation
Mortality rate	Overall 10%	19% to 41%
Future recurrence	Unlikely	Probably likely

It is important, however, to prepare not just for future MERS outbreak but for other zoonotic coronaviruses that may "jump" the species barrier to produce another novel unexpected zoonosis epidemic. In order to prevent future coronavirus zoonosis emerging, we need more basic research to fully understand the mechanisms of cross-species virus transmission. Current investigations indicate that the surface S protein of the coronavirus and the host proteases that cleave the protein before membrane fusion are key factors for interspecies transmission [86]. It is believed that two mutations may have allowed the bat coronavirus HJKU4 to enter human cells, enabling the S protein to be activated by human proteases [87]. Could there be environmental or extrinsic factors that facilitate key mutations to enable cross-species transmission from bats to camels to humans, and are these modifiable? These are areas for future research.

Rather than developing specific agents to treat MERS-CoV or vaccines for prevention, which may be after the fact, it would be more prudent to develop new treatment and prevention that could be effective against all zoonotic coronaviruses that may emerge in the future. All coronaviruses require proteolytic activity of nsp 5 protease [3CL-pro] during replication, and this has been identified as a common target for development of a general anti-coronavirus agent [88]. Development of a universal antiviral agent for animal coronaviruses may be feasible, but could be difficult, as screening of a peptidomimetic library identified 43 compounds with good to excellent inhibitory potency against a bat coronavirus [HKU4-CoV] [89]. Another target for multiple coronaviruses is the coronavirus helicase [nsp 13], which is also important in viral replication. A replication inhibitor of the viral helicases of SARS-CoV, mouse hepatitis virus, and MERS-CoV, SSYA10-001, may be a suitable candidate as a broad spectrum coronavirus inhibitor [90].

Designing a universal vaccine for current and future zoonotic coronaviruses may be a very difficult undertaking. Neutralizing antibodies against the spike glycoprotein were shown to be strain specific with very little cross-reactivity within or across subgroups [91]. In addition, the nucleocapsid proteins do not share cross-reactive epitopes across subgroups of coronaviruses. It has been proposed that vaccine designed for emerging animal coronaviruses should include chimeric spike proteins containing neutralizing epitopes from multiple strains across subgroups [91]. Vaccine manufacturing companies would likely not be enticed on such seemingly nonprofitable enterprise, but scientist should still pursue such a venture, even for proof of concept with animal model experiments, as the need for a universal coronavirus vaccine may arise sometime in the future.

4.9.2 Conclusion

MERS-CoV is the most lethal of the six known human coronaviruses and produces a higher mortality than SARS-CoV. Moreover, it is likely to continue to afflict humans for the foreseeable future unlike SARS-CoV which has not reappeared since its first appearance over a decade ago. MERS-CoV uses various methods to evade the host innate antiviral immunity, which may explain its high pathogenic capability. Recent advances of our understanding of the immunopathogenesis of MERS may lead to more effective therapy. Although the overall mortality of clinical recognizable cases of MERS globally is about 41%, treatment in South Korea reduced the death rate to 19% [45].

References

1. Yuen KY, Wong SSY, Peiris JSM (2007) The severe acute respiratory syndrome. In: Fong IW, Alibek K (eds) New and evolving infection of the 21st century. Springer, New York, pp 163–193
2. Guan Y, Zheng BJ, He YQ et al (2003) Isolation and characterization of viruses related to SARS coronavirus from animals in southern China. Science 302:276–278
3. Zhong NS, Zhong BJ, Li YM et al (2003) Epidemiology and cause of severe acute respiratory syndrome [SARS] in Guangdong, People's Republic of China, in February, 2003. Lancet 362:1353–1358
4. Zaki AM, van Boheemen S, Besteboer TM, Osterhaus AD, Fouchier RA (2012) Isolation of a novel coronavirus from a man with pneumonia in Saudi Arabia. N Engl J Med 367:1814–1820
5. Cotten M, Watson SJ, Zumla AI et al (2014) Spread, circulation and evolution of the Middle East syndrome coronavirus. MBio 5(1). doi:10.1128/mBio.01062-13
6. World Health Organization. Middle East respiratory syndrome coronavirus [MERS-CoV] in Republic of Korea as of 22 July, 2015. http://www.wpro.who.int/outbreaks emergencies/wpro coronavirus/en/.
7. Woo PC, Lau SK, Huang Y, Yuen KY (2009) Coronavirus diversity, phylogeny and interspecies jumping. Exp Biol Med 234:1117–1127
8. Birch CJ, Clothier HJ, Secull A et al (2005) Human coronavirus OC43 causes influenza-like illness in residents and staff of aged-care facilities in Melbourne, Australia. Epidemiol Infect 133:275–277

9. El-Sahly HM, Atmar RL, Glezen WP, Greenberg SB (2000) Spectrum of clinical illness in hospitalized patients with "common cold" virus infections. Clin Infect Dis 31:96–100
10. Van der Hook L, Pyrc K, Jebbink MF et al (2004) Identification of a new coronavirus. Nat Med 10:368–373
11. Woo PC, Lau SK, Chu CM et al (2005) Characterization and complete genome sequence of a novel coronavirus, coronavirus HKU1 from patients with pneumonia. J Virol 79:884–895
12. Chan JF, To KK, Tse H, Jin DY, Yuen KY (2013) Interspecies transmission and emergence of novel viruses: lessons from bats and birds. Trends Microbiol 21:544–555
13. Lau SK, Li KS, Tsang AK et al (2013) Genetic characterization of betacoronavirus lineage C virus in bats reveals marked divergence in the spiked protein pipistrellus bat coronavirus HKU5 in Japanese pipistrellae: implications for the origin of the novel Middle East respiratory syndrome coronavirus. J Virol 87:8638–8650
14. Huyah J, Li S, Yount B et al (2012) Evidence supporting a zoonotic origin of human coronavirus strain NL63. J Virol 86:12816–12825
15. Glozu-Rausch F, Ipsen A, Seebens A et al (2008) Detection and prevalence of group 1 coronaviruses in bats, northern Germany. Emerg Infect Dis 14:626–631
16. Ge XY, Yang XL, Chmura AA et al (2013) Isolation and characterization of a bat SARS-like coronavirus that uses ACE 2 receptor. Nature. doi:10.1038/nature12711
17. Li W, Moore MJ, Vasilieva N et al (2003) Angiotensin-converting enzyme 2 is a functional receptor for the coronavirus. Nature 426:450–454
18. Gu J, Gong E, Zhong B et al (2005) Multiple organ infection and the pathogenesis of SARS. J Exp Med 202:415–424
19. Raj VS, Mou H, Smits SL et al (2013) Dipeptidyl peptidase 4 is a functional receptor for the emerging human coronavirus-EMC. Nature 495:251–254
20. Scobey T, Yount BL, Sims AC et al (2013) Reverse genetics with a full-length infectious cDNA of the Middle East respiratory syndrome coronavirus. Proc Natl Acad Sci USA 110:16157–16162
21. Lu G, Wang Q, Gu OB (2015) Bat-to-human: spike features determining 'host jump' of coronaviruses, SARSCoV, MERS CoV, and beyond. Trends Microbiol. doi:10.1016/j.tim.2015
22. Gralinski LE, Bankhead A III, Jeng S et al (2013) Mechanisms of severe acute respiratory syndrome coronavirus-induced acute lung injury. MBio 4:e00271–e00213
23. Zhou J, Chu H, Li C et al (2014) Active replication of Middle East respiratory syndrome coronavirus and aberrant induction of inflammatory cytokines and chemokines in human macrophages: implication for pathogenesis. J Infect Dis 209:1331–1342
24. Chu H, Zhou J, Wong BH et al (2015) Middle East respiratory syndrome coronavirus efficiently infects human primary T lymphocytes and activates both the extrinsic and intrinsic apoptosis pathways. J Infect Dis 213(6):904–914
25. Yang Y, Zhang L, Geng H et al (2013) The structural and accessory proteins M, ORF 4a, ORF 4b, and ORF 5 of Middle East respiratory syndrome coronavirus [MERS-CoV] are potent interon antagonists. Protein Cell 4:951–961
26. Zielecki F, Weber M, Eickmann M et al (2013) Human cell tropism and innate immune system interaction of human respiratory coronavirus EMC compared to those of severe acute respiratory syndrome coronavirus. J Virol 87:5300–5304
27. Faure E, Poissy J, Goffard A et al (2014) Distinct immune response in two MERS-CoV infected patients: can we go from bench to bedside? PLoS One 9:e88716
28. Josset L, Menachery VD, Gralinski LE et al (2013) Cell host response to infection with novel human coronavirus EMC predicts potential antivirals and important difference with SARS coronavirus. MBio 4:e00165–e00113
29. Rockx B, Corti D, Donaldson E et al (2008) Structural basis for potential cross-neutralizing human monoclonal antibody protection against lethal human and zoonotic severe acute respiratory syndrome coronavirus challenge. J Virol 82:3220–3235
30. Chan RW, Chan MC, Agnihothram S et al (2013) Tropism and innate immune responses of novel betacoronavirus lineage C virus in human ex vivo respiratory organ cultures. J Virol 87:6604–6614

31. Bolles M, Donaldson E, Baric R (2011) SARS-CoV and emergent coronaviruses: viral determinants of interspecies transmission. Curr Opin Virol 1:624–634
32. Deming D, Sheehan T, Heise M et al (2006) Vaccine efficacy in senescent mice challenged with recombinant SARS-CoV bearing epidemic and zoonotic spike variants. PLoS Med 3:e525
33. Yu IT, Li Y, Wong TW, Tam W, Chan AT, Lee JH, Leung DY, Ho T (2004) Evidence of airborne transmission of the severe acute respiratory syndrome virus. N Engl J Med 350:1731–1739
34. van Boheemen S, de Graaf M, Lauber C et al (2012) Genomic characterization of a newly discovered coronavirus associated with acute respiratory syndrome in humans. MBio 3:e0043–e00412
35. Ithete NL, Stoffberg S, Corman VM et al (2013) Close relative of human Middle East respiratory syndrome coronavirus in bat, South Africa. Emerg Infect Dis 19:1819–1823
36. Memish ZA, Mishra N, Olival KJ et al (2013) Middle East respiratory syndrome coronavirus in bats, Saudi Arabia. Emerg Infect Dis 19:1819–1823
37. Perera RA, Wanmg P, Goman MR et al (2013) Seroepidemiology for MERS coronavirus using microneutralization and pseudo-particle virus neutralization assays reveal a high prevalence of antibody in dromedary camels in Egypt, June 2013. Euro Surveill 18:20574
38. Reusken CB, Haagmans BL, Miller MA et al (2013) Middle East respiratory syndrome coronavirus neutralizing antibodies in dromedary camels: a comparative serological study. Lancet Infect Dis 13:859–866
39. Haagmans BL, Al Dhahiry SH, Reusken CB et al (2014) Middle East respiratory syndrome coronavirus in dromedary camels: an outbreak investigation. Lancet Infect Dis 14:140–145
40. Azhar EI, El-Kafrawy SA, Faarj SA, Hassan AM, Al-Saeed MS, Hashem AM, Madani TA (2014) Evidence for camel-to-human transmission of MERS coronavirus. N Engl J Med 370:2499–2505
41. Farag EA, Reusken CB, Haagmans BL et al (2015) High proportion of MERS-CoV shedding dromedaries at slaughterhouse with potential epidemiological link to human cases, Qatar 2014. Infect Ecol Epidemiol 5:28305
42. Drosten C, Meyers B, Muller MA et al (2014) Transmission of MERS-coronavirus in household contacts. N Engl J Med 37:828–835
43. Assiri A, Mc Greer A, Perl TM et al (2013) Hospital outbreak of Middle East respiratory syndrome coronavirus. N Engl J Med 369:407–416
44. Oboho I, Tomczyk SM, Al-Asmari AM et al (2015) 2014 MERS-CoV outbreak in Jeddah a link to healthcare facilities. N Engl J Med 372:846–854
45. Kim S-H, Chang SY, Sung M et al (2016) Extensive viable Middle East respiratory syndrome [MERS] coronavirus contamination in air and surrounding environment in MERS isolation wards. Clin Infect Dis 63:363–369
46. Breban R, Riou J, Fontanet A (2013) Interhuman transmissibility of Middle East respiratory syndrome coronavirus: estimation of pandemic risk. Lancet 382:694–699
47. Al-Abdallat MM, Payne DC, Alqasrawi S et al (2014) Hospital-associated outbreak of Middle East respiratory syndrome coronavirus: a serologic, epidemiologic, and clinical description. Clin Infect Dis 589:225–233
48. Assiri A, Al-Tawfiq JA, Al-Rabeeah AA et al (2013) Epidemiogical, demographic, and clinical characteristics of 47 cases of Middle East respiratory syndrome coronavirus disease from Saudi Arabia: a descriptive study. Lancet Infect Dis 13:752–761
49. Saad M, Omrani AS, Baig K et al (2014) Clinical aspects and outcomes of 70 patients with Middle East respiratory syndrome coronavirus infection: a single-center experience in Saudi Arabia. Int J Infect Dis 29:301–306
50. Poon LL, Lau SK, Wong BH et al (2003) Early diagnosis of SARS coronavirus infection by real time RT-PCR. J Clin Virol 28:233–238
51. Woo PC, Lau SK, Wong BH et al (2004) False positive results in a recombinant severe acute respiratory syndrome associated coronavirus [SARS-CoV] nucleocapsid enzyme-linked immunosorbent assay due to HCOV-OC43 and HCOV-229E rectified by Western blotting with recombinant SARS-CoV spike polypeptide. J Clin Microbiol 42:5885–5888

52. Memish ZA, Al-Tawfiq JA, Mckhdoom HQ et al (2014) Respiratory tract samples, viral load, and genome fraction yield in patients with MERS. J Infect Dis 210:1590–1594
53. Pass D, Patel DE, Reusken C et al (2015) First international external quality assessment of molecular diagnosis for MERS-CoV. J Clin Virol 69:81–85
54. Ding Y, Wang H, Shen H et al (2003) The clinical pathology of severe acute respiratory syndrome [SARS]: a report from China. J Pathol 200:282–289
55. Cui W, Fan Y, Wu W, Zhang F, Wang JY, Ni AP (2003) Expression of lymphocytes and lymphocyte subset in patients with severe acute respiratory syndrome. Clin Infect Dis 37:857–859
56. Li T, Qiu Z, Zhang L et al (2004) Significant changes of peripheral T lymphocytes subsets in patients with severe acute respiratory syndrome. J Infect Dis 189:648–651
57. Faire E, Poissy J, Goffard A et al (2014) Direct immune response in two MERS-CoV-infected patients; can we go from bench to bedside? PLoS One. doi:10.1371/journal.pone.0088716
58. Tomlinson B, Cockram C (2003) SARS experience at Prince of Wales Hospital, Hong Kong. Lancet 361:1486–1487
59. Seto WH, Conly JM, Pessoa-Silva CC, Malik M, Eremin S (2013) Infection prevention and control measures for acute respiratory infections in healthcare settings: an update. East Mediterr Health J 19(suppl. 1):S39–S47
60. Center for Disease Control and Prevention (2014) Interim infection prevention and control recommendation for hospitalized patients with Middle East respiratory syndrome coronavirus [MERS-CoV]. http://www.cdc.gov/coronavirus/meis/infection-prevention-control.htm#infection-prevention.
61. Cinatl J Jr, Morgenstern B, Bauer G, Chandra P, Rabenau H, Doerr HW (2003) Treatment of SARS with human interferons. Lancet 362:293–294
62. Haagmans BL, Kuiken T, Martina BE et al (2004) Pegylated interferon-alpha protects type 1 pneumocytes against SARS coronavirus in macaques. Nat Med 10:290–293
63. Soo YO, Cheng Y, Wong R et al (2004) Retrospective comparison of convalescent plasma with continuing high-dose methylprednisolone treatment in SARS patients. Clin Microbiol Infect 10:676–678
64. Omrani AS, Saad MM, Baig K et al (2014) Ribavirin and interferon alpha-2a for severe Middle East respiratory syndrome coronavirus infection: a retrospective study. Lancet Infect Dis 14:1090–1095
65. Shalhoub S, Farahat F, Al-Jiffri A, Simhairi R, Shamma O, Siddiqi N, Mushtaq A (2015) INFα-2a or IFN-β-1a in combination with ribavirin to treat Middle East respiratory syndrome coronavirus pneumonia: a retrospective study. Antimicrob Chermother 70(7):2129–2132
66. Kuiken T, Fouchier RA, Schutten M et al (2003) Newly discovered coronavirus as the primary cause of severe acute respiratory syndrome. Lancet 362:263–270
67. Bukreyer A, Lamirande EW, Buchholz UJ et al (2004) Mucosal immunization of African green monkey [*Cercopithecus aethiops*] with an attenuated parainfluenza virus expressing SARS coronavirus spike proteins for the prevention of SARS. Lancet 363:2122–2127
68. Mc Auliffe J, Vogel L, Roberts A et al (2004) Replication of SARS coronavirus administered into the respiratory tract of African green, rhesus and cynomolgus monkeys. Virology 330:8–15
69. van den Brand JMA, Smits SL, Haagmans BL (2013) Pathogenesis of the Middle East respiratory syndrome coronavirus. J Pathol 235:175–184
70. Yao Y, Bao LO, Deng W et al (2014) An animal model of MERS produced by infection of rhesus macaques with MERS coronavirus. J Infect Dis 209:236–242
71. de Wit E, Ramussen AL, Falzarano D et al (2013) Middle East respiratory syndrome coronavirus [MERS-CoV] causes transient lower respiratory infection in rhesus macaques. Proc Natl Acad Sci USA 110:16598–16603
72. Falzarano D, de Wit E, Feldman F et al (2014) Infection with MERS-CoV causes lethal pneumonia in the common marmosets. PLoS Pathog 10:e1004431
73. Alagaili AN, Briese T, Mishra N et al (2014) Middle East respiratory syndrome coronavirus infection in dromedary camels in Saudi Arabia. MBio 5:e01002–e01014

74. van Doremalen N, Munster VJ (2015) Animal models of Middle East respiratory syndrome coronavirus. Antiviral Res 122:28–38
75. Hartr BJ, Dyall J, Postnikova E et al (2014) Interferon-β and mycophenolic acid are potent inhibitors of Middle East respiratory syndrome coronavirus in cell-based assays. J Gen Virol 95:571–577
76. Dyall J, Coleman CM, Hart BJ et al (2014) Repurposing of clinically developed drugs for treatment of Middle East respiratory syndrome coronavirus infection. Antimicrob Agents Chemother 58:4885–4893
77. Chan JF, Yao Y, Yeung M et al (2015) Treatment with lopinavir/ritonavir and interferon-β1b improves outcome of MERS-CoV infection in a nonhuman primate model of marmoset. J Infect Dis 212(12):1904–1913
78. Du L, Zhao G, Yang Y et al (2014) A conformation-dependent neutralizing monoclonal antibody specifically targeting receptor-binding-domain in Middle East respiratory syndrome coronavirus spike protein. J Virol 88:7045–7053
79. Corti D, Zhao G, Pedotti M et al (2015) Prophylactic and postexposure efficacy of a potent monoclonal antibody against MERS coronavirus. Proc Natl Acad Sci USA 112(33):10473–10478
80. Lee SS, Wong NS Probable transmission chains of MERS-CoV and the multiple generations of secondary infections in South Korea. Int J Infect Dis. doi:10.1016/j.ijid.29015.07.014
81. Zhao J, Perera RA, Kayali G, Meyerholz D, Perlman S, Peiris M (2015) Passive immunotherapy with dromedary immune serum in an experimental animal model of Middle East respiratory syndrome coronavirus infection. J Virol 89:6117–6120
82. Ma C, Wang L, Tao X et al (2014) Searching for an ideal vaccine candidate among different MERS coronavirus receptor-binding fragments the importance of immunofocusing in subunit vaccine design. Vaccine 32:6170–6176
83. Kim E, Okada K, Kenniston T et al (2014) Immunogenicity of an adenoviral-based Middle East respiratory syndrome coronavirus vaccine in BALB/c mice. Vaccine 32:5975–5982
84. Barlan A, Zhao J, Sankar MK, Li K, Mc Cray PB Jr, Perlman S, Gallagher T (2014) Receptor variation and susceptibility of Middle East respiratory syndrome coronavirus. J Virol 88:4953–4961
85. van Doremelan N, Miazgowicz KL, Milne-Price S et al (2014) Host species restriction of Middle East syndrome coronavirus through its receptor, dipeptidyl peptidase 4. J Virol 88:9920–9932
86. Lu G, Wang Q, Gao GF (2015) Bat-to-human: spike features determining 'host jump' of coronaviruses, SARS-CoV, MERS-CoV, and beyond. Trends Microbiol. doi:10.1016/j.tim.2015.06.003
87. Yang Y, Liu C, Du L, Jiang S, Shi Z, Baric R, Li F (2015) Two mutations were critical for bat-to-human transmission of MERS coronavirus. J Virol 89(17):9119–9123
88. St. John SE, Tomar S, Stauffer SR, Mesecar AD (2015) Targeting zoonotic viruses: structure-based inhibition of the 3C-like protease from bat coronavirus HKU4—the likely reservoir host to the human coronavirus that cause Middle East respiratory syndrome [MERS]. Bioorg Med Chem. doi:10.1016/j.bmc.2015.06.039
89. Tomar S, Johnston ML, St. John SE et al (2015) Ligand-induced dimerization of MERS coronavirus NSP 5 protease [3CL pro]: implication for NSP 5 regulation and the development of antivirals. J Biol Chem 290:19403–19422
90. Adedeji AO, Singh K, Kassim A et al (2014) Evaluation of SSYA10-001 as a replication inhibitor of severe acute respiratory syndrome, mouse hepatitis, and Middle East respiratory syndrome coronaviruses. Antimicrob Agents Chemother 58:4894–4898
91. Agnihothram S, Gopal R, Yount BL et al (2014) Evaluation of serologic and antigenic relationships between Middle Eastern respiratory syndrome coronavirus and other coronaviruses to develop vaccine platforms for the rapid response to emerging coronaviruses. J Infect Dis 209:995–1006

Chapter 5
Emergence of New Tickborne Infections

5.1 Introduction

Ticks are responsible for the most diverse range of microbial pathogens transmitted by vectors and are second only to mosquitoes as the most frequent vector of human infectious diseases. These hematophagous arthropods parasitize every class of vertebrates in nearly all areas of the world [1] and represent one of the most important mediators of zoonoses to humans. In the past decade or more, there have been global expansion and emergence of several tickborne diseases: Lyme disease, ehrlichiosis, anaplasmosis, babesiosis, and others. Moreover, new tickborne diseases continue to be discovered in recent years, such as novel phleboviruses of the *Bunyaviridae* family emerging separately in Asia, Europe, and North America. One of these viruses was first recognized in China in 2010, presenting in patients with fever, leucopenia, thrombocytopenia, and organ dysfunction, called severe fever with thrombocytopenia syndrome virus [SFTSV] [2]. A closely related phlebovirus, called Heartland virus, was subsequently discovered in 2012 to cause a similar clinical syndrome in 2012 in Missouri, United States [US] [3].

Recently described new emerging tickborne infections also include previously unrecognized species of rickettsia causing human diseases, such as *Rickettsia slovaca*, first clinically recognized in 1997 in Europe as causing tickborne lymphadenopathy [TIBOLA] [4], and newly described spirochete causing relapsing fever-type illness, first appearing in humans in Russia in 2011 [5], and subsequently reported in the US in 2013, known as *Borrelia miyamotoi* [6]. In 2010 a new member of the *Anaplasmataceae* family, *Candidatus neoehrlichia mikurensis*, was recognized to cause a septicemic illness in Europe [7].

I.W. Fong, *Emerging Zoonoses*, Emerging Infectious Diseases of the 21st Century,
DOI 10.1007/978-3-319-50890-0_5

5.2 Ticks

There are at least 869 species or subspecies of ticks recognized [1]. These comprise of two major families: the Ixodidae or hard-ticks because of their sclerotized, hard dorsal plate, which represent the most important group medically and numerically, and the Argasidae or soft-ticks because of their flexible cuticle. A third family consisting of a single species, the Nuttalliellidae, is confined to southern Africa [1]. Ticks have three stages in their life cycle: the larval, nymphal, and adult forms. Hard-ticks and soft-ticks differ morphologically and in their life cycle and ecology. Ixodid ticks have several advantageous attributes for transmitting infectious pathogens as vectors more than soft-ticks or argasids. Ixodid ticks feed for long periods [several days], firmly attached to the host, and remain unnoticed as their bite is painless, and their feeding hosts include a large variety of vertebrates in diverse habitats [1]. Argasids, conversely, feed briefly and often on a single host species, and they live in dry areas, mostly in sheltered habitat near their hosts [1]. It is believed that ticks evolve about 225 million years ago in the Paleozoic or early Mesozoic eras, initially parasitizing reptiles [8].

Ixodids or hard-tick adult size averages 20–30 mm, females longer than males, and the larva is about 2 mm with progressive increase in size to the nymph, which is under 1 cm, and then to the adult form [1]. Each stage of the tick attaches and feeds on a single host over several days; then once satiated it detaches and finds a resting place to molt to the next stage. The larva and nymph are the main transmitters of diseases, as the adult feeds only briefly and the males may not feed at all. Ixodid ticks actually spend >90% of their life cycle unattached from the host and mainly live in open areas of meadows but are transported to woods by their hosts [9]. These ticks exhibit active host-seeking behavior by climbing plants and attaches to passing animal hosts or emerge from their habitat and attack nearby animals [1].

The life cycle of the ixodid ticks usually extends to 2 years but may vary from 6 months to 6 years, depending on the environmental conditions such as temperature and humidity [1]. Some tick species are host specific, feeding only a limited variety of animals, but this may vary with the stage, and some ticks have different host for each feeding stage. In general the animals in their habitat influence host selection and diverse species of ticks have different affinities for attacking humans [10]. The relationship with ecology or environment, natural animal inhabitants and host specificity or range, and geographic distribution are all interrelated with respect to the life cycle of these ticks. The brown dog tick, *Rhipicephalus sanguineus*, for instance, is well adapted to the vegetation and climate in the Mediterranean region and other areas with similar conditions such as Mexico, Arizona, California, Texas, and parts of Brazil, all regions where the brown dog tick can transmit rickettsial pathogens. Each feeding stage of *R. sanguineus* has high specificity, and it readily feeds on dogs and is important vectors for Mediterranean spotted fever in the Mediterranean region, and occasionally Rocky Mountain spotted fever in southwestern US and Mexico. In contrast, *Ixodes scapularis* [black-legged deer tick] in the US and *Ixodes ricinus* [European sheep tick] in Europe are adapted to the woods and forests with relative high humidity and are sensitive to desiccation from dry places. Ticks inhabiting wooded areas, *I. scapularis*, *I. ricinus*, *Amblyomma american* [lone star tick], and others, feed on different host species, such as small and large mammals and birds [1, 11].

The life cycle of ixodid wooded ticks, as exemplified by *I. ricinus*, starts with the gravid female laying 100–1000 of eggs on the grass, and the larvae hatch in 4–6 weeks, which attach and feed on small mammals [rodents] and birds, become engorged, detach in grass, and molt *to adults in 10–20 weeks* and which then attach and feed on small or large mammals and birds, then the engorged female mates with males to continue the cycle [11]. Soft-bodied ticks [argasids] are inhabitants of sheltered environments, nests, caves, burrows, and primitive man-made shelters. Unlike ixodids, they lack cement from the salivary glands for firm attachment but produce anticoagulants and cytolytic substances to facilitate brief multiple feeding times [1]. Thus, the time spent on the host is relatively short, up to a few hours, and after meals they can be found in cracks and crevices or below the soil surface in their home environment.

5.3 Historical Aspects

Ticks have been recognized to inflict bites on humans for thousands of years by ancient Greek scholars [1]. Their ability to transmit infectious disease was first recognized in animals; cattle fever caused by tickborne protozoan [*Babesia bigemina*] was described in Texas at the end of the nineteenth century [12]. Subsequently, within the first decade of the twentieth century, ticks were implicated in the transmission of microbes to humans. *Borrelia duttoni*, transmitted by a soft-bodied tick [*Ornithodoros moubata*], was found to cause relapsing fever in 1905 [13], and in 1909 Rocky Mountain spotted fever agent, *Rickettsia rickettsii*, was shown to be transmitted by the wood hard-tick *Dermacentor andersoni* [14]. Although Mediterranean spotted fever was first described in Tunis in 1910 [15], it was not until the 1930s that the brown dog tick [*Rhipicephalus sanguineus*] was recognized as the vector of the causative *Rickettsia conorii* [16].

Tularemia was recognized in squirrels in 1911 and the first human case may have been described in Japan in 1837 [11], but the first documented human case was reported in 1914 [17]. However, the epidemiology and the role of diverse ticks in the transmission of *Francisella tularensis* were not reported until 1929 [18]. In the era following World War II, many tickborne diseases of animal and humans were discovered of bacterial, viral, and protozoan etiologies [19]. Q-fever, caused by *Coxiella burnetii*, was first described in Australia in 1935 [20], and although the organism can be found in >40 species, it rarely is a vector-transmitted disease [most commonly transmitted by inhalation or ingestion], and tick transmission was initially reported in 1947 [21].

Lyme borreliosis, named after the town Old Lyme in Connecticut [US] in 1995, was discovered to be caused by *Borrelia burgdorferi* in 1982 and transmitted by ixodid ticks [22], the black-legged deer tick *I. scapularis* in the northeastern and upper midwestern US and the western black-legged tick [*I. pacificus*] along the Pacific coast. A similar condition was described in Europe more than 100 years before, with a rash of "erythema chronicum migrans," which was subsequently found to be caused by *Borrelia garinii*, *B. burgdorferi*, and *Borrelia afzelii* and transmitted by ixodid ticks [*I. ricinus*] [23]. Lyme borreliosis is the most common vector-transmitted disease in the US and Europe, but also occurs widely in other countries of the former Soviet Union, China, Japan, Australia, and probably in North Africa [11]. The main reservoir of Lyme borreliosis are small mammals, especially rodents such as the white-footed mouse [*Peromyscus leucopus*] in northeastern US and *Apodemus* species in Europe [24, 25]. Although deer are hosts for the black-legged ticks, they are not reservoirs for Lyme borreliosis. In Europe *B. afzelii* life cycle is maintained in rodents and *B. garinii* in avian reservoirs [26].

Rickettsiosis, caused by intracellular bacteria, has been recognized to be expanding globally for several decades and represent one of the major zoonoses [27]. Prior to 1974 only four tickborne rickettsiosis were known with one tickborne spotted fever group identified in separate regions: Rocky Mountain spotted fever due to *R. rickettsii* was first described as a clinical entity in 1899 in the Americas; *R. conorii* present in Europe, southeast Asia, and North Africa causing Mediterranean spotted fever; *R. siberia* recognized in Siberia and western Russia; and *R. australis* found in Australia [11]. The members of the *Rickettsia* genus are classified into four groups: the spotted fever group rickettsiae, typhus group rickettsiae, *Rickettsia bellii* group, and *Rickettsia Canadensis* [27]. The recognition and expansion of pathogenic *Rickettsia* have been facilitated by molecular techniques in recent times. Rickettsiae once considered nonpathogenic to humans and new species have been identified in the past 25 years. There are now 26 *Rickettsia* species validated in addition to subspecies mentioned. Some new species identified since 2005 include *Rickettsia asiatica*, *Rickettsia heilongjiangesis*, *Rickettsia hoogstraalii*, *Rickettsia raoulti*, and *Rickettsia tamourae* [27]. Tickborne rickettsiosis is now present in all continents of the world and some Caribbean and Pacific islands. *Rickettsia africae* which causes African tick-bite fever is believed to have been exported from Africa during the slave trade of the 1800s to several Caribbean islands, Guadeloupe, St. Kitts, Nevis, Dominica, US Virgin Islands, Montserrat, St. Lucia, Martinique, and Antigua [27].

5.3.1 Tickborne Zoonoses: General Background

Tickborne infections of humans, farm, and domestic animals are primarily associated with wildlife animal reservoirs. Humans and domestic animals are incidental hosts that result from infringement of the usual circulation between wildlife and tick vectors in their natural habitats. The risks of tickborne diseases vary geographically and are determined by the climate, environment, presence of rodents and other vertebrate reservoirs, and the species of ticks parasitizing wild and domestic animals [28]. These zoonoses can emerge in previously non-endemic areas when climate conditions and circumstances favorable to the maintenance and transmission arise.

This may include displacement of the hosts of vectors and reservoir animals by human expansion and development. Birds are important reservoirs for several pathogens and act as vehicles for infected ticks and probably play an important role in the dispersal of tickborne zoonoses [29]. Although for each pathogen one or more tick vectors and several animal reservoirs may exist, various combinations of coinfection with different microbes may coexist, and studies have demonstrated mixed infection in ticks of 7–10.9% [29, 30]. For example, the black-legged deer tick [*I. scapularis*], from the northeastern and upper midwestern US, can be coinfected and transmit multiple pathogens to humans such as Lyme and myamotoi borreliosis, anaplasma, babesia, and Powassan virus. Concurrent infection by two or more tickborne microbes is fairly common and frequently detected in clinical and veterinary practice [31]. Coinfection with multiple pathogens may be transmitted by ticks simultaneously during the same blood meal or at separate times. The natural reservoir hosts of many microbes [*B. burgdorferi*, *Anaplasma phagocytophilum*, and *Babesia microti*] transmitted by Ixodes ticks are rodents, which are often coinfected with multiple organisms and may transfer the microbes to the larvae or nymphs and then to animals and humans [32]. Simultaneous coinfection of several pathogens may result in diverse host responses, which can conflict and antagonize each other, allowing the pathogens to synergistically infect the host more successfully, with more prolonged and severe disease with multitude of clinical manifestations, than normally reported with a single pathogen infection [33]. Sequential infection with a new microbe may allow the establishment of a new infection which otherwise may not have occurred because of elimination by the host immune defense mechanisms [31].

The majority of tickborne infections are transmitted during the course of the blood meal from contaminated tick saliva [i.e., *B. burgdorferi*, relapsing fever borreliae, and spotted fever rickettsiae], regurgitated midgut contents [*B. burgdorferi*], feces [*Coxiella burnetii*], or coxal fluid in argasid ticks with relapsing fever borreliae [27]. Transmission from contamination of abraded skin or mucosa of the eye following crushing of the ticks with the fingers is possible. Transmission by accidental ingestion of the arthropods is well known in animals and is possible to occur in humans. Transmission of several tickborne zoonoses can occur by blood transfusion and sharing of needles and syringes from an unsuspected blood donor or drug abuser. Different *Babesia* species, for instance, have been reported to be transmitted from blood transfusions in both animals [dogs] and humans [31].

5.4 New Tickborne Bunyaviruses

Tickborne viruses that cause human diseases belong to three main families: *Bunyaviridae*, *Flaviviridae*, and *Reoviridae*. The most common tickborne viruses belong to the *Flaviviridae* family and are endemic in Europe and Asia: tickborne encephalitis virus [TBEV] being the most common in Europe, louping illness virus [LIV], Omsk hemorrhagic fever virus [OHFV], and Kyasanur Forest disease virus [KFDV]; in the Middle East Alkhurma hemorrhagic fever virus [AHFV]; and in

North America only the Powassan virus from this group is present [34]. TBEV is widely distributed in Europe and the vector is *I. ricinus*, the sheep dog tick. Prior to 2009–2010, the only recognized members of the *Bunyaviridae* family causing tickborne infection in humans were the Crimean-Congo virus and the Bhanja virus. Since then two similar bunyaviruses of the *Phlebovirus* genus have emerged in China in 2009, severe fever with thrombocytopenia syndrome virus [SFTSV], and the Heartland virus in the US in 2012.

The *Bunyaviridae* family of viruses constitutes the largest group of RNA viruses with more than 350 identified [35]. They are enveloped, spherical virions with a diameter of 80–120 μm, containing three single-stranded RNA segments [negative sense], with no matrix proteins [35]. The family of *Bunyaviridae* contains five genera: *Hantavirus* genus, type species Hantaan virus; *Nairovirus* genus, type species Crimean-Congo hemorrhagic fever virus [CCHFV]; *Orthobunyavirus* genus, type species Bunyamwera virus; *Phlebovirus* genus, type species Rift Valley fever virus [RVFV]; and *Tospovirus* genus, type species Tomato spotted wilt virus [36, 37]. Presently there are at least 40 viruses of the *Bunyaviridae* family not assigned a genera or species, including Bhanja virus, Forecariah virus, and the Kismayo virus [38]. The Bhanja serogroup viruses are most closely related to SFTSV and the Heartland virus, and members of this group have been isolated from ticks on livestock and wild animals in India, Southern Europe, and Africa [38].

The *Bunyaviridae* are generally found in arthropods and wild animals, especially rodents, and some members can infect humans. The Hantavirus is the only member of this family not transmitted by vectors, but by rodent excreta [36]. The *Orthobunyavirus* [i.e., La Crosse virus] infects only mosquito vectors, the *Nairovirus* [i.e., CCHFV] is largely limited to ticks, and the *Phlebovirus* [i.e., sandfly fever] is transmitted by sandflies and midges, but RVFV which is a member of this group can infect a wide range of arthropods and is primarily transmitted by mosquitoes [36].

5.4.1 Severe Fever with Thrombocytopenia Syndrome [SFTS]

SFTS was first recognized as a distinct clinical entity in Henan province of China in 2007. Patients presented with high fever, gastrointestinal bleeding, abdominal pain, bloating, nausea, and vomiting, with low platelets and white blood count, elevated alanine and aspartate transaminases, and proteinuria [39]. In 2010, a novel bunyavirus designated Huaiyangshan virus and subsequently changed to SFTSV was isolated from patients with this syndrome [40].

It is believed that the SFTSV actually originated in China about 50–150 years ago [41]. The SFTSV and Heartland virus are assigned in the *Phlebovirus* genus, but they are different from other known phleboviruses. The phleboviruses consist of about 70 antigenically distinct serotypes classed into two groups. The *Phlebovirus* fever group is transmitted by the phlebotominae sandflies or mosquitoes and the Uukuniemi group by ticks [42]. Although SFTSV have limited sequence similarities to other viruses of the Uukuniemi group, it was assigned to this subtype due to serological similarities and lack of small nonstructural protein on the M segment, and ticks are also the common vector [43]. Isolates of SFTSV from different geographical regions share 90% genetic sequence similarity and are grouped into five sublineages A–E [44]. Lineage A constitutes isolates from animals, dogs, cats, goats, buffaloes, and cattle, with no geographical clustering pattern [41].

Segmented-genome viruses such as the bunyaviruses are capable of rapid recombination, which is associated with pathogenicity and transmissibility among vectors and hosts and increases the risk for new outbreaks [45]. Recently two strains of SFTSV were found to have reassortment in the small RNA segment, which may drive rapid changes in the in the virus [46]. The basis of genetic diversity of SFTSV, although not completely understood, may be explained by lack of proofreading function of its RNA-dependent RNA polymerase, with high mutation rate [about 10^{-4} substitution per site each year] during its replication [41].

5.4.1.1 Epidemiology of SFTSV Infection

SFTS as a distinct syndrome was first reported in rural regions in Henan and Hubei provinces of Central China in 2009 [44]. Since then the disease has been found in 11 province of China with over 2500 reported cases. There is evidence that the microbe is widely distributed in China and only a small proportion of infected subjects develop clinical disease. Serosurveillance studies of populations in hilly regions of China showed that 1.0–3.8% of people had SFTSV antibodies [47–49]. There is also evidence that the virus has been circulating naturally in some regions of China showing seasonal variance with most cases occurring from May to July, and clinical disease occurred mainly in older subjects [92%] [50]. It has been estimated that the annual incidence of disease is about 5 per 100,000 of the rural population [49]. Seroprevalence and asymptomatic viremia of SFTSV in blood donors from endemic regions have also been reported. In a study of 17,208 blood donors, seropositivity ranged from 0.27% to 0.54%, but very low-grade viremia was detected in only two subjects [51]. Overall, SFTS is mainly reported from rural areas of central and northeastern China from May to September, targeting farmers >50 years of age [52]. Subjects were predominantly affected from farming-related exposures and numerous domestic and wild animals were infected by SFTSV [53].

SFTS was first reported outside China in North Korea in 2009 [54] and a subsequent fatal case was reported from South Korea in 2012 [55]. During 2013 SFTS was diagnosed in 35 patients in South Korea, and phylogenic analysis of SFTSV isolates from South Korea and China was closely related [56]. Locally transmitted cases have also been identified in Japan with 11 cases recognized by 2013, all living in western Japan, and were >50 years of age, and there were six fatalities [57]. Phylogenetic analysis of the Japanese isolates indicated that the genotype was independent from those in China. It has been estimated that SFTSV was likely circulating unrecognized in animals and humans in Korea and Japan for sometimes before these reports. In 2014, there were 108 suspected cases of SFTS in Japan and 41 were confirmed cases by PCR [58].

5.4.1.2 Vector and Ecology

The tick vector of SFTSV is considered to be a widely distributed hard-tick of the Ixodidae family, *Haemaphysalis longicornis*, present in China and other countries of Asia [59]. In endemic areas of China, the prevalence of SFTSV in these ticks collected from domestic animals ranged from 2.1% to 5.4%, and the RNA sequences of viral isolates were very closely related to those in patients [39, 59]. The virus is also detected at a lower rate [0.6%] of another widely distributed tick, *Boophilus (Rhipicephalus) microplus*, in endemic and non-endemic areas [44]. Viral RNA of SFTSV was also detected in *H. longicornis* ticks at all stages of its life cycle, indicating both transstadial and transovarial routes of transmission and the ability of ticks to play a role as a vector and a reservoir of the virus [60]. In a study from South Korea, 13,000 ticks were examined for SFTSV from nine provinces, and the minimum detection rate in *H. longicornis* was 0.46%, with the highest prevalence in the southern region [61]. In South Korea SFTSV was also detected in other hard-ticks at lower rates, *Ixodes nipponensis* and *Amblyomma testudinarium* [62], and in China the virus was also detected in mites from field mice and goats in endemic regions [44]. Common with many tickborne infections, not all patients with clinical disease have a history of tick bites, and in patients from China only 52% with SFTS recall tick exposure [48].

5.4.1.3 Reservoir Hosts

A variety of domestic and wild animals have been found to carry SFTSV, at different rates in various endemic regions of China. Seroprevalence in domestic animals ranged from in cattle 32–80%, goats 57–95%, chickens 1–36%, dogs 6–55%, and pigs 2–6% [44, 52, 63, 64]. Low levels of viral RNA were found in a small fraction [1.7–5.3%] of the animals studied [48, 65]. The viruses from animals and those from patients and ticks shared 95.4% of the genomic sequences [65]. Many wild animals such as deer, hedgehog, weasel, possum, and some bird species are hosts for the vector ticks and could carry the virus. However, rodents are known reservoirs of many bunyaviruses, but the rates of infection of rats [3.03–8%] with SFTSV are lower than in livestock [48, 66]. Hence, it appears that domestic animals may be the amplifying hosts of SFTSV and are probably the main reservoir of the virus.

Transmission of SFTSV is considered mainly from tick bites, but there is also evidence from multiple reports that the virus can be transmitted from human to human by direct contact with blood of infected patients [67–71]. A cluster of cases in families/households have been reported to be transmitted by blood contact, and blood transfusion and laboratory accidents from handling infected blood are potential means of transmission of SFTSV.

5.4.1.4 Pathogenesis and Immunity of SFTSV Infection

Although the pathogenesis of SFTS is not fully understood, major strides have been made in the few years since its description in understanding the mechanisms of the disease. Like other severe viral infections, cytokine and chemokine imbalance appears to be important in the pathogenesis, and severe, fatal cases of SFTS usually demonstrate "cytokine storm." Increased levels of tumor necrosis factor-alpha [TNF-α], interferon-gamma [IFN-γ], and IFN-induced protein-10 were associated with disease severity [72]. SFTSV gain entry to many human and animal cells, including macrophages and dendritic cells, by binding of the virus glycoprotein [GN/GC] to a receptor, the C-type lectin DC-SIGN [73].

Dynamic changes in viral load, T-cell subsets, and cytokines have been measured and analyzed in patients with SFTS and correlated with outcome. High levels of peak viral RNA load, serum liver enzymes, and serum interleukin [IL]-6 and IL-10 were associated with higher fatality rates [74]. These markers declined within 2 weeks of onset in survivors, and CD69+ T-cells were elevated early after infection, while HLA-DR+ and CTLA4+ T-cells elevated during the recovery phase of survivors [74]. Hence, high SFTSV viral load, very low platelets, high transaminases, marked elevation of proinflammatory and anti-inflammatory cytokines, and activation of CD 69+ T-cells were markers of severe disease and poor outcome. Cytokine storm with suppression of CD4+ and CD3+ lymphocytes with progressive decline but higher B lymphocytes has also been reported in severe and fatal cases of SFTS [75, 76].

Analysis of the immune response during the course of illness may also assist in understanding the pathogenesis of disease. In a study of 298 confirmed cases of SFTS and 55 followed after convalescence, during the first week of illness, there was a loss of T, B, and NK lymphocytes which were subsequently restored, but severe disease was associated with slower recovery and lower humeral immunity [77]. SFTSV-specific IgM antibody could be detected within 9 days, peaked at 4 weeks, and persisted for 6 months. IgG antibody could be detected in most patients within 6 weeks, peaked at 6 months, and persisted for at least 3 years [77]. There is also evidence that SFTSV is capable of infecting monocytes and suppresses IFN-beta and NF-kappa B promoter activities, facilitating the virus replication in human monocytes by restricting the innate immune response [78]. The virus can also disrupt type 1 interferon signaling by the nonstructural protein-mediated sequestering of signal molecules [STATS I and II] into inclusion bodies [79]. IFN-β production is an important host defense mechanism against viral pathogens, and inhibition of this response has been reported with other bunyaviruses [80, 81].

Limited pathological studies of fatal SFTS cases have been reported and the characteristic feature is the presence of necrotizing lymphadenitis of lymphoid tissues [82]. Leucopenia and thrombocytopenia which are hallmarks of the disease were not due to bone marrow suppression or aplasia, but are likely related to peripheral destruction or sequestration [83].

5.4.1.5 Animal Models of SFTSV Infection

Animal models are useful for elucidating the pathogenic mechanisms and development of new therapies for many human diseases. Newborn mice and rats, especially Kunming mice, are highly susceptible to SFTSV, and infected mice demonstrate pathological changes with large areas of necrosis only in the liver, but the virus can be detected in numerous organs [84]. Interferon-α/interferon-β knockout mice are also highly susceptible to infection with the virus, with 100% mortality in 3–4 days after inoculation [85]. The virus is found in numerous organs with heavy viral burden in mesenteric lymph nodes and spleen but no detectable histological changes. It appears that C57/BL6 mouse may be a better model for SFTSV infection, as the animals demonstrate leucopenia and thrombocytopenia similar to humans [86]. Moreover, histopathological changes were found in the spleen, liver, and kidney, but the spleen appears to be the primary target as viral replication could be demonstrated in this organ. The thrombocytopenia is probably caused by virus-bound platelets that underwent phagocytosis by splenic macrophages [86]. Nonhuman primates are considered the gold standard animal models for studying human disease pathogenesis, and a recent study has been reported by infecting rhesus macaques. SFTSV infection of *Macaca mulatta* did not result in severe disease symptoms or death but caused fever, thrombocytopenia, leucopenia, and elevated transaminases and cardiac enzymes [87]. Minor pathological lesions were found in the liver and kidney during the late stages of infection, and elevation of inflammatory cytokines was present in the blood. Thus, infection of this primate model resembles mild SFTS in humans.

Other animal studies assessed potential therapeutic and preventative interventions. The nonstructural protein of the S segment [SFTSV/NSs] fraction of SFTSV appears to antagonize interferon and suppress the host's innate immunity to facilitate infection. However, immunization with recombinant SFTSV/NSs was ineffective in promoting virus clearance in infected C57L/6J mice [88]. In another study using a mouse model, antiserum from a recovered patient with SFTS prevented lethal infection with the virus and improved clinical signs in nonlethal infection [89]. Other agents tested including steroids, ribavirin, and a site-I protease inhibitor were ineffective.

5.4.1.6 Clinical Aspects of SFTS

The incubation period of SFTS after a tick bite is estimated to be 5–14 days [44, 90] and after exposure to infected blood as 7–12 days [67]. The major stages of the disease recognized include an initial febrile flu-like illness, a second stage of multiorgan failure, and a subsequent convalescent phase [44, 52]. The first stage of the disease is characterized by sudden onset of fever from 38 to 41°C, headache, fatigue, myalgia, nausea, vomiting, and diarrhea, associated with high viral load, leucopenia, thrombocytopenia, lymphadenopathy, and elevation of transaminases and creatine phosphokinase. This stage may last for 5–11 days but may resolve after a week and enter a convalescent phase in patients with mild disease. The second stage occurs after 5 days and in more severe disease progress to develop multiorgan failure, first involving the liver, then the heart, lungs, and kidneys, which can persist for

7–14 days or lead to death [44, 90]. Coma may occur in about 6% but as high as 27% and confusion is seen in 22–36% [44, 52]. A recent report of 538 patients with SFTS described development of encephalitis in 19% of cases, with a fatality rate of 44.7% in this subgroup [91]. In nonfatal cases the biomarkers start to decrease with decline of the viral load and improvement of the platelet count by day 9–11, whereas in fatal cases the viral load, biomarkers, and thrombocytopenia continue to increase. The overall case fatality rate of 2500 reported cases in China averages at 7.3% [44], but the initial report was 12–30% [39].

Convalescence in survivors varies from 11 days to 19 days after onset of illness but in most severe cases occurred after 14 days [40, 52]. Clinical symptoms improve and the laboratory abnormalities gradually return to normal after 3–4 weeks [89]. Prognostic factors for disease severity and outcome of SFTS are related to host factors, clinical manifestations, and laboratory parameters. Age is a key factor in the severity of disease and outcome, with older age carrying a worst prognosis [44, 52, 90]. Children rarely become infected [possibly from less tick exposures and more robust immunity] and those who become infected manifest mild symptoms with fever, malaise and gastrointestinal symptoms, and minor laboratory abnormalities [92]. Clinical features associated with adverse prognosis include neurological manifestations, acute respiratory distress syndrome, and disseminated intravascular coagulopathy [93]. Host immune responses and viral replication are important factors in determining clinical severity and outcome, and high viral load in blood at admission is associated with a worst prognosis and fatality [93]. Laboratory parameters associated with poor prognosis and fatality include hypoalbuminemia, hyponatremia, coagulation disturbance, elevated transaminases, elevated creatinine, decreased lymphocyte count, and elevated lactic dehydrogenase levels in the late stage [94, 95].

5.4.1.7 Diagnosis of SFTS

Diagnosis of SFTS is based on epidemiological exposure risk and clinical features with presence of fever, thrombocytopenia, and leucopenia. Laboratory diagnosis can be confirmed by viral nucleic acid test with polymerase chain reaction [PCR] or by serology. Rapid diagnosis in acute illness is best accomplished with a reverse transcriptase [RT] PCR which is highly sensitive and specific [44]. Potential detection limit of 10 viral RNA copies/μL was achieved using quantitative real-time PCR with 98.6% sensitivity and over 99% specificity [96]. Moreover, the quantitative PCR at acute diagnosis can assist in the prognosis based on the viral load [58].

Serological methods for diagnosis include conventional immunofluorescence and serum neutralization assays which are not helpful in acute cases or early diagnosis, and they are costly and require well-trained personnel [52]. Indirect enzyme immunoassay [EIA] and double-antigen sandwich EIA have been used to detect total antibodies or viral-specific IgM and IgG [52]. A highly sensitive and specific EIA utilizing a glycoprotein N from the nucleocapsid has been developed [63]. In serological diagnosis acute and convalescent sera are best tested, thus providing a delayed diagnosis, although the presence of specific IgM antibodies in acute disease can be diagnostic.

5.4.1.8 Treatment of SFTS

Management of patients with SFTS is largely supportive with correction of fluid and metabolic disturbances and blood transfusion if necessary for significant blood losses. Platelet transfusion may be required for severe thrombocytopenia [$<30 \times 10^{-9}$/L] with significant bleeding, and appropriate antibiotics for suspected or proven secondary bacterial infections [44]. Mechanical ventilation may be needed in severe cases with ARDS or coma for airway protection and hemodialysis for severe renal failure.

No specific therapy has been shown to be effective for SFTSV infection. Ribavirin was of interest as it had been considered effective or approved for treatment of other bunyavirus infections, such as Rift Valley fever virus and Crimean-Congo hemorrhagic fever virus [97, 98]. While ribavirin has some in vitro activity against SFTSV, it did not effectively reduce virus replication in pre-infected cells [99]. Moreover, ribavirin treatment of patients with SFTS produced no significant clinical benefit and had no effect on platelet counts or viral loads [100]. Convalescent sera from recovered patients infected with SFTSV have high neutralizing antibodies against the virus and may have a role in treatment of severe cases or postexposure prevention after contact with infected blood [89, 101].

5.4.2 *Heartland virus*

Heartland virus infection was first described from a patient in Missouri [US] in 2012 and the agent is a phlebovirus closely related to SFTSV [3]. Phylogenetic and serological analysis revealed that Heartland virus should belong to the Bhanja group of viruses in the *Phlebovirus* genus [38]. Bhanja viruses have been isolated from various species of hard-ticks and are divided into the African and Eurasian lineages. *Amblyomma americanum* [lone star tick] is the vector of Heartland virus [HRTV]. Investigations had revealed the presence of viable virus in *A. americanum* nymphs in a patient's farm and nearby conservation area, with >97.6% sequence identity to human strains [102]. Ticks probably become infected by feeding on viremic hosts during their larval stage and may transmit the virus to humans during the spring and early summer when nymphs are plentiful.

Serological investigations in wild and domestic animals in Missouri have detected high antibody prevalence to HRTV in raccoon [42.6%], horse [17.4%], and white-tailed deer [14.3%], which suggest that these species are possible candidate reservoir hosts [103]. Antibodies were also found in dogs [7.7%], but no HRTV was isolated from any animal sera or ticks.

5.4.3 *Clinical Features of Heartland virus Infection*

The original two index cases were male farmers, aged 57 and 67 years, who presented with fever, leucopenia, and thrombocytopenia [3]. Both patients survived without hemorrhagic complications or multiorgan failure. An additional five nonfatal cases were subsequently identified through active surveillance in Missouri [104]. More recently a fatal case of HRTV infection was described in 80-year-old male on a farm in Tennessee [100]. The patient had a history of multiple tick bites, with detectable tick on his body 2 weeks prior to onset of illness, and had comorbid conditions of chronic obstructive lung disease and alcoholism. He presented with weakness, fever, and altered mental status, and tests showed persistent leucopenia, progressive thrombocytopenia, anemia, and elevated transaminases. His clinical course deteriorated on broad-spectrum antibiotics and subsequent hypotension, hypoxia, and renal dysfunction occurred before his death. Autopsy findings were largely nonspecific and the bone marrow revealed myeloid hyperplasia and trilineage hematopoiesis. Although this fatal case resembles severe cases of SFTS, there was no evidence of necrotizing lymphadenitis as described in fatal cases of infection with SFTSV, but the spleen demonstrated white-pulp depletion and scattered immunoblasts [105]. The differential diagnosis of HRTV infection would include infections with borrelia, rickettsia, ehrlichia, anaplasma, viruses, thrombotic thrombocytopenia, and hematological malignancies.

A subsequent fatal case has been described in Oklahoma [105], bringing the total number of HRTV infection detected to date to nine cases with two fatalities from three states. However, more cases are likely to be recognized in the future from other regions of the US, as wildlife serological studies have determined that HRTV is widespread within the central and eastern US [106]. Of 1428 animals tested, 103 were seropositive for the HRTV including 55 deer, 33 raccoon, 11 coyote, and 4 moose. Thirteen states had seropositive animals from Florida, Georgia, Illinois, Indiana, Kansas, Kentucky, Maine, Missouri, New Hampshire, North Carolina, Tennessee, Texas, and Vermont [106].

Similar to SFTS, there is no known specific therapy for HRTV infection. Management is primarily supportive, but doxycycline should be used initially in severe cases for treatable differential diagnoses [borreliosis, rickettsiosis, ehrlichiosis, etc.] until these conditions are excluded. There is no data yet on pathogenesis and animal models of HRTV infection. However, the mechanisms in disease pathogenesis maybe similar to that of SFTSV since the two viruses are closely related.

5.4.4 **Borrelia miyamotoi** *Disease*

Borrelia miyamotoi is a spirochete that is closely related to species that cause relapsing fever. It was first discovered in *Ixodes persulcatus* ticks in Japan in 1994 and subsequently documented in ticks and rodents in Europe and North America, but was not recognized to cause human infection until reported from Russia in 2011 [5]. In 2013, the first case of *B. miyamotoi infection* was described in the US with clinical manifestation of meningoencephalitis [6]. Subsequently two more cases were reported in the US presenting like human granulocytic anaplasmosis [107]. Cases of *B. miyamotoi* infection were also reported from New England [108], the Netherlands [109], and Japan [110].

The largest case series of *B. miyamotoi* disease [BMD] was recently reported from northeastern US. Blood samples from acutely febrile patients presenting to primary care offices, emergency departments, or urgent care clinics in 2013–2014, requesting testing for tickborne infections, were sent to IMUGEN [Norwood, Massachusetts] [111]. Whole blood PCR for presence of specific DNA sequences of common tickborne pathogens and BMD were performed. Among 11,515 patients tested, 97 BMD cases were identified by PCR. Most of the patients presented with fever, chills, marked headache, myalgia, or arthralgia, and 24% required hospitalization [111]. Elevated liver enzymes [82%], leucopenia, and thrombocytopenia were common, and symptoms resolved after doxycycline and patients treated with amoxicillin or ceftriaxone also improved. Serology for the acute diagnosis of BMD was poor, as only 1 of 39 cases with circulating DNA had *B. miyamotoi* IgM and none with IgG, but convalescent sera were reactive in 78%. In this study of acute febrile patients, BMD was confirmed less frequently than babesiosis and human granulocytic anaplasmosis.

B. miyamotoi infection is transmitted by the black-legged deer ticks [*I. scapularis*] and has a similar range and distribution as Lyme borreliosis. Most cases of BMD occurred in July and August, suggesting transmission by larval ticks which have their peak activity in these months. Lyme disease, babesiosis, and human anaplasmosis mainly occur in June and early July, when the nymphal deer ticks are most abundant [112]. *B. miyamotoi* had been shown to undergo transovarial transmission and can be transmitted experimentally by larvae [113].

In regions where tickborne infections are possible, with or without known tick bites/exposure such as in many parts of northeastern, north-central, and far western US, empiric treatment with doxycycline may be reasonable for acutely sick, febrile patients for possible borrelia, rickettsia, ehrlichia, and anaplasma infections until a diagnosis is confirmed.

5.5 Conclusion and Future Perspectives

Novel tickborne infections continue to emerge in distant unrelated regions of the world with worrisome frequency. New, previously unrecognized, pathogens from tick exposure will continue to be discovered in the future. This is further exemplified by the recent detection of a novel *Orthomyxovirus* of the genus *Thogotovirus*, labeled Bourbon virus, associated with a fatal infection in a previously healthy male from Kansas after tick bites in 2014 [114]. The differential diagnosis of acute febrile illness after tick exposure is quite diverse and largely depends on the local region or geography at the time of exposure. For instance, in the US, there are 14 tickborne diseases listed by CDC, and now there will be 15 with the addition of Bourbon virus [see Table 5.1]

Tickborne diseases are theoretically preventable with appropriate clothing [long sleeve shirts and pants], avoidance of high-risk exposures, use of insecticides or acaricides to control tick population, or use of personal insect repellants. Despite these known preventative measures, for decades tickborne diseases have continued to

Table 5.1 Tickborne diseases of North America, mainly United States

Disease	Pathogen	Vector	Distribution
Anaplasmosis	*Anaplasma phagocytophilum*	*I. scapularis*	Eastern and upper-middle US
		I. pacificus	Pacific coast/Western US
Babesiosis	*Babesia microti*	*I. scapularis*	Northeast/upper midwestern US
BMD	*Borrelia miyamotoi*	*I. scapularis*	Northeastern/midwestern US
Bourbon virus disease	*Thogotovirus*	Unconfirmed tick	Kansas
Colorado tick fever	Colorado tick fever virus	*Dermacentor andersoni*	Rocky Mountain states
Ehrlichiosis	*Ehrlichia chaffeensis*/*E. ewingii*	*Amblyomma americanum*	Southcentral/Eastern US
Heartland virus	*Phlebovirus*/*Bunyavirus*	Lone star tick [*A. americanum*]	Missouri, Tennessee, Oklahoma
Lyme disease	*Borrelia burgdorferi*	*I. scapularis*	Northeast/upper midwestern US
		I. pacificus	Pacific Coast
Powassan disease	Powassan virus	*I. scapularis* and Groundhog tick	Northeast, Great Lakes area
R. parkeri rickettsiosis	*Rickettsia parkeri*	Gulf Coast tick	Gulf Coast, Southwestern US
RMSF	*Rickettsia rickettsii*	Am. dog tick	Eastern US [most common]
		RM wood tick	Western US
		Brown dog tick	Arizona, Texas, and Mexico
364D rickettsiosis	*Rickettsia phillipi*	Pacific Coast tick	California
STARI [Southern tick-associated rash illness]	Unknown agent	Lone star tick	Eastern, Southeastern US
Tickborne relapsing fever	*Borrelia hermsii*	*O. hermsii* [soft-tick] of squirrel, chipmunk	Western US [15 states]
Tularemia	*Francisella tularensis*	Dog tick, wood tick, and lone star tick	Throughout US

BMD Borrelia miyamotoi disease, *I* ixodes, *Am* American, *O Ornithodoros*, *RM* Rocky Mountain, *SF* spotted fever

flourish with worldwide expansion and emergence. Thus, new approaches for prevention of tickborne infections are needed. Directed mass education of targeted at-risk populations [farmers, outdoor campers, hikers, etc.] in endemic areas may be of benefit to inform the public of the risk of these diseases and preventative measures, including large visible signs in these areas. Development of specific vaccines for these various pathogens is not feasible nor would be cost-effective due to their relatively low frequency. However, it may be worthwhile to evaluate new strategies such as a vaccine to prevent tick bites that could be used to prevent many tickborne diseases. Anti-tick vaccines could potentially reduce tickborne infections by reducing the tick burden or interference in host-tick-human transmission by vaccination of the vertebrate hosts [115]. Anti-tick vaccines would be environmentally safe, unlikely to select resistance as compared to insecticides, and can include multiple antigens to target a broad range of ticks [116].

References

1. Sonenshire DE (1991) Biology of ticks, vol 1. Oxford University Press, New York
2. Xu B, Liu L, Huang X et al (2011) Metagenetic analysis of fever, thrombocytopenia and leucopenia syndrome [FTLS] in Henan Province, China: discovery of a new bunyavirus. PLoS Pathog 7:e1002369
3. Mc Mullan LK, Folk SM, Kelly AJ et al (2012) A new phlebovirus associated with severe febrile illness in Missouri. N Engl J Med 367:834–841
4. Ibarra V, Oteo JA, Portillo A et al (2006) *Rickettsia slovaca* infection: DEBONEL/TIBOLA. Ann N Y Acad Sci 1078:206–214
5. Platonov AE, Karan LS, Kolyasnikova NM et al (2011) Humans infected with relapsing fever spirochete *Borrelia miyamotoi*, Russia. Emerg Infect Dis 17:1816–1823
6. Gugliotta JL, Goethert HK, Berardi VP et al (2013) Meningoencephalitis from *Borrelia miyamotoi* in an immunocompromised patient. N Engl J Med 368:240–245
7. Fehr JS, Bloemberg GV, Ritter G et al (2010) Septicemia caused by tickborne bacterial pathogen *Candidatus Neoehrlichia mikurensis*. Emerg Infect Dis 16:1127–1129
8. Klompen JSH, Blank WC IV, Kierens JE, Oliver JH Jr (1996) Evolution of ticks. Annu Rev Entomol 41:141–161
9. Needham GR, Peel PD (1991) Off-host physiological ecology of ixodid ticks. Annu Rev Entomol 36:659–681
10. Estrada-Pena A, Jongejan E (1999) Ticks feeding on humans: a review of records on human biting Ixodoidea with special reference to pathogen transmission. Exp Appl Acarol 23:685–715
11. Parola P, Raoult D (2001) Ticks and tickborne bacterial diseases in humans: an emerging infectious threat. Clin Infect Dis 32:897–928
12. Smith T, Kilbourne FL (1893) Investigations into the nature, causation and prevention of Texas or southern cattle fever. Bull Bur Anim Ind US Dept Agric 1:301
13. Dutton JE, Todd JL (1905) The nature of tick fever in the eastern part of Congo Free State, with notes on the distribution and bionomics of the tick. Br Med J 2:1259–1260
14. Ricketts HT (1909) Some aspects of Rocky Mountain spotted fever as shown by recent investigations. Med Rec 16:1843–1855
15. Connor A, Bruch A (1910) Une fievre eruptive observe en Tunisie. Bull Soc Pathol Exot Filial 8:492–496
16. Brumpt E (1932) Longevite du virus de la fievre boutonneuse [*Rickettsiae conori*, n.sp.] chez a la tique *Rhinicephalus sanguineus*. CR Soc Biol 110:1197–1199
17. Wherry WB, Lamb BH (1914) Infection of man with *Bacterium tularense*. J Infect Dis 15:331–340
18. Francis E (1929) A summary of the present knowledge of tularemia. Medicine (Baltimore) 7:411–432
19. Sonenshire DF (1993) Biology of ticks, vol 2. Oxford University Press, New York
20. Mauvin M, Raoult D (1999) Q-fever. Clin Microbiol Rev 12:518–553
21. Ecklund CM, Parker RR, Lackman DB (1947) Case of Q-fever probably contracted by exposure to ticks in nature. Public Health Rep 62:1413–1416
22. Johnson RC, Schmid GP, Hyde FW, Steigerwalt AG, Brenner DJ (1984) *Borrelia burgdorferi* sp. nov.: etiological agent of Lyme disease. Int J Syst Bacteriol 34:496–497
23. Rijpkema SGT, Tazelaar D, Molkenboer M et al (1997) Detection of *Borrelia afzeli*, *Borrelia burgdorferi* sensu stricto, *Borrelia garinii*, and group VS 116 by PCR in skin biopsies of patients with erythema migrans and acrodermatitis chronicum atrophicans. Clin Microbiol Infect 3:109–116
24. Cook V, Barbour AG (2015) Broad diversity of host responses of the white-footed mouse *Peromyscus leucopus* to Borrelia infection and antigens. Ticks Tick Borne Dis 6:549–558
25. Gern L, Estrada-Pena A, Frandsen F et al (1998) European reservoir hosts of *Borrelia burgdorferi* sensu lato. Zentralbl Bakteriol 287:196–204

26. Gern L, Humair PF (1998) Natural history of *Borrelia burgdorferi* sensu lato. Wien Klin Wochenschr 110:856–858
27. Parola P, Paddock CD, Socolovschi C et al (2013) Update on tickborne rickettsiosis around the world: a geographic approach. Clin Microbiol Rev 26:657–702
28. Fritz CL (2009) Emerging tickborne diseases. Vet Clin North Am Small Anim Pract 39:265–278
29. Franke J, Fritzsch J, Tomaso H, Straube E, Doprn W, Hiodebrandt A (2010) Coexistence of pathogens in host seeking and feeding ticks within a single natural habitat in Central Germany. Appl Environ Microbiol 76:6829–6836
30. Eshoo MW, Crowder CD, Carolan HE, Rounds NA, Ecker DJ, Jaag H, Mottes B (2014) Broad-range survey of tickborne pathogens in southern Germany reveals a high prevalence of Babesia microti and a diversity of other tickborne pathogens. Vector Borne Zoonotic Dis 8:584–591
31. Banneth G (2014) Tickborne infections of animals and humans: a common ground. Int J Parasitol 44:591–596
32. Johnson RC, Kodner C, Jarnefeld J, Eck DK, Xu Y (2011) Agents of human anaplasmosis and Lyme disease at Camp Ripley, Minnesota. Vector Borne Zoonotic Dis 11:1529–1534
33. Krause PJ, Mc Kay K, Thompson CS et al (2002) Disease–specific diagnosis of co-infecting tickborne zoonoses: babesiosis, human granulocytic ehrlichiosis, and Lyme disease. Clin Infect Dis 34:1184–1191
34. Lani R, Mopghaddam E, Haghani A, Chang LY, Abubaker S, Zandi K (2014) Tick-borne viruses: a review from the perspective of therapeutic approaches. Ticks Tick-borne Dis 5:457–465
35. Plyusnin A, Beaty BJ, Elliott RM et al (2011) Family *Bunyaviridae*. In: Ninth report of the International Committee on Taxonomy of Viruses, 1st edn. Elsevier, London, pp 725–741
36. Mertz GJ (2009) Bunyaviridae: Bunyavirus, Phlebovirus, Nairoviruses, and Hantaviruses. In: Richman DD, Whitley RJ, Hayden RG (eds) Clinical virology, 3rd edn. ASM Press, Washington, DC, pp 977–1007
37. Anonymous. *Bunyaviridae*. Wikipedia. https://en.wikipedia.org/wiki/Bunyaridae
38. Matsuno K, Weiserd C, Travassos de Rosa APA et al (2013) Characterization of the Bhanja serogroup viruses *[Bunyaviridae*]: a novel species of the genus *Phlebovirus* and its relationship with other emerging tickborne phleboviruses. J Virol 87:3719–3727
39. Yu XJ, Liang MF, Zhang SY et al (2011) Fever with thrombocytopenia associated with novel bunyavirus in China. N Engl J Med 364:1523–1532
40. Xu B, Liu L, Huang X et al (2011) Metagenomic analysis of fever, thrombocytopenia and leucopenia syndrome [FTLS] in Hennan Province, China: discovery of a new bunyavirus. PLoS Pathol 364:e1002369
41. Lam TT, Liu W, Bowden TA et al (2013) Evolutionary and molecular analysis of the emergent severe fever with thrombocytopenia syndrome virus. Epidemics 5:1–10
42. Bishop DH, Calisher CH, Casals J et al (1980) Bunyaviridae. Intervirology 14:125–143
43. Palacios G, Savji N, Travassos de Rosa A et al (2013) Characterization of the Uukuiemi virus group [Phlebovirus: Bunyaviridae]: evidence of seven distinct species. J Virol 87:3187–3195
44. Liu Q, He B, Huang SY, Wei F, Zhu XQ (2014) Severe fever with thrombocytopenia syndrome, an emerging tickborne zoonosis. Lancet Infect Dis 14:763–772
45. Horne KM, Vanlandingham DL (2014) Bunyavirus—vector interactions. Viruses 6:4373–4397
46. Ding NZ, Luo ZF, Niu DD et al (2013) Identification of two severe fever with thrombocytopenia syndrome virus strains originating from reassortment. Virus Res 178:543–546
47. Cui F, Cao HX, Wang L et al (2013) Clinical and epidemiological study on severe fever with thrombocytopenia syndrome in Yiyuan County, Shandong Province, China. Am J Trop Med Hyg 88:510–512
48. Zhao L, Zhai S, Wen H et al (2012) Severe fever with thrombocytopenia syndrome virus, Shandong Province, China. Emerg Infect Dis 18:963–965
49. Zhang X, Liu Y, Zhao L et al (2013) Emerging hemorrhagic fever in China caused by a novel bunyavirus SFTSV. Sci China Life Sci 56:697–700
50. Zhang L, Ye L, Ojcias DM et al (2014) Characterization of severe fever with thrombocytopenia syndrome in rural regions of Zhejiang, China. PLoS One 9:e111127

51. Zeng P, Ma L, Gao Z et al (2015) A study of seroprevalence and rates of asymptomatic viremia of severe fever with thrombocytopenia syndrome virus among Chinese blood donors. Transfusion 55:965–971
52. Liu S, Chai C, Wang S et al (2014) Systemic review of severe fever with thrombocytopenia syndrome: virology, epidemiology, and clinical characteristics. Rev Med Virol 24:90–102
53. Li Z, Hu J, Bao C et al (2014) Seroprevalence of antibodies against SFTS virus infection in farmers and animals, Jiangsu, China. J Clin Virol 60:185–189
54. Denic S, Janbeith J, Nair S, Conca W, Tariq WU, Al-Salam S (2011) Acute thrombocytopenia, leucopenia, and multiorgan dysfunction: the first case of SFTS Bunyavirus outside of China? Case Rep Infect Dis 2011:1–4
55. Kim KH, Yi J, Kim G et al (2013) Severe fever with thrombocytopenia syndrome, South Korea, 2012. Emerg Infect Dis 19:1892–1894
56. Park SW, Hun MG, Yun SM, Park C, Lee WJ, Ryou J (2014) Severe fever with thrombocytopenia virus, South Korea. Emerg Infect Dis 20:1880–1882
57. Takahashi T, Maeda K, Suzuki T et al (2014) The first identification and retrospective study of severe fever with thrombocytopenia syndrome in Japan. J Infect Dis 209:816–827
58. Yoshikawa T, Fukushi S, Tani H et al (2014) Sensitive and specific PCR systems for detection of both Chinese and Japanese severe fever with thrombocytopenia syndrome virus strains and prediction of patient survival based on viral load. J Clin Microbiol 52:3325–3333
59. Zhang YZ, Zhou DJ, Qin XC et al (2012) The ecology, genetic diversity, and phylogeny of Huaiyangshan virus in China. J Virol 86:2864–2868
60. Wang S, Li J, Niu G et al (2015) SFTS virus in ticks in endemic areas of China. Am J Trop Med Hyg 92:648–649
61. Park SW, Song BG, Shin EH et al (2014) Prevalence of severe fever with thrombocytopenia syndrome virus in *Haemaphipalis longicornis* ticks in South Korea. Ticks Tick-borne Dis. 5:975–977
62. Yun SM, Lee WG, Ryou J et al (2014) Severe fever with thrombocytopenia syndrome virus in ticks collected from humans, South Korea, 2013. Emerg Infect Dis 20:1358–1361
63. Jiao Y, Zeng X, Guo X et al (2012) Preparation and evaluation of recombinant severe fever with thromboctypenia syndrome virus nucleocapsid protein for detection of total antibodies in human and animal sera by double-antigen sandwich enzyme-linked immunosorbent assay. J Clin Microbiol 50:372–377
64. Ding S, Yin H, Xu X et al (2014) A cross–sectional survey of fever with thrombocytopenia syndrome virus infection of domestic animals in Laizhou City, Shandong Province, China. Jpn J Infect Dis 67:1–4
65. Niu G, Li J, Liang M et al (2013) Severe fever with thrombocytopenia syndrome virus among domesticated animals, China. Emerg Infect Dis 19:756–763
66. Ge HM, Wang QK, Li Z et al (2012) Investigation of severe fever with thrombocytopenia syndrome [SFTS] virus carrying situation in rodents in SFTS epidemic area of Donghai county, Jiangsu. J Prev Med 23:12–14 [in Chinese]
67. Tang X, Wu W, Wang H et al (2013) Human-to-human transmission of severe fever with thrombocytopenia syndrome Bunyavirus through contact with infectious blood. J Infect Dis 207:736–739
68. Gai Z, Liang M, Zhang Y et al (2012) Person-to-person transmission of severe fever with thrombocytopenia syndrome Bunyavirus through blood contact. Clin Infect Dis 54:249–252
69. Bao CJ, Guo XL, Qi X et al (2011) A family cluster of infection by a newly recognized Bunyavirus in eastern China, 2007; further evidence of person-to person transmission. Clin Infect Dis 53:1208–1214
70. Liu Y, Li Q, Hu W et al (2012) Person-to-person transmission of severe fever with thrombocytopenia syndrome virus. Vector Borne Zoonotic Dis. 12:156–160
71. Chen H, Hu K, Zou J et al (2013) A cluster of cases of human-to-human transmission caused by severe fever with thrombocytopenia syndrome Bunyavirus. Int J Infect Dis 17:e206–e208
72. Deng B, Zhang S, Geng Y et al (2012) Cytokine and chemokine levels in patients with severe fever with thrombocytopenia syndrome virus. PLoS One 7:e41365

73. Hofmann H, Li X, Zhang X et al (2013) Severe fever with thrombocytopenia syndrome virus glycoproteins are targeted by neutralizing antibodies and can use DC-SIGN as a receptor for pH-dependent entry into human and animal cell lines. J Virol 87:4384–4394
74. Li J, Han Y, Xing Y et al (2014) Concurrent measurement of dynamic changes in viral load, serum enzymes, T cell subsets, and cytokines in patients with severe fever with thrombocytopenia syndrome. PLoS One 9:e91679
75. Sun Y, Jin C, Zhang F et al (2012) Host cytokine storm associated with disease severity of severe fever with thrombocytopenia syndrome. J Infect Dis 206:1085–1094
76. Sun L, Hu Y, Niyansaba A, Tong Q, Lu L, Li H, Jie S (2014) Detection and evaluation of immunofunction of patients with severe fever with thrombocytopenia syndrome. Clin Exp Med 14:389–395
77. Lu QB, Cui N, Hu JG et al (2015) Characterization of immunological responses in patients with severe fever with thrombocytopenia syndrome: a cohort study in China. Vaccine 33:1250–1255
78. Qu B, Qi X, Wu X et al (2012) Suppression of the interferon and NF-Kappa B responses by severe fever with thrombocytopenia syndrome virus. J Virol 86:8388–8401
79. Ning YJ, Feng K, Min YQ, Cao WC, Wang M, Deng F, Hu Z, Wang H (2015) Disruption of type 1 interferon signaling by the nonstructural protein of severe fever with thrombocytopenia syndrome virus via hijacking of STAT2 and STAT1 into inclusion bodies. J Virol 89:4227–4236
80. Billecocq A, Spiegel M, Vialat P et al (2004) NSs protein of Rift Valley fever virus blocks interferon production by inhibiting host gene transcription. J Virol 78:789–806
81. Van Knippenberg I, Carlyon-Smith C, Elliott RM (2010) The N-terminus of *Bunyamwera orthobunyavirus* nss protein is essential for interferon antagonism. J Gen Virol 91:2002–2006
82. Hiraki T, Yoshi M, Suzuki T et al (2014) Two autopsy cases of severe fever with thrombocytopenia syndrome [SFTS] in Japan: a pathognomonic histological feature and unique complication of SFTS. Pathol Int 64:569–575
83. Quan Tai X, Feng Zhe C, Xiu Guang S, Dong GC (2013) A study of cytological changes in the bone marrow of patients with severe fever with thrombocytopenia syndrome. PLoS One 8:e83020
84. Chen XP, Cong ML, Li MH, Kang YJ, Feng YM, Plusnin A, Xu J, Zhang YZ (2012) Infection and pathogenesis of Huaiyanshan virus [a novel tickborne bunyavirus] in laboratory rodents. J Gen Virol 93:1288–1293
85. Liu Y, Wu B, Paessler S, Walker DH, Tesh RB, Yu XJ (2014) The pathogenesis of severe fever with thrombocytopenia syndrome virus infection in alpha/beta interferon knockout mice: insights into the pathogenic mechanisms of a new hemorrhagic fever. J Virol 88:1781–1786
86. Jin C, Liang M, Ning J et al (2012) Pathogenesis of emerging severe fever with thrombocytopenia syndrome virus in C57/BL6 mouse model. Proc Natl Acad Sci U S A 109:10153–10158
87. Jin C, Jiang H, Liang M et al (2015) SFTS virus infection in nonhuman primates. J Infect Dis 211:915–925
88. Liu R, Huang DD, Bai JY et al (2015) Immunization with recombinant SFTSV/NSs protein does not promote viral clearance in SFTSV-infected C57BL/6J mice. Viral Immunol 28:113–122
89. Shimada SS, Posadas-Herrera G, Aoki K, Morita K, Hayasaka D (2015) Therapeutic effect of post-exposure treatment with antiserum on severe fever with thrombocytopenia syndrome [SFTS] in a mouse model of SFTS virus infection. Virology 482:19–27
90. Gai ZT, Zhang Y, Liang MF et al (2012) Clinical progress and risk factors for death in severe fever with thrombocytopenia syndrome patients. J Infect Dis 206:1095–1102
91. Cui Liu R, Wang LY et al (2015) Severe fever with thrombocytopenia syndrome bunyavirus-related human encephalitis. J Infect 70:52–59
92. Wang LY, Cui N, Lu QB, Wo Y, Wang HY, Liu W, Cao WL (2014) Severe fever with thrombocytopenia syndrome in children: a case report. BMC Infect Dis 14:366
93. Deng B, Zhou B, Zhang S et al (2013) Clinical features associated with severe fever with thrombocytopenia syndrome Bunyavirus infection in Northeast China. PLoS One 6:e80802

94. Zhang YZ, He YW, Dai YA et al (2012) Hemorrhagic fever caused by a novel Bunyavirus in China: pathogenesis and correlates of fatal outcome. Clin Infect Dis 54:527–533
95. Cui N, Bao X-L, Yang ZD et al (2014) Clinical progression and predictors of death in patients with severe fever with thrombocytopenia syndrome in China. J Clin Virol 59:12–17
96. Sun Y, Liang M, Qu J et al (2012) Early diagnosis of novel SFTS Bunyavirus infection by quantitative real-time PCR assay. J Clin Virol 53:48–53
97. Debing Y, Jochmans D, Neyts J (2013) Intervention strategies for emerging viruses: use of antivirals. Curr Opin Virol 3:217–224
98. Tasdelen FN, Ergonul O, Doganci I, Tulex N (2009) The role of ribavirin in the therapy of Crimean-Congo hemorrhagic fever early use is promising. Eur J Clin Microbiol Infect Dis 28:929–933
99. Shimojima M, Fukushi S, Tani H et al (2014) Effects of ribavirin on severe fever with thrombocytopenia syndrome virus in vitro. Jpn J Infect Dis 67:423–427
100. Lu QB, Chui N, Li H et al (2013) Case fatality and effectiveness of ribavirin in hospitalized patients with severe fever with thrombocytopenia syndrome in China. Clin Infect Dis 57:1292–1299
101. Guo X, Zhang I, Zhang W et al (2013) Human antibody neutralizes severe fever with thrombocytopenia syndrome virus, an emerging hemorrhagic fever virus. Clin Vaccine Immunol 20:1426–1432
102. Savage HM, Godsey MS Jr, Lambert A et al (2013) First detection of Heartland virus [*Bunyaviridae: Phlebovirus*] from field collected arthropods. Am J Trop Med Hyg 89:445–452
103. Bosco-Lauth AM, Panella NA, Root JJ et al (2015) Serological investigation of Heartland virus [*Bunyaviridae: Phlebovirus*] exposure in wild and domestic animals adjacent to human case sites in Missouri 2012-2013. Am J Trop Med Hyg 92:1163–1167
104. Pastula DM, Turabelidze G, Yates KF et al (2014) Notes from the field: Heartland virus disease—United States, 2012-2013. MMWR Morb Mortal Wkly Rep 63:270–271
105. Oklahoma State Department of Health. Oklahoma State Health Department confirms first case and death of Heartland virus. http:/www.ok.gov/health/Organization/Office-of-Communication-News-releases/2014-New-Releases/Oklahoma-State–Health Department-Confirms-First-Case-and-Death-of –Heartland-virus.htm/
106. Riemersma KK, Komar N (2015) Heartland virus neutralizing antibodies in vertebrate wildlife, United States, 2009-2014. Emerg Infect Dis 21:1830–1833
107. Chowdri HR, Gugliotta JL, Beradi VP, Goethert HK, Molloy PJ, Sterling SL, Telford SR III (2013) *Borrelia miyamotoi* infection presenting as human granulocytic anasplasmosis. Ann Intern Med 159:21–27
108. Krause PJ, Narasimham S, Worser GP et al (2013) Human *Borrelia miyamotoi* infection in the United States. N Engl J Med 368:291–293
109. Hovius JW, Wever B, Sohne M et al (2013) A case of meningoencephalitis by the relapsing fever spirochete *Borrelia miyamotoi* in Europe. Lancet 382:658
110. Sato K, Takano A, Konnai S et al (2014) Human infections with *Borrelia miyamotoi*, Japan. Emerg Infect Dis 20:1391–1393
111. Molloy PJ, Telford SR III, Chowdri HR et al (2015) *Borrelia miyamotoi* disease in northeastern United States. A case series. Ann Intern Med 163:91–98
112. Bacon RM, Kugeler KJ, Mead PS, Center for Disease Control and {Prevention [CDC] (2008) Surveillance for Lyme disease United States, 1992-2006. MMWR Surveill Summ 57:1–9
113. Scoles GA, Papers M, Beati L, Fish D (2001) A relapsing fever group transmitted by *Ixodes scapularis* ticks. Vector Borne Zoonotic Dis 1:21–34
114. Kosoy OI, Lambert AJ, Hawkinson DJ, Pastula DM, Goldsmith CS, Hunt DC, Staples JD (2015) Novel Thogotovirus associated febrile illness and death, United States, 2014. Emerg Infect Dis 21:760–764
115. Liu XY, Bonnet SI (2014) Hard tick factors implicated in pathogen transmission. PLoS Neglected Trop Dis 8:e2566
116. Nuttall PA, Trimmell AR, Kazimirova M, Labuda M (2006) Exposed and concealed antigens as vaccine targets for controlling ticks and tick-borne diseases. Parasite Immunol 28:155–163

Chapter 6
Chikungunya Virus and Zika Virus Expansion: An Imitation of Dengue Virus

6.1 Introduction

Chikungunya virus [CHIKV] and dengue viruses [DENV] have many similarities in clinical manifestations and epidemiology, as well as means of transmission by *Aedes* species of mosquitoes. Moreover, the emergence and pattern of spread of dengue virus disease [DVD] from tropical and subtropical regions of the world to a global dispersal more than 50 years ago is being repeated by CHIKV in the last few years. Although both conditions are largely self-limited, systemic viral infections transmitted primarily between humans and mosquitoes, they evolve from Africa or Asia as zoonotic infections between nonhuman primates and other small animals with humans as secondary hosts. Presently, endemic and epidemic outbreaks of dengue viruses and CHIKV are through human-to-human transmission by mosquitoes, rather than enzootic or epizootic means. The spread of these viruses globally is directly related to the dispersal and adaptation of the vectors, *Aedes aegypti* and more recently *Aedes albopictus*. The primary vector, the urban adapted *Ae. aegypti mosquito,* is widely distributed in tropical and subtropical regions of the world. It is believed that this vector emerged from Africa during the slave trade in the fifteenth to the nineteenth centuries to the Americas and Caribbean, spread through Asia during commercial trading in the eighteenth and nineteenth centuries, and spread more globally in the past 50–60 years with increased international travel and trade [1]. *Ae. aegypti* was initially introduced in Europe during the seventeenth to nineteenth centuries where it existed in southern Europe until its disappearance during the twentieth century and has since returned on Madeira and the Black Sea Coast [2]. *Ae. aegypti* eggs may have been transmitted to the Americas by water containers on slave ships, and now it is believed that larvae and eggs of the Asian tiger mosquito, *Ae. albopictus,* were imported into the western hemisphere from Japan through trade of used car tires, where the mosquitoes lay its drought-resistant eggs in collected rainwater [3]. Although the burden global of dengue is estimated between 50 and 100 million cases a year affecting about 100

I.W. Fong, *Emerging Zoonoses*, Emerging Infectious Diseases of the 21st Century,
DOI 10.1007/978-3-319-50890-0_6

countries with over 2.5 billion people at risk [4], there is no similar estimate of the global burden or annual incidence of CHIKV disease as outbreaks are more variable, and small sporadic outbreaks are often attributed to dengue. Zika virus, also transmitted by *Aedes* species of mosquitoes, was recognized to cause mild sporadic disease in Africa and Asia for decades until its emergence in the southwestern Pacific Ocean in 200 and sudden explosive outbreak throughout the Americas and Caribbean in 2015–2016.

6.2 Historical Aspects

6.2.1 CHIKV

CHIKV was first isolated from febrile patients during an outbreak in the southern province of Tanzania [Makonde Plateau] in 1952–1953 [5]. The name chikungunya was derived from the Swahili or Makonde word meaning "to become contorted" or "that which bend up," due to the severe muscle and joint pains that can continue for years and may be so severe that some patients adopt a bent or stooping position [5]. *Ae. aegypti* is the primary mosquito vector in Africa and other tropical or subtropical countries, but multiple other *Aedes* species have been implicated in Africa, where CHIKV is maintained in a sylvatic cycle among mosquitoes, wild primates, and other small mammals [rodents, bats, and squirrels] and birds [6]. However, urban transmission from virus circulating in eastern Africa in nonhuman primates to humans had occurred on multiple occasions [7]. Outbreaks of CHIKV infection across Africa were associated with heavy rainfall in rural forested areas with increased mosquito population and spillover from the enzootic forest cycle to epizootic savannah cycle, with exposure to nonimmune populations. During epidemics the CHIKV can circulate between human populations without the presence of animal reservoirs [8]. Subsequent emergence and spread of the virus beyond Africa are estimated to have occurred in the eighteenth century, with carriage of infected mosquitoes aboard sailing ships along with barrels of stored water and with humans to maintain the propagation of the mosquitoes and a local cycle [9]. CHIKV of African lineage is believed to have been introduced in Asia in the nineteenth century or sometimes between 1879 and 1956, after adaptation into urban cycle [10].

Outbreaks of CHIKV in Asia were first recognized in 1954 from the Philippines with subsequent outbreaks in 1956 and 1968 [8]. During the 1970s and early 1980s, the virus spread throughout southern and Southeast Asia, with frequent epidemics and sporadic activity in many Asian countries, but eventually the virus activity dwindled and ceased for many years with only localized outbreaks. After several decades of inactivity, CHIKV reemerged in 2000 as an urban epidemic in Kinshasa [Democratic Republic of the Congo] after 39 years of absence [11] and reappeared in Indonesia after 20 years of inactivity in 2001–2003 [12]. In 2004, an outbreak involving the ECSA [eastern, central, and southern African] lineage progenitor began in the coastal towns of Kenya and spread to several Indian Ocean Islands and onto India where it caused massive outbreaks of several million people [13]. Infected air travelers from the Indian Ocean basin outbreaks retuned to Europe, Australia, the Americas and other parts of Asia, and secondary local transmission resulted in Italy, urban France, South and Southeast Asia [10]. The extent of the explosive outbreaks in the Indian Ocean basin has been attributed partly due to the ease of air travel which facilitated rapid spread, large dense urban populations previously naïve to CHIKV exposure and vulnerable to infections, dense populations of *Aedes* mosquitoes with introduction of *Ae. albopictus* mosquitoes into this region from native Asia islands in 1985, and adaptive mutations of the new Indian lineage CHIKV strains which resulted in enhanced transmission by the new mosquito vector [14–16]. *Ae. albopictus* mosquito species were not implicated as a major vector in prior Asian epidemics and older Asian lineage of CHIKV was not well adapted to this mosquito [17].

During the outbreaks of the Indian Ocean lineage [IOL] of CHIKV between 2006 and 2009, many infected travelers returned to the Americas without provoking local transmission, despite many imported cases [18]. However, in October–December 2013, a single case of an Asia-lineage CHIKV infection was introduced in the island of St. Martin and rapidly spread across the western hemisphere. It is estimated that more than 1.2 million cases of CHIKV infection occurred locally from 44 countries and territories throughout the Americas, except for Canada [19]. Local transmission was reported throughout almost all Caribbean islands, all countries in Central America, several South American countries, and parts of Mexico and the USA [20]. In 2014, a total of 2811 CHIKV cases were reported from US states with 12 locally transmitted in Florida; all other cases were from returning travelers from affected countries [21]. ArboNET [the national surveillance system for arthropod-borne disease] reported 4710 cases from US territories for 2014, mainly locally transmitted cases from Puerto Rico, US Virgin Islands, and American Samoa [21].

6.2.2 DENV

The site of origin of DENV is controversial as some consider an African origin and being the same as the principal vector, *Ae. aegypti*, while others contend an Asian origin [22]. Early Chinese literature described dengue-like illness during the Chin Dynasty [Common Era 265–420] and other subsequent dynasties, and seven centuries later, a similar illness appeared in the French West Indies and Panama in the seventeenth century [22]. Possible dengue pandemic occurred in the latter part of the eighteenth century with widespread geographic distribution from present-day Jakarta to Philadelphia. Benjamin Rush has been attributed to provide the first detailed clinical description of DENV infection, as well as the as the application of the term "breakbone fever," in an epidemic in Philadelphia in 1879 [22]. The discovery that dengue fever was due to a virus was from experiments in human volunteers at Fort McKinley in the Philippines in 1907 [23]. *Ae. aegypti* mosquito was confirmed to be the primary vector by 1926 [24] and subsequently *Ae. albopictus* was incriminated as well in 1931 [25].

Records suggest that there were at least five dengue pandemics from Africa to India/Oceana and to the Americas from 1823–1916, each lasting 3–7 years and propagated by the slave trade and commerce with entry via seaports of coastal regions [22]. After the World War 1, the pattern of DENV infections changed in Southeast Asia, Indian subcontinent, and the Philippines from episodic epidemics to persistent endemic state, but the Caribbean and the Americas remained intermittently active. By the late 1920s, major epidemics occurred throughout the Gulf and Atlantic states, the Caribbean, South Africa, Egypt, and Greece, affecting millions of people. With eradication of *Ae. aegypti*, DENV was eliminated from the Mediterranean region by the mid-1940s [22]. By the 1960s and 1970s, there was dramatic increase in dengue activity in tropical and subtropical regions of the world, and by the 1990s there was a global distribution of all DENV serotypes to urban populations. Since the arrival of the twenty-first century, the annual incidence of dengue fever and intensity of epidemics have dramatically increased around the world. Several major outbreaks have occurred in this century in the Americas, Southeast Asia, and Asia, and all DENV serotypes have global hyperendemicity, and it is expected that recurrent outbreaks will continue possibly in cycles of 3–5 years, with annual mortality of 20,000 or more [22].

Although most dengue infections arise from human reservoirs, the ancestral DENVs are believed to have circulated in tropical forests to maintain the cycle among nonhuman primates and mosquitoes. The sylvatic DENVs are genetically and ecologically different from the urban viruses responsible for most endemic and epidemic outbreaks. Sylvatic cycles of DENV have been demonstrated in Asia and Africa between canopy-dwelling mosquitoes and tree-dwelling animals, such as monkeys, slow lorises, civets, and squirrels [26, 27].

6.2.3 Zika Virus

Zika virus [ZIKV] was first discovered in 1947 after a rhesus monkey developed a febrile illness, while caged as a sentinel animal on a tree platform in the Zika forest of Uganda, for research on jungle yellow fever [28]. ZIKV was subsequently isolated from *Aedes africanus* mosquitoes from the same forest in 1948 [29]. Although serological studies in the early 1950s indicated that humans could be infected with ZIKV, the virus was first isolated from humans in 1968 in Nigeria and from 1971 to 1975 [30, 31]. Subsequent studies from 1951 to 1981 found serological evidence of ZIKV infection throughout Africa [Uganda, Tanzania, Egypt, Central African Republic, Sierra Leone, and Gabon] and parts of Asia [India, Malaysia, the Philippines, Thailand, Vietnam, and Indonesia] [32]. ZIKV was isolated from *Ae. aegypti* mosquitoes in Malaysia in 1968–69 [33], and since then it has been considered the primary vector for the virus. Infection with ZIKV first emerged outside Africa and Asia in 2007, when it caused an outbreak in the southwestern Pacific Ocean Yap Island, of the Federated States of Micronesia [34]. Since then it has spread to other Pacific islands, French Polynesia in 2013–2014, Caledonia, Cook Islands, Cape Verde, and then Easter Island [Chile], and subsequently to Brazil and Columbia [35]. Since 2015–early 2016, an explosive outbreak has occurred in the Americas and Caribbean with over 30 countries affected, and it is predicted that by the end of 2016, four million people could be affected [Amedeo Zika Virus, http://amedeo.com/medicine/zik.htm]. By February 1, 2016, the WHO declared ZIKV a global public health emergency because of the reported association of microcephaly in newborns of infected pregnant women in Brazil.

As of August 2016, there have been 29 locally acquired mosquito-borne cases of ZIKV infection in Florida and 2487 travel-associated cases in the USA, of which 22 were sexually transmitted and 7 Guillain-Barre syndrome-associated cases [CDC. Case counts in the US. Accessed August 24, 2016 at http://www.cdc.gov/ncezid]. The ZIKV progression globally is circling back to Asia, as there is now an outbreak in Singapore with over 200 cases and a few cases in Malaysia [CNN broadcast, September 5, 2016]. In the past year, 2015–2016, Puerto Rico had experienced a ZIKV outbreak with at least 28,219 confirmed cases [36]. Moreover, recently locally transmitted cases of ZIKV were reported in Brownsville, Texas by CDC on Dec 14, 2016 [https://www.cdc.gov/media/releases/2016/s1214-brownsville-texas-zikaguidance.html].

6.3 Virology

6.3.1 CHIKV

Phylogenetic studies indicate that there are three CHIKV genotypes: Asian, East/Central/South African, and West African [37]. It has been estimated that the African genotypes emerged between 100 and 840 years ago, while the Asian genotype between 50 and 310 years ago [38]. CHIKV belong to the genus *Alphavirus* of the *Togaviridae* family and is grouped together with the Semliki Forest antigenic complex alphaviruses such as Ross River virus and others, which are also mosquito-borne [38]. CHIKV and other alphaviruses have single-stranded, positive-sense RNA genome which encodes four nonstructural proteins and three main structural proteins, the capsid and two envelope glycoproteins [10]. Unlike DENV there is limited antigenic diversity between the three genotypes of CHIKV, and previous infection seems to provide lifelong immunity [8].

6.3.2 DENV

DENV belong to the genus *Flavivirus*, family *Flaviviridae,* with four distinct but antigenically related serotypes [DENV-1–DENV-4], and each serotype has closely related multiple genotypes [22]. A potential fifth serotype [DENV-5] was isolated from a patient in Borneo, but it is unclear if this strain can cause sustained transmission between humans [39]. Infection with a particular serotype will usually produce lifelong immunity to that serotype and associated genotypes, but cross protection against other serotypes is generally weak and short-lived, lasting for 2–3 months [40]. Dengue virions are spherical, about 50 nm in diameter with a host-derived lipid bilayer containing a single-stranded, positive-sense RNA genome coding for three structural and seven nonstructural proteins [40]. Ancestral dengue virus arose 1000–2000 years ago among monkeys in either Africa or Asia, and it is estimated that DENV-1 and DENV-2 emerged more recently, within the past three centuries [41]. Current evidence supports the hypothesis that all four endemic DENVs evolved from sylvatic progenitors in the forest of Asia and only DENV-2 could have emerged from African jungles [22]. The emergence of endemic DENV with humans as the only reservoir may have occurred within the past 2000 years and theoretically should be controllable with an effective vaccine. However, sylvatic DENV still exist in the forests of Asia and Africa, and fitness studies of sylvatic DENV-2 compared with human-circulating endemic/epidemic strains suggest that little genetic adaptation was required for the initial emergence of endemic strains. Hence sylvatic reservoirs in Asia and Africa will remain as a source of reemergence [22].

6.3.3 ZIKV

ZIKV is a flavivirus related to DENV, yellow fever, West Nile, and the Japanese encephalitis viruses, and it is a positive-sense single-stranded RNA virus with 10,794 nucleotides encoding 3419 amino acids [31]. There is only one strain recognized to date with three to five clades based on site and place of isolation, i.e., Zika Uganda 47, Senegal 84 [a, b, c], and Yap 2007. Based on phylogenetic analysis of ZIKV sequences, there are only two main virus lineages [African and Asian], and the strain responsible for the Yap Island and Pacific Ocean epidemics and the outbreak in the Americas likely originated in Southeast Asia [42, 43]. It has been estimated that ZIKV emerged in East Africa around 1920, and in the late 1940s, the seropositivity was 6.1% of the population in Uganda, and by the late 1960s, Kenya demonstrated seropositivity of 52% [44]. ZIKV has been isolated from several species of *Aedes* mosquitoes and wild monkeys, which are considered the natural hosts. In Africa ZIKV still maintains a sylvatic transmission cycle between nonhuman primates by species of *Aedes* mosquitoes, but in Asia this has not been demonstrated but possibly exists. It was demonstrated in the laboratory that infected *Ae. aegypti* mosquitoes could transmit ZIKV to mice and monkeys and that the extrinsic incubation period in mosquitoes was about 10 days [45]. The virus can be isolated from monkeys 9 days after inoculation and in humans the first day of illness and can be detected up to 11 days after onset [31, 33]. *Ae. aegypti* mosquitoes are the main vectors of ZIKV, but it is likely the virus can also be transmitted by *Ae. albopictus*. Most of the ZIKV transmission is from human to mosquito-human spread of the virus, but transmission of infections has been documented through intrauterine infection of the developing fetus, sexual transmission, potential blood transfusion, and laboratory exposure [46–49]. Sexual transmission from male returning travelers who acquired infection in South America has been reported, and this may occur before, during, and after development of symptoms [50, 51]. High viral RNA or replicative viral particles have been detected in sperm up to 62 days after onset of symptoms [52, 53]. Preliminary data [one case] indicate that ZIKV can be transmitted by body fluids such as tears or saliva, when the viral load is extremely high [54]. MMWR Morb. Mortal. Wkly Rep. 2016 Sep. 13; e-pub [http://www.cdc.gov/mmwr/volumes/65/wr/mm653e4.htm].

6.4 Vectors of CHIKV, ZIKV, and DENV

Ae. aegypti, the primary mosquito vector for DENV, ZIKV, and CHIKV, is distributed worldwide in tropical and subtropical regions between 35°N and 35°S [latitude], with the lowest winter temperature of 10 °C in regions where the species can survive overwintering [55]. Expansion of *Ae. aegypti* population globally appears to have occurred after the end of eradication programs in the 1970s with return of the pre-eradication levels by 1995 [56]. The mosquito was previously widespread in southern Europe where it caused local outbreaks of yellow fever in the nineteenth century. It is well adapted to the urban environment and breed in small puddles of water and water collected in cans, empty containers, or old car tires in yards or homes. The female mosquitoes are the vectors of these viruses, and they feed on human hosts from early morning to just before sunset. After acquiring the virus from an infected host, there is an incubation period of 4–10 days and the mosquitoes remain infectious for their life-span [56]. Although the mosquitoes have limited range of flight, about 500 m, they are capable of feeding on multiple persons in a short period of time, and the movement of the infected individuals throughout and between the communities may help drive epidemics [56]. *Ae. albopictus,* another urban or domestic mosquito, is considered a secondary vector for DENV and CHIKV and has been implicated in previous epidemics of both viruses. *Ae. aegypti* had been eradicated by DDT in some regions and *Ae. albopictus* emerged to cause outbreaks of DENV and even larger epidemics of CHIKV in the islands of the Indian Ocean between 2006 and 2006 [57]. It appears that CHIKV had undergone adaptive mutations to *Ae. albopictus* to allow for greater transmissibility with more efficient crossing of the mosquito gut membrane barrier [15]. *Ae. albopictus* tolerates a wider range of temperature variation and environment than *Ae. aegypti*, from tropical to temperate climate, due to diapause or mechanism for tolerating harsh winters [57]. Although *Ae. aegypti* is endemic in some southern US states [Florida, Texas, and California], *Ae. albopictus* has spread to 36 states with its northernmost boundary in the northeastern USA including parts of New Jersey, southern New York, and Pennsylvania [58]. This mosquito is endemic in southern Europe where it has caused local outbreaks of CHIKV infection and colonized almost all Mediterranean countries [59]; besides its widespread distribution includes throughout the Americas [excluding Canada], Europe, Asia, Africa, Australia, and the Pacific [41]. In dengue endemic areas, *Ae. aegypti* is considered the major vector for transmission, and *Ae. albopictus* is considered less efficient at transmitting the virus; however, some studies suggest that the latter may be more effective at dengue transmission because of the longer life-span of this species [60]. It has also been suggested that *Ae. albopictus* is less adapted to the urban environment than *Ae. aegypti* and may be involved in maintaining the sylvatic and rural transmission in endemic areas [61]. In a recent unique study, the direct susceptibility of these two species of mosquito to dengue infection after feeding on viremic humans was compared, *Ae. albopictus* was significantly less likely than *Ae. aegypti* to develop an infectious phenotype 14 days after direct feeding for DENV-2 and DENV-4 but not for serotypes 1 and 3 [62].

Various *Aedes* species had been found to be infected with ZIKV and are potential vectors of the virus [63], although transmission is believed to be mainly by *Ae. aegypti.* However, *Ae. albopictus* [which is capable of transmitting over 20 arboviruses] has been shown in the laboratory to be capable of transmitting ZIKV [64]. Furthermore, there is evidence that *Ae. albopictus* carried ZIKV in the wild and likely transmitted the virus to humans in Africa [65].

6.5 Pathogenesis of Disease

6.5.1 CHIKV Disease Pathogenesis

Currently the mechanisms involved in the pathogenesis of CHIKV disease are incompletely understood. The virus is capable of infecting and replicating in human endothelial cells, fibroblasts, and macrophages but cannot replicate in lymphocytes, monocytes, and dendritic cells, and replication was associated with cytopathic effect and induction of apoptosis in infected cells [66]. In subjects with severe CHIKV infection, increased proinflammatory cytokines, interleukin [IL]-1 and IL-6, are associated with severity of disease [67], and dysregulation of the inflammatory response may be inducing the disease manifestations. Moreover, the virus can infect and replicate in human muscle cells with inflammatory reaction which can explain the symptoms of CHIKV-induced myositis [68]. Studies in mice and nonhuman primates showed that the virus replicates in high concentration in joint tissues with associated influx of inflammatory cells, mainly monocytes, macrophages, and natural killer cells, and foot swelling in the mouse was reduced with macrophage depletion [69]. Furthermore, in the primate model during the acute phase of CHIKV infection, the virus disseminates to the lymphoid tissue, liver, central nervous system, joints, and muscle, and chronically [44 days postinfection], there is low-grade persistence of the virus in splenic macrophages and endothelial cells of the liver sinusoids [70]. In the mouse model, type 1 interferons [IFNs] are upregulated and are important in controlling the infection, and type 1 IFN receptor-deficient mice are more susceptible to severe disease [71]. Neuroinvasion and replication in the brains of infected mice are also seen with pathological changes in the brain parenchyma and meningeal inflammation and associated neurological signs [72].

6.5.1.1 Immune Response to CHIKV

Type-1 IFNs, as part of the innate immunity, are central to the control of viral infections including CHIKV [73]. Recent studies indicate that CHIKV does not activate leukocytes or dendritic cells for the induction of IFNs, but the type-1 IFNs are produced by infected fibroblasts [74]. Clinical studies indicate that the innate immunity efficiently clears circulating CHIKV within 4–7 days [73]. Normally adaptive immune response with activation of specific B and T lymphocytes occurs after a week and likely does not play a role in controlling the acute infection, but the relative role in the chronic phase of the disease is unknown. Generation of CHIKV-specific neutralizing antibodies to protect against reinfection occurs during convalescence, and recovery and prophylactic administration of specific antibodies can protect against infection or disease in mice, but not 24 h after inducing infection [75]. The role of cellular immunity in CHIKV infection is undetermined, but there is likely a beneficial response, as alphavirus-specific cytotoxic T lymphocytes can eliminate persistently infected macrophages with Old World alphaviruses [76]. Many patients recovering from CHIKV infection have chronic joint disease for several months to years, and the mechanism remains unclear. A single report suggests possible induction of autoimmune response by lymphocytes caused by cross-reactivity between viral and host antigens [77]. It has also been suggested that chronic joint symptoms are due to persistence of the virus with reactive inflammation with persistent virus-specific IgM [78], and persistence of viral RNA has been demonstrated in a patient with chronic arthralgia for 18 months after CHIKV infection [79]. Current evidence indicates that CHIKV infection results in lifelong immunity and there is no significant antigenic variation between genotypes to result in immune escape [8].

6.5.2 DENV Disease Pathogenesis

The pathophysiology of DENV infection is not well understood because of the lack of a suitable animal model. Most studies have focused attention on severe dengue disease or the dengue hemorrhagic syndrome [DHS]. Clinical epidemiologic studies have identified several risk factors for severe disease: young age, female sex, high body mass index, virus strain, and genetic variants of the MHC-1-related sequence B and phospholipase C epsilon 1 genes [80]. Reinfection or two sequential infection with different serotypes is linked to severe dengue disease through antibody-dependent enhancement of the immune response, with the infecting virus forming an immune complex that enter Fc receptor-bearing cells [81]. The concept of antibody-dependent enhancement by weakly cross-reactive antibody from another serotype does not explain adequately that in most dengue endemic countries, the populations at large have been exposed to multiple serotypes with cross-reactive antibodies to all four serotypes by age 14 years and above, yet severe dengue syndromes are relatively rare [82–84]. Studies in Thailand found that only 0.5–2% of secondary infection resulted in shock [84]. This suggests that there is a genetic predisposition to severe dengue syndromes. Another aspect of the hypothesis of severe disease mechanism suggests that high viral load in the blood with infection of endothelial cells results in stimulation of high levels of cytokines and soluble mediators, resulting in vascular fragility and bleeding. However, there is no good evidence to support the correlation of high viral load and severity of disease. Hence, it has been proposed that an aberrant immune response with an imbalance between proinflammatory and anti-inflammatory cytokines results in a cytokine storm [85]. Several cytokines may induce bleeding and increase vascular permeability, through induction of apoptosis [cell death] of platelet precursors and endothelial cells, to produce thrombocytopenia and affect endothelial cell adherent junctions [85].

There is also evidence that the sequence of DENV serotypes infection plays a role in the risk of development of severe dengue syndromes. The highest risk for severe disease has been associated with DENV-1/DENV-2 followed by DENV-1/DENV-3 and lowest for DENV-2/DENV-3 [86]. Studies have found that young children have the highest risk of developing severe disease after a second infection [85] and infants born to dengue-immune mothers are at the greatest risk of severe disease [85]. It was initially opined that risk of severe dengue syndromes would occur only within the first 5 years after primary infection, but attack rates of these syndromes in the same population with primary DENV-1 and secondary DENV-2 were significantly higher 20 years versus 4 years after repeat infection [84]. A recent review and meta-analysis of factors associated with dengue shock syndrome [DSS] confirmed the association of several recognized risk factors, young age, secondary dengue infection, female sex, DENV-2, and several manifestations of neurological disorders and liver damage [the latter two factors likely represent severity of the disease], but there have been a sustained decline in DSS and dengue hemorrhagic fever [DHF] in Southeast Asia for the last 40 years [87]. It is postulated that declining rates of severe dengue disease may be related to educational programs, early rehydration treatment, and the increasing age of the average population.

6.5.2.1 DENV Immunity

Following a mosquito bite with DENV, the virus enters and replicates in Langerhans cells and then migrates to lymphoid tissue for further amplification, and viremia can be detected in 5–7 days with primary infection and only lasts 2–3 days during secondary infection [88]. During primary viremia, the virus infects tissue macrophages of several organs, especially the liver and spleen where the virus may continue to replicate after the cessation of the viremia [88]. The initial innate response to the infection consist of the combined effect of type-1 IFNs, tumor necrosis factor alpha [TNF-α], and cell surface receptor-ligand interactions in stimulating the anti-dengue response of primary human natural killer [NK] cells [89]. The NK cells produce cytokines and lysis of target infected cells. There is also recent evidence that activated and apoptotic platelets aggregate with monocyte and signal-specific cytokine [IL-1β, IL-8, IL-10, and MCP-1] responses that may contribute to the pathogenesis of dengue [90]. During primary infection, IgM antibodies become detectable about day 5 after onset of symptoms, peaks at day 10, and may last for 6 months [86]. Neutralizing IgG antibodies appear after the first week with primary infection and peak during convalescence, 14 days or after, and provide specific serotype protection for life and cross protection for heterologous virus for at least 2–3 months after infection and for up to 1–3 years [88]. During secondary infection, IgG antibody is produced faster and to higher titers and IgM if produced is diminished and may last for four months.

It is believed that secondary infection of a heterologous virus can induce antibody-dependent enhancement [ADE] of infection. The exact mechanism of ADE is still unclear. It is postulated that cross-reactive non-neutralizing antibodies enhance the uptake of virus by monocytic cells and this results in extreme immune activation that results in cytokine storm [91]. Other hypotheses include that virus-antibody complexes activate the complement pathway to enhanced immune response resulting in cytokine storm or dengue-specific antibodies cross-react with host proteins to activate the coagulation pathway and alter endothelial cell function [91]. There is evidence that CD4+ and CD8+ T-cell activation plays a role in protection against dengue infection and aberrant CD8+ cell response may be involved in the cytokine storm in severe disease [88, 92]. Occasionally severe dengue disease [DHF] can be seen in primary infection in older children and adults [92].

6.5.3 ZIKV Disease Pathogenesis

There is sparse data on the pathogenesis of ZIKV infection, but once the virus enters the human body by a mosquito bite, it infects dermal fibroblasts and dendritic cells through adhesion factors and induces transcription of Toll-like receptor 3 [TLR3] and other molecules, to stimulate several IFN genes and strongly enhance IFNβ gene expression [93]. ZIKV gains entry to skin fibroblasts and immune cells by the phosphatidylserine receptor AXL and upregulates the autophagy pathway, which enhances replication in autophagosomes [93]. There is evidence that flavivirus replication and pathogenesis involve several cellular pathways such as endoplasmic reticulum stress, cellular signaling response termed unfolded protein response, and autophagy [94]. Limited studies in patients had demonstrated polyfunctional T-cell activation, Th1, Th2, Th9, and Th17 response, during the acute phase with respective cytokine level increases, followed by a decrease in the recovery phase [95]. ZIKV is sensitive to the antiviral effects of both type-1 and type-2 IFNs [95], which likely result in clearance of the virus. However, flaviviruses in general surmount IFN type 1 signaling in order to proliferate intracellularly. ZIKV accomplish this by degradation of the STAT2 signaling molecule downstream of the IFN receptor in human cells [96].

In utero transmission and congenital abnormalities from ZIKV infection have attained global attention since the large epidemic in Brazil and the Americas, such as microcephaly, spontaneous abortion, and intrauterine growth restriction. Recently, four studies using the mouse pregnant model have investigated the effect of ZIKV on the fetuses [97–100]. Inoculation of different isolates of ZIKV was via peripheral routes or directly into the fetal brain, and all studies demonstrated infection of the fetuses with brain cell damage. One of the studies [97], using deficient-type IFN-signaling mouse model and cutaneous inoculation of an Asian ZIKV strain in the first trimester [human equivalent], showed placental infection, injury, and impairment, with fetal brain injury and neuronal cell death. These findings would explain the human experience with ZIKV in the first and second trimester of pregnant women. These studies indicate that ZIKV can cross the placental barrier to gain access to the fetus and infect placental cells and fetal endothelial cells to cause vascular damage and impair growth and development. Moreover, seeding to the fetal brain can injure or cause death of neuroprogenitor cells and inhibit cell differentiation to explain the cortical thinning, microcephaly, and brain structure abnormalities in neonates of infected pregnant women [101]. Although ZIKV has been documented in maternal blood 5 days after acute symptoms, it also has been rarely reported 8 weeks and even up to 107 days after the onset of symptoms, suggesting persisting viral replication in the fetus or placenta [102].

6.5.4 ZIKV Immunity

Most ZIKV infections are asymptomatic. Viremia is believed to occur from several days before onset of illness to a week after illness and has the potential for transmission by blood transfusion [103]. ZIKV-specific IgM antibodies develop during the first week of illness, but data on duration are limited [33]. IgG neutralizing antibodies develop shortly after IgM antibodies and probably persist for many years after infection to confer prolonged and possibly lifelong immunity [104]. However, this presumption is largely based on data from other related flaviviruses such as West Nile virus and yellow fever virus. There is no known antigenic variation of different genotypes; thus, immunity should be cross protective for isolates from different regions of origin, as there is no strain subtypes of ZIKV.

6.6 Clinical Manifestations

6.6.1 CHIKV Clinical Disease

The acute manifestation of CHIKV infection is abrupt with high fever, chills, headaches, myalgia, and arthralgia and clinically indistinguishable from dengue fever. The incubation period ranges from 1 to 12 days but averages 2–4 days, but unlike dengue asymptomatic infection is very uncommon, about 3–25% of people with serological evidence of infection [8]. A maculopapular, erythematous rash on the trunk with possible involvement of the face, limbs, palms, and soles can be present in 20–80% of patients and is rarely bullous in children [10]. Severe polyarthralgia, which can be disabling, is the hallmark of this disease, and debilitating polyarthralgia in endemic countries even with circulating DENV have a positive predictive value of 80% for CHIKV viremia [8, 105]. The arthralgia in dengue fever is usually milder and of shorter duration during first few days with fever. In CHIKV infection, polyarthralgia is usually symmetrical and involves joints of the arms and lower limbs [about 90%] and less often the spine, and large joints and small joints are frequently affected [8, 10]. Joints with previous damage from other disease such as osteoarthritis are often more severely affected with disabling arthritis. Swelling and periarticular edema may be seen especially in the interphalangeal joints, wrists, and ankles and represent active arthritis, but joint effusions have not been a notable feature. Nonspecific symptoms of vomiting, diarrhea, abdominal pain, confusion, and weakness have been reported in large outbreaks of hospital referred patients [8].

Although CHIKV is not considered a neurotropic virus, various neurological complications have been reported in multiple large outbreaks from India and the Indian Ocean Islands, both in adults requiring hospitalization and in young children and neonates with mother-to-child in utero transmission [8, 106]. In adults requiring admission to hospital in these outbreaks, neurological manifestations have been reported in 15–25% of cases, and these include encephalopathy, seizures, encephalitis, Guillain-Barre syndrome, and encephalomyelitis [72]. Neurological signs were less frequent and severe in hospitalized children but also included seizures, encephalitis, meningeal symptoms, and acute encephalopathy. Infected neonates were especially prone to CNS complications with 50% showing abnormalities on magnetic resonance imaging of the brain including white matter lesions, cerebral hemorrhage, and edema, which may lead to death and severe disabilities in 10–20% [72]. Autopsy of infected neonates and murine model of CHIKV infection have demonstrated brain pathological changes from the infection. Death rate following CHIKV infection was estimated to be 1:1000 cases in La Reunion outbreak [106] but is usually reported rarely in other outbreaks, and mortality is generally associated with multiple comorbid illness in the elderly or neonatal infection. These complicated cases may be associated with rare complications such as hemorrhage, liver dysfunction, renal failure, heart failure, and myocarditis [8, 10, 66]. Other rare manifestations of CHIKV infections include conjunctivitis, uveitis, retinitis, iritis, and chondritis of the ear [10].

Most patients improve 1–2 weeks after the onset of acute illness, but a high proportion of patients have persistent arthralgia from several months to years. In a study from the Reunion Island outbreak, 57% of infected patients had persistent or recurrent polyarthralgia after 15 months [107]. In this study and others, persistent arthralgia was related to age and being greater in those older than 45 years of age. Even 3 years after acute infection of CHIKV 12% of patients may still have residual joint pain, stiffness, and swelling [108]. Chronic arthralgia is more common in the distal joints and may mimic rheumatoid arthritis, with increased inflammatory markers and erosive changes on imaging [rarely deforming polyarthritis] in up to 50% of patients followed for 36 months [109]. Patients with joint pains up to 36% met the criteria for rheumatoid arthritis but with negative rheumatoid factor, and MRI findings may include joint effusion, bony erosions, marrow edema, synovial thickening, tendinitis, and tenosynovitis [110]. Treatment of acute and chronic arthralgia of CHIKV infection is primarily the use of nonsteroidal anti-inflammatory agents [NSAIDs], and although methotrexate has been used for more disabling arthritis, there is no clinical trial with immunosuppressive agents [111].

CHIKV infection is most often confirmed by serological methods. Indirect immunofluorescence and ELISA tests are sensitive and rapid techniques for diagnosis and differentiate the presence of IgM or IgG antibodies [8]. Specific IgM antibody can be detected between 2 and 7 days after onset of illness by ELISA and immunofluorescence and even after 1 day by lateral-flow rapid test. IgG antibodies are most often detected after 5–7 days and occasionally after 2 days. IgM antibody persists for 3–4 months, and reports of persistent IgM after 2 years may suggest persistent virus in the joints [78]. IgG antibody specific for CHIKV persists for many years. CHIKV can be isolated on mammalian cell cultures such as Vero cells, and RT-PCR can be used to detect the virus from 1 to 7 days after onset of symptoms [8]. Antigen capture ELISA technique has also been used to detect the virus in serum and cerebrospinal fluid after 2 days of illness [10].

6.6.2 DENV Disease

Most patients infected with the DENV in endemic regions are asymptomatic or experience minimal symptoms [112], a major difference from CHIKV infection. The incubation period after inoculation is 3–7 days and starts suddenly, and the illness is divided into three phases: an acute febrile phase, a critical phase around the time of defervescence, and a recovery phase [80]. The initial phase is usually heralded by high fever, headaches, retro-orbital pain, myalgia, arthralgia, nausea and vomiting, and sometimes a transient macular rash [80, 88]. Mild bleeding, bruising, and petechiae may be evident and an enlarged liver may be palpable. Although children have high fever, the other symptoms may be less severe than adults in this phase. Common laboratory abnormalities often include mild to moderate leukopenia, thrombocytopenia, and liver enzyme elevations, similar to disturbances that can be seen with CHIKV infection. Majority of patients recover without complications after 3–7 days.

A small proportion of susceptible patients, mainly children and young adults, may progress to the critical phase at the time of resolving fever [80, 88]. This stage of illness is characterized by worsening symptoms of weakness with systemic signs attributable to increased vascular permeability and leaky vessels. These consist of signs of hypotension and shock, associated with worsening and persistent vomiting, abdominal pain with tender hepatomegaly, development of pleural and peritoneal effusions, and subsequently bleeding from the mucosa, restlessness, and lethargy. Failure to institute adequate medical therapy may result in refractory shock, severe gastrointestinal bleeding, hepatorenal failure, encephalopathy, rarely myocarditis, and death. Results of laboratory tests at this phase usually reveal moderate to severe thrombocytopenia, with platelet counts often below 20×10^{-9} per liter, and increase in activated partial-thromboplastin time, and low fibrinogen may exhibit levels resembling but not typical for disseminated intravascular coagulopathy [80].

Up to two-thirds of patients may have hypokalemia, and hypokalemic paralysis is an underemphasized neuromuscular complication of severe DENV infection [113]. Elevated creatine kinase from rhabdomyolysis is seen occasionally [114], and acute kidney failure occurs in 2–5% of patients and carries a high mortality [115]. In a recent study of 796 pediatric patients, the most common findings included thrombocytopenia [96%], abdominal pain [71%], and vomiting [59%], but the most important factors associated with severe dengue were rash, severe thrombocytopenia, and anemia [116]. This critical phase is short-lived and lasts for 48–72 h with subsequent rapid recovery in the recovery phase in most cases. The mortality in patients with refractory shock, usually from late medical attention, can be up to 20% but with early appropriate treatment is usually very low, from 0.1% to 0.5% [88]. The critical phase is characterized by very low viremia, peak T-cell activation, and vasculopathy. Current evidence suggests that platelet and complement activation through antibody-mediated immune complex results in the combined effect of anaphylatoxins and inflammatory molecules and platelet sequestration results in vasculopathy [117]. Since 2009, the World Health Organization has replaced the 1997 classification of dengue/dengue hemorrhagic fever/dengue shock syndrome with dengue and severe dengue [118].

During the recovery phase, there is gradual improvement of any organ dysfunction, and a second mild maculopapular rash may appear that eventually resolves in 1–2 weeks, but fatigue may persist in adults for several weeks [80]. Unlike CHIKV infection, chronic sequelae from dengue have been unrecognized. However, neurological complications such as encephalitis, myositis, myelitis, Guillain-Barre syndrome, and neuropathies can occur with long-lasting effect [119, 120].

Laboratory diagnosis of dengue is usually accomplished by serological methods. IgM antibodies can be detected as early as 4 days after onset of fever by ELISA or lateral-flow rapid test [80]. Although IgM antibodies from a single specimen provide a presumptive diagnosis, seroconversion from paired sample is confirmatory. In secondary dengue infection, dengue-specific IgG antibodies may predominate over IgM due to the rapid anamnestic antibody response. In the initial febrile phase rapid, early diagnosis can be made by detection of DENV RNA by RT-PCR or detection of the virus-expressed soluble nonstructural protein 1 [NS1] by ELISA or lateral-flow rapid test [80]. In primary infection during the febrile phase, detection of NS1 antigen is >90% but is 60–80% in secondary infection [121, 122]. Antigenemia may persist for several days after resolution of the fever, and combined antigen and antibody testing improves the diagnostic sensitivity [123].

Management of dengue fever is largely symptomatic and supportive as with CHIKV infection. Mild to moderate disease can be managed with NSAIDS for pain control, but aspirin is best avoided because of increased risk of bleeding. Severe disease requires prompt fluid and electrolyte disturbance correction with intravenous crystalloids and oral replacement therapy. Blood and platelet transfusions may be required for bleeding complications, but there is a lack of guidelines for transfusion support in patients with severe dengue and no studies on the benefit of prophylactic platelet transfusion [124]. Patients in shock may require vasopressors to support an adequate blood pressure, hemodialysis for those in renal failure, treatment for heart

Table 6.1 Comparison of CHIKV, ZIKV, and DENV infections

	CHIKV	DENV	ZIKV
Virology	*Togaviridae*	*Flaviviridae*	*Flaviviridae*
Virus	Enveloped, single-stranded RNA	Same	Same
Origin	Africa	Africa or Asia	Africa
Strains	1 strain, 3 genotypes	4–5 strains	1 strain, 2 lineages
Vector	*Aedes aegypti, Ae. albopictus*	Same	Same
Distribution	Tropical and subtropical globally	Same	Africa, SE Asia, Americas, Pacific
Clinical aspects			
Incubation period	2–4 days	3–7 days	3–12 days
Subclinical infection	Rare [3–<20%]	Common	80%
Symptoms: acute	Severe polyarthralgia	Flu-like illness, bruising	Mild flu-like illness
Duration of acute	7–14 days	5–7 days	3–5 days
Severe disease	Very rare [neonates]	Children [2%]	Rare Guillain-Barre syndrome
Critical phase	None	Shock and bleeding [48–72 h]	None
Recovery phase	None	7–14 days	None
Chronic disease	Arthralgia/arthritis [12–50% at 3 years]	None	None
Congenital	Rare CNS disorder	None	Microcephaly
Treatment	NSAIDS	NSAIDS, blood and platelets	Acetaminophen, NSAIDS
Mortality	<1: 1000	Severe dengue—0.2–20%	Stillbirth [rare]
Promising drug	None	Prochlorperazine at onset	None
Vaccine	None	Dengvaxia approved in Mexico	None

failure in subjects with myocarditis and fluid overload, and mechanical ventilation to manage ARDS and comatose patients. Table 6.1 shows a comparison of the clinical and virological features of DENV, ZIKV and CHIKV infections.

6.6.3 ZIKV Disease

Prior to the Yap Island outbreak of 2007, ZIKV caused only sporadic infections with mild flu-like illness resembling a mild form of dengue fever in Africa and Southeast Asia. Thus cases of ZIKV infections were probably misdiagnosed and under recognized and under reported. In the Yap Island outbreak, it was estimated that 73% of the population were infected, but most patients were asymptomatic [about 80%], and clinical symptoms were mild with mainly fever, rash, arthralgia, myalgia, and conjunctivitis [125]. Although *Aedes hensilli* was the predominant mosquito circulating at the time, the virus was never detected in this species. A large epidemic of ZIKV occurred in French Polynesia in 2013–2014 before spreading throughout the Pacific region [126]. This unusual epidemic was unprecedented as there were concurrent outbreaks of DENV, CHIKV, and ZIKV around the same time [127]. In the French Polynesia, there were reports of increased severe neurological complications [Guillain-Barre syndrome] first time associated with ZIKV [127], but possibly related to concomitant infection with CHIKV or DENV. The Polynesia outbreak occurred rapidly soon after the index case, with 28,000 cases of ZIKV infection in the first four months and 42 cases of Guillain-Barre syndrome [44]. Based on phylogenetic analysis, the ZIKV introduction to Polynesia was independent of the Yap Island outbreak and came separately from Southeast Asia. [44]. Transmission to the Americas appears to have originated from the Pacific islands.

ZIKV spread from the South Pacific islands to South America by February 2014, where it was first recognized to be locally transmitted on Easter Island off the coast of Chile, and locally transmitted cases appeared in May 2015, in the northeastern region of Brazil [WHO; Weekly epidemiological record. November 6, 2015. http://www.who.int/wer]. By December 2015, there were 440,000–1,300,000 cases of ZIKV infection in Brazil with a 20-fold increase in infants born with microcephaly and with detection of the virus RNA in the amniotic fluid of affected newborns [128]. Microcephaly has never been associated with previous ZIKV outbreaks, but the virus was detected in tissues of a baby with microcephaly that died shortly after birth [129]. Lack of standardization of the definition of microcephaly may have affected the reporting of this condition, but CDC has recommended a definition of an occipitofrontal circumference below the third percentile for age and sex [130]. Microcephaly can lead to seizures, developmental delays, learning disabilities, and impaired motor function. Infections associated with microcephaly would usually occur during the early weeks of fetal development to result in loss of brain cells in the developing brain. However, microcephaly can result from fetal brain disruption sequence, with normal brain development in early pregnancy and subsequent collapse of the skull with fetal brain destruction [131]. Reports of infants with microcephaly related to ZIKV infection in Brazil are consistent with fetal brain disruption, and there is some evidence that this could occur even in the second or third trimester of pregnancy, although the greatest risk is in the first trimester [132]. Preliminary report from Brazil has noted the presence of fetal abnormalities in 29% of pregnant women infected with ZIKV on ultrasonography [133]. There is increased fetal death with infection between 6 and 32 weeks of gestation [133]. The possible explanations for this newly recognized complication of ZIKV infection are the following: (1) previous outbreaks were too small to detect a rare complication; (2) the virus may have mutated to become more neurovirulent; and (3) synergistic effect of coinfection with another circulating virus such as CHIKV or DENV. However, a recent retrospective study of the French Polynesian outbreak in 2013–2015 suggests that the association of microcephaly and ZIKV infection was missed but is evident on recent analysis [134]. In this study, the baseline prevalence of microcephaly was two cases per 10,000 neonates, and the risk associated with ZIKV infection was 95 cases per 10,000 women infected in the first trimester. Other neurological complications of ZIKV recently reported include meningoencephalitis, acute myelitis, and optic abnormalities [especially in neonates with or without microcephaly] [135, 136].

Diagnosis of ZIKV infection is often assessed by serological methods such as ELISA for IgM or IgG antibodies, which can be difficult to interpret due to cross-reaction with other flavivirus such as DENV or West Nile virus, and antibodies may not be present in early infection. Fetal cord blood at birth with ZIKV-IgM would be specific as IgM would not cross the placenta barrier. Antibodies appear about 5–6 days after onset of symptoms, and confirmation of a positive serology may require the ZIKV plaque neutralization test [PRNT], which is laborious and time consuming [137]. For the acute diagnosis with symptoms less than 10 days, serology and RT-PCR should be performed. A one-step rRT-PCR appears to be very sensitive and specific [138]. ZIKV is present in the blood for the first 3–5 days of illness but can be detected in urine >10 days after onset of illness [139], and in saliva even when the blood was negative by RT-PCR [140].

There is no specific treatment for ZIKV infection and only symptomatic or supportive therapy is available. For pregnant female with headaches or pain, acetaminophen can be used; otherwise, aspirin or NSAIDS may be used when dengue fever is not likely. Serial ultrasounds every 3–4 weeks are recommended in pregnant women with confirmed or suspected ZIKV infection. Fetal abnormalities may be detected by 18–20 weeks of gestation but usually later and may include abnormalities other than microcephaly such as intrauterine growth retardation, hydrops fetalis, anhydramnios, cerebral calcification, hydrocephalus, brain atrophy, absent corpus callosum, and hydranencephaly [141]. For suspected congenital infection amniocentesis and testing by RT-PCR could be considered with appropriate counseling of the parents of available options. At birth testing of the infant serum or umbilical cord blood, and cerebrospinal fluid, should be tested for ZIKV by serology and RT-PCR, plus PCR of the placenta [137]. Neurological assessment including head ultrasound, ophthalmologic examination, and hearing evaluation should be performed and neurological monitoring throughout infancy to assess for long-term sequelae. A recent report documented ocular abnormalities in infants with microcephaly presumably related to intrauterine ZIKV infection [142]. These findings included macular alterations, gross pigment mottling or choreoretinal atrophy, and optic nerve abnormalities in a high percentage [35%] of affected infants [143].

6.6.3.1 Congenital ZIKV Syndrome

The largest and most comprehensive study on congenital ZIKV syndrome, including microcephaly, was recently published based on data from the Brazilian ministry of health surveillance system for microcephaly. A total of 7830 suspected cases were reported to the Brazilian Ministry of Health by June 2016. Among a total of 5554 live-born infants with suspected microcephaly, investigation was completed on 1501 suspected cases [27%], of which 602 [40%] were considered definite or probable cases of congenital ZIKV neurological disorder [based on head circumference for gestational age below 2SD and specific neuroimaging findings [141]]. Rashes in the third trimester of pregnancy were associated with brain abnormalities despite normal head size [present in about 20% of definite or probable cases], but rash during pregnancy did not occur in a third of the affected cases. Distinctive neuroimaging findings included brain calcifications, ventricular enlargement, or both with negative serology for syphilis, toxoplasmosis, and cytomegalovirus; however, some of the probable cases had incomplete data to exclude these other congenital infections. Also reported at the same time is a case series [$N = 5$] of the pathological findings of congenital ZIKV syndrome [144]. In all three fatal cases with microcephaly, [two also had severe arthrogryposis] viral antigens were localized to glial cells and neurons and associated with microcalcification. ZIKV antigens were detected in chorionic villi of one of the first trimester placentas [with spontaneous abortion], and tissue samples from all five cases were positive for ZIKV RNA by RT-PCR [144]. Thus, these findings provide strong evidence for ZIKV causing congenital central nervous system malformations, including microcephaly, arthrogryposis [flexion contracture of the limb/joint], and spontaneous abortion.

6.7 Potential Treatments and Vaccines

6.7.1 CHIKV Treatment

Although in vitro studies show that IFN-α have activity against CHIKV and combined with ribavirin there is synergistic antiviral effect [145], they are unlikely to have a significant role in future treatment. The major debilitation is from chronic severe arthralgia or arthritis, and novel or existent therapies are needed. Chloroquine or its derivative hydroxychloroquine have anti-inflammatory properties and have been used for rheumatological diseases for decades. However, clinical studies with chloroquine in CHIKV infection have produced mixed results. Early, small open clinical trial suggested that chloroquine might be beneficial for chronic arthralgia after CHIKV infection, but a later study of 27 patients with acute CHIKV treated with chloroquine compared to 27 controls showed no benefit and increased incidence of chronic arthralgia [146, 147]. Administration of specific CHIKV immunoglobulins harvested from recovered patients has high neutralizing activity and has been shown to have protective and therapeutic activity [within 24 h after inoculation] in the mouse model [75]. However, the potential use in an outbreak would be limited to specific high-risk scenarios such as neonates born to viremic mothers or adults with underlying rheumatological conditions such as rheumatoid arthritis. Development of an effective CHIKV vaccine should be simpler than producing an effective DENV vaccine because of limited antigenic variation among genotypes and lack of antibody enhancement. Although several vaccines in phase 1 or preclinical studies are in development [148], marketing a commercially viable vaccine will be challenging. CHIKV outbreaks have been unpredictable, and to perform a phase 3 clinical trial would be difficult as disease surveillance declines after epidemics. Usually afflicted population would have long-lasting protection, and new outbreaks require a susceptible population at risk.

6.7.2 DENV Vaccines

Development of an effective and safe DENV vaccine has been a major goal and priority by the scientific and research community for over two decades. The major concern of vaccine-induced disease from antibody enhancement has not been realized from vaccines that have been tested to date, likely from development of tetravalent vaccines to provide antibodies to all serotypes at the same time. Several DENV vaccines are in development at various stages, but the Sanofi Pasteur vaccine [Dengvaxia], developed over 20 years and the first dengue vaccine to be marketed and approved in Mexico in December 2015, for an initial phase with 40,000 people [children] to be inoculated. [htp://www.sanofipasteur.com/en/articles/dengvaxia-world-s-first-dengue-vaccine-approved-in-mexico.aspx]. The vaccine is a recombinant, live-attenuated, tetravalent dengue vaccine [CYD-TDV] composed of four chimeric live flaviviruses, each derived from the yellow fever virus genome with gene segments of each of the four DENV serotypes [142]. The vaccine was given to >35,000 children between the ages of 2 and 16 years in Asian-Pacific and Latin American countries as three subcutaneous injections given at months 0, 6, and 12 and assessed in three clinical trials [148–151]. The vaccine efficacy after a 3 year assessment was shown to reduce dengue due to all four serotypes in nearly two-thirds of participants, and pooled analysis showed the vaccine prevented 9 out of 10 cases of severe dengue and 8 out of 10 hospitalizations in ages 9–16 years, but there was an unexplained higher incidence of hospitalization in children younger than 9 years of age [151]. This latter observation has to be carefully monitored during long-term follow-up to assess for antibody enhancement-induced disease. The vaccine was found to be safe and serious adverse event in the first 28 days of vaccination in the vaccine and control group was similar [each 1%] [151].

There are several drawbacks of the CYD-TDV vaccine including the need for multiple dosing; modest protective efficacy of only 60% overall at 3 years, with even lower efficacy [35.5%] in flavivirus-naïve subjects [150, 151]; and evidence of unequal and waning immunogenicity long after the third dose of the vaccine [152]. Another potential concern would be development of severe dengue disease in vaccine recipients much later in life, a trend which is already appearing in the pre-vaccine era.

Currently there is no effective agent for treatment of dengue. Chloroquine, a cheap and widely available drug, has modest in vitro activity against DENV, but a randomized controlled trial failed to show any clinical benefit [153].

6.7.3 ZIKV Vaccine

Research in developing an effective vaccine for ZIKV is occurring at a rapid unprecedented pace. Preliminary studies of a purified inactivated ZIKV [PIV] vaccine and a DNA vaccine expressing an optimized premembrane and envelope [prM-Env] immunogen showed protective efficacy in mice challenged with both ZIKV strains from Brazil and Puerto Rico [154]. Further studies by the same investigators have recently shown that three different vaccine platforms protect against ZIKV challenge in rhesus monkeys [155]. The purified inactivated virus vaccine induced ZIKV-specific neutralizing antibodies and completely protected against ZIKV challenge with strains from both Brazil and Puerto Rico. In addition, adoptive transfer studies demonstrated that purified immunoglobulin from vaccinated monkeys conferred passive protection. A plasmid DNA vaccine and a single-shot recombinant rhesus adenovirus vector expressing ZIKV prM-Env also completely protected monkeys against ZIKV challenge. Hence, rapid clinical development of a ZIKV vaccine for humans is a reality.

6.8 Future Prospects

The most vexing challenge which would provide the most cost-effective benefit against CHIKV, ZIKV, and DENV infections is effective vector control. Although simple measures have been available to communities for many decades to decrease the risk of mosquito bites, they are poorly adhered to over an indefinite period. Two novel approaches to control the mosquito population appear to be of some benefit from preliminary studies. The release of genetically modified male mosquitoes that sterilize the circulating female population to reduce egg output and subsequent mosquito population had been shown to be feasible in a localized region such as a small island [156]. However, it is difficult to envision that this technique could be highly effective in large continents in Asia, Africa, and the Americas with numerous adjoining countries. An alternative strategy is the induction of widespread biologic resistance to DENV and CHIKV in *Aedes* mosquito population. *Wolbachia* species are bacterial endosymbionts of insects, helminths, and crustaceans that are transmitted by transovarial and transstadial [between stages of development passage] means and are important in the pathogenesis of filariasis [157]. *Wolbachia*-infected *Ae. aegypti* are partially resistant to DENV and CHIKV infections [158] and the bacteria can invade and establish infection in *Ae. aegypti* populations [158, 159]. Wolbachia infected mosquitoes are also highly resistant to ZIKV [Duttra et al., 2016; Cell Host & Microbe; 19: 771–4]. Unlike the use of chemical insecticides, there would be minimal risk to humans and less chance of resistant mutants developing. *Wolbachia* species are nonpathogenic to humans [160]. Laboratory reared male mosquitoes infected with Wolbachia when released in the environment mate with multiple females, rendering them infertile and eventually reduce the pest population. Preliminary studies in targeted areas in Australia, Indonesia, Vietnam, and China indicate that this is feasible and 90% of local mosquitoes can become infected within weeks. Plans are in progress to widely release Wolbachia infected mosquitoes to fight ZIKV in Rio de Janeiro and Medellin [Columbia] over the next 2 years [161]. The USA is reviewing and considering similar strategy and a biotechnology firm [MosquitoMate] is seeking approval to market Wolbachia infected mosquitoes as a pesticide [http://www.nature.com/news/us-reviews-plan-to-infect-to-stop–12/19/2016].

The most promising of the dengue vaccines in development [TV005] is a live-attenuated tetravalent vaccine designed by the Laboratory of Infectious Diseases at the National Institutes of Health [Bethesda, Maryland]. All the vaccine components have a DENV genetic background and share a core attenuating, 30-nucleotide deletion in the viral genome of each strain, yielding replication-deficient attenuated viruses [162]. The precursor vaccine [TV003] DENV-2 component was less immunogenic than other serotypes, but the TV005 vaccine contained a highe-dose component of DENV-2. A recent preclinical study has shown that a single subcutaneous injection of the TV005 vaccine elicited a tetravalent response in 90% of vaccines at 3 months after vaccination and a trivalent response in 98% [163]. The vaccine was well tolerated. Further studies such as a phase 3 clinical trial in a large at-risk population may take several years to accomplish but are eagerly awaited.

Development of vaccines for CHIKV is feasible and in the very early stages. Two live-attenuated CHIKV vaccine candidates have been shown to be safe and effective in preventing viremia and clinical disease in nonhuman primates after a single dose [164]. It is also possible to develop a DNA vaccine to initiate replication of live vaccine CHIKV in vitro and in vivo that can elicit neutralizing antibodies and protect mice from a neurovirulent CHIKV [165]. The main challenges facing development of an effective CHIKV vaccine will be completion of a large phase 3 clinical trial, as disease outbreaks are sporadic and unpredictable, and to have a viable commercial market to make it cost-effective.

Drug development of antiviral agents for DENV and CHIKV is in its infancy, and progress toward drug candidates has been very slow. A not so novel but practical approach and cost efficient is the repurposing of existent marketed drugs for a new indication. Prochlorperazine [PCZ] is readily available and approved for treatment of nausea and vomiting as a dopamine D2 receptor antagonist. PCZ can block DENV infection by targeting viral binding and viral entry through D2R, and clathrin-associated mechanisms and administration soon after infection can protect against lethality in Stat1-deficient mouse model [166]. Thus, prophylactic and early administration in the acute, febrile phase of dengue preclinical studies of this drug would be unnecessary and may improve clinical symptoms and outcome. Hence, a large randomized, controlled, clinical trial is warranted with PCZ for any new outbreak of dengue, as further preclinical studies of this drug would be unnecessary.

ZIKV vaccine development is being explored by pharmaceuticals, and the US government has promised to invest hundreds of millions of dollars toward an effective vaccine. The obvious candidates for an effective ZIKV vaccine would be uninfected pregnant women or women of child-bearing age in an outbreak. A major challenge is to conduct a phase three efficacy trial before termination of any ongoing outbreak. While separate vaccines for DENV,CHIKV, and ZIKV is the first step for effective prevention of these emerging zoonoses, a multiple virus vaccine for all three viruses should be explored in experimental animal models. Further research is needed to unravel the mechanism of ZIKV infection and microcephaly and to explore potential preventative measures in pregnant nonhuman primates. Another laboratory model to study microcephaly is being initiated by scientists in Brazil by infecting cerebral organoids with ZIKV. Cerebral organoids are tiny models of the human brain grown from stem cells in laboratory dish that already has been used to study the mechanisms of genetic mutations causing microcephaly in Austria [167].

Perhaps more research and funding would be better served to investigate development of effective means of eradicating or controlling *Aedes* mosquitoes or explore the potential of a vaccine to protect against mosquito bites. This would be one solution to protect against four emerging viruses of public health concern, as yellow fever virus is transmitted by the same vector. Currently, the fight against the vector in Brazil and other affected countries includes a combination of education of the public to reduce mosquito bites [protective clothing, screened doors and windows, insect repellant with DEET], methods to discourage breeding sites in and around the home [such as discarding empty containers and old tires] or use of larvicides for collected ponds of water, and insecticide spraying in and around the home. There is little evidence to support the efficacy of various mosquito abatement programs, but measures used in most countries have been of temporary benefit. Thus, it is reasonable for governments to employ a combination of methods: behavioral education campaigns to manage water containers to reduce breeding sites, biological methods such as predatory organisms or bacteria [i.e., *Wolbachia* species] and genetic methods to reduce the population by sterilizing male mosquitoes, and chemical control techniques [insecticide sprays, larvicides, cleaning water containers with household chemicals, or adding paraffin to standing water].

Novel vector control measures are being explored to reduce the mosquito density in local communities as conventional measures often fail. The CDC developed an autocidal gravid ovitrap [AGO] to attract and capture female *Ae. aegypti* mosquitoes responsible for disease transmission. The AGO trap uses wet hay as attractant, is of low cost with no pesticides, and can be used for extended period. In a recent study in Puerto Rico, communities using the AGO traps had tenfold lower mosquito densities than nonintervention communities. Moreover, the proportion of chikungunya virus infection in the intervention communities was one-half that of nonintervention communities [168]. Insectivorous bats may serve as an alternative approach to achieve mosquito control in communities. Under laboratory conditions, bats can consume up to 600 mosquitoes per hour, and a 32% reduction in oviposition [egg deposits] by *Culex* spp. with bat predation had been observed [169].

Addendum Two recent studies have provided new insights on the incidence of congenital abnormalities in maternal ZIKV infection. In a US study of 442 completed pregnancies with recent ZIKV infection, birth defects were identified in 6% and microcephaly in 4% of live births. Birth defects were higher [11%] in first trimester infection with ZIKV [170]. Similar findings were reported from a Brazilian study, with microcephaly in 4 of 117 [3.4%] live births and adverse outcome in 55% of first trimester infection and 29% after the third trimester [171].

References

1. Mousson L, Dauga C, Garrigues T, Shaffner F, Vazeille M, Failloux AB (2005) Phylogeography of Aedes [Stegomyia] aegypti [L.] and Aedes [Stegomyia] albopictus [Skuse] [Diptera: Culicidae] based on mitochondrial DNA variations. Genet Res 86:1–11
2. Medlock JM, Hansford KM, Schaffner F et al (2012) A review of the invasive mosquitoes in Europe: ecology, public health risks, and control options. Vector Borne Dis 12:435–447
3. Hawley WA, Reiter P, Copeland RS, Pumpuni CB, Craig GB (1987) *Aedes albopictus* in North America: probable introduction in used tires from northern Asia. Science 236:1114–1115
4. World Health Organization (2012) Global strategy for dengue prevention and control: 2012–2020, pp. 1–43

5. Robinson MC (1955) An epidemic of virus disease in Southern Province, Tanganyika Territory, in 1952-53---1: clinical features. Trans R Soc Trop Med Hyg 49:28
6. Diallo M, Thonnon J, Traore-Lamizana M, Fontenille D (1999) Vectors of Chikungunya virus in Senegal: current data and transmission cycles. Am J Trop Med Hyg 60:281–286
7. Volk SM, Chen R, Tsetsarkin KA et al (2010) Genome-scale phylogenetic analyses of Chikungunya virus reveal independent emergences of recent epidemics and various evolutionary rates. J Virol 84:6497–6504
8. Burt FJ, Ralph MS, Rulli NE, Mahalingam S, Heise MT (2012) Chikungunya: a re-emerging virus. Lancet 379:662–671
9. Carey DE (1971) Chikungunya and dengue: a case of mistaken identity? J Hist Med Allied Sci 26:243–262
10. Weaver SC, Lecuit M (2015) Chikungunya virus and the global spread of a mosquito-borne disease. N Engl J Med 372:1231–1239
11. Pastorino B, Muyembe-Tamfun JJ, Bessaud M et al (2004) Epidemic resurgence of Chikungunya virus in Democratic Republic of the Congo: identification of a new central Africa strain. J Med Virol 74:277–282
12. Laras K, Sukri NC, Larasati RP et al (2005) Tracking the re-emergence of epidemic Chikungunya virus in Indonesia. Trans R Soc Trop Med Hyg 99:128–141
13. Weaver SC, Osorio JE, Livengood JA, Chen R, Stinchcomb DT (2012) Chikungunya virus and prospects for a vaccine. Expert Rev Vaccines 11:1087–1101
14. Schuffenecker I, Iteman I, Michault A et al (2006) Genome microevolution of Chikungunya viruses causing the Indian Ocean outbreak. PLoS Med 3:e263
15. Vazeille M, Moutailler S, Coudrier D et al (2007) Two Chikungunya viruses from the outbreak of La Reunion [Indian Ocean] exhibit different patterns of infection in the mosquito, *Aedes albopictus*. PLoS One 2:e1168
16. Tsetsarkin KA, Weaver SC (2011) Sequential adaptive mutations enhance efficient vector switching by Chikungunya virus and its epidemic emergence. PLoS Pathog 7:e1002412
17. Tsetsarkin KA, Chen R, Leal G et al (2011) Chikungunya virus emergence is constrained in Asia by lineage-specific adaptive landscapes. Proc Natl Acad Sci U S A 108:7872–7877
18. Lanciotti RS, Kosoy OL, Laven JJ et al (2007) Chikungunya virus in US travelers returning from India, 2006. Emerg Infect Dis 13:764–767
19. Gaines J. Chikungunya update for clinicians. CDC Medscape, May 4, 2015. http://www.medscape.com/viewarticle/843623.
20. Leparc-Goffart I, Nourgairede A, Casadou S, Prat C, de Lamballerie X (2014) Chikungunya in the Americas. Lancet 383:514
21. Centers for Disease Control and Prevention. Laboratory confirmed chikungunya virus cases reported to ArboNET by territory—United States, 2014. http://www.salud.gov.pr/Chikungunya/Pages/default.aspx
22. Vasilakis N, Weaver SC (2008) The history and evolution of human dengue emergence. Adv Virus Res 72:1–76
23. Ashburn PM, Craig CF (1907) Experimental investigations regarding the etiology of dengue fever. J. Infect. Dis. 4:440–475
24. Siler JF, Hall MW, Hitchens AP (1926) Dengue: Its history, epidemiology, mechanism of transmission, etiology, clinical manifestations, immunity and prevention. Philipp J Sci 29:1–252
25. Simmonds JS, St. John JH, Reynolds FH (1931) Experimental studies of dengue. Philipp J Sci 44:1–252
26. Rudnick A, Marchette NJ, Garcia R (1967) Possible jungle dengue—recent studies and hypotheses. Jpn J Med Sci Biol 20:69–74
27. Smith CE (1956) The history of dengue in tropical Asia and its probable relationship to the mosquito *Aedes aegypti*. J Trop Med Hyg 59:243–251
28. Dick GW, Kitchen SF, Haddow AJ (1952) Zika virus. 1. Isolation and serological specificity. Trans R Soc Trop Med Hyg 46:509–520
29. Macnamara FN (1954) Zika virus: a report on three cases of human infection during an epidemic of jaundice in Nigeria. Trans R Soc Trop Med Hyg 48:139–145

30. Moore DL, Causey OR, Carey DE et al (1975) Arthropod-borne viral infections of man in Nigeria, 1964–1970. Ann Trop Med Parasitol 69:49–64
31. Hayes EB (2009) Zika virus outside Africa. Emerg Infect Dis 15:1347–1350
32. Marchette NJ, Garcia R, Rudnick A (1969) Isolation of Zika virus from *Aedes aegypti* mosquitoes in Malaysia. AmJTrop Med Hyg 18:411–415
33. Lanciotti RS, Kosoy OL, Laven JJ et al (2008) Genetic and serological properties of Zika virus associated with an epidemic. Yap State, Micronesia, 2007. Emerg Infect Dis 14:1232–1239
34. Thiboutot MM, Kannan S, Kawalekar OU et al (2010) Chikungunya: a potential emerging epidemic? PLoS Neglected Trop Dis 4:e623
35. Cherian SS, Walimbe AM, Jadhav SM et al (2009) Evolutionary rates and timescale comparison of Chikungunya viruses inferred from the whole genome/E1 gene with special reference to the 2005–07 outbreak in the Indian subcontinent. Infect Genet Evol 9:16–23
36. Lozier M et al (2016) Incidence of Zika virus disease by age and sex - Puerto Rico, November 1, 2015-October 20, 2016. MMWR Morb Mortal Wkly Rep 65(44):1219–1223
37. Normile D (2013) Tropical medicine. Surprising new dengue virus throws a spanner in disease control efforts. Science 342:415
38. Rodenhuis-Zybert IA, Wishut J, Smit JM (2012) Dengue virus life cycle: viral and host factors modulating infectivity. Cell Mol Life Sci 67:2773–2786
39. Holmes E, Twiddy S (2003) The origin, emergence and evolutionary genetics of dengue virus. Infect Genet Evol 3:19–28
40. Fredericks AC, Fernandez-Sesma A (2014) The burden of dengue and Chikungunya worldwide: implications for the southern United States and California. Annals Glob Hlth 80:466–475
41. World Health Organization (2011) Comprehensive guidelines for prevention and control of dengue and dengue hemorrhagic fever. WHO Regional Office for South-East Asia, New Delhi
42. Haddow AD, Schuh AJ, Yasuda CY et al (2012) Genetic characterization of Zika virus strain: geographic expansion of the Asian lineage. PLoS Negl Trop Dis 6:e1477
43. Enfissi A, Codrington J, Roosblad J, Kazanji M, Rousset D (2016) Zika virus genome from the Americas. Lancet 387:227–228
44. Gatherer D, Kohl A (2015) Zika virus: a previously slow pandemic spreads rapidly through the Americas. J Gen Virol 97(2):269–273
45. Boorman JP, Poprterfield JS (1956) A simple technique for infection of mosquitoes with viruses: transmission of Zika virus. Trans R Soc Trop Med Hyg 50:238–242
46. Besnard M, Lastere S, Teissier A, Cao-Lormeau VM, Musso D (2014) Evidence of perinatal transmission of Zika virus, French Polynesia, December 2013 and February 2014. Eur Surveill 19:20751
47. Oliveira Melo AS, Malinger G, Ximenes R, Szejnfeld PO, Alves Sampaio S, Bispo De Filippis AM (2016) Zika virus intrauterine infection causes fetal brain abnormality and microcephaly: tip of the iceberg? Ultrasound Obstet Gynecol 47:6–7
48. Musso D, Nhan T, Robin E et al (2014) Potential for Zika virus transmission through blood transfusion demonstrated during an outbreak in French Polynesia. November 2013 to February 2014. Eur Surveill 19:20751
49. Musso D, Roche C, Robin E, Nhan T, Teissier A, Cao-Lorrmeau VM (2015) Potential sexual transmission of Zika virus. Emerg Infect Dis 21:359–361
50. Hills SL, Russell K, Hennessey M, Williams C, Oster AM, Fischer M, Mead P (2016) Transmission of Zika virus through sexual contact with travelers to areas to areas of ongoing transmission continental United States, 2016. MMWR Morb Mortal Wkly Rep 65:215–216
51. Venturi G, Zammarchi L, Fortuna C et al (2016) An autochthonous case of Zika virus due to possible sexual transmission, Florence, Italy, 2014. Euro. Surveill. 21(8). doi: 10.2807/1560-7917
52. Atkinson B, Hearn P, Afrough B et al (2016) Detection of Zika virus in semen. Emerg Infect Dis 22:940
53. Mansuy JM, Dutertre M, Mengelle C et al (2016) Zika virus: high infectious viral load in semen, a new sexually transmitted pathogen? Lancet Infect Dis 16:405

54. Brent C et al (2016) Preliminary findings from an investigation of Zika virus infection of a patient with no known risk factors – Utah. MMWR Morb Mortal Wkly Rep 65(36):981–982
55. Mammen MP, Pimgate C, Koenraadt CJM et al (2008) Spatial and temporal clustering of dengue virus transmission in Thai villages. PLoS Med 5:e205
56. Randolph SE, Rogers DJ (2010) The arrival, establishment and spread of exotic diseases: patterns and prediction. Nat Rev Microbiol 8:361–371
57. Hanson S, Craig GB (1994) Cold acclimation, diapause, and geographic origin affect cold hardiness in eggs of *Aedes albopictus* [Diptera: Culicidae]. J Med Entomol 31:192–201
58. Rochlin I, Ninivaggi DV, Hutchinson ML, Farajollahi A (2013) Climate change and range expansion of the Asian tiger mosquito [*Aedes albopictus*] in northeastern USA: implication for public health practitioners. PLoS One 8:e60874
59. Schaffner F, Medlock JM, Van Bortel W (2013) Public health significance of invasive mosquitoes in Europe. Clin Microbiol Infect 19:685–692
60. Brady OJ, Golding N, Pigott DM et al (2014) Global temperature constraints on *Aedes aegypti* and *A. albopictus* persistence and competence for dengue virus transmission. Parasit Vectors 7:338
61. Gratz NG (2004) Critical review of the vector status of *Aedes albopictus*. Med Vet Entomol 18:215–227
62. Whitehorn J, Kien DTH, Nguyen NM et al (2015) Comparative susceptibility of *Aedes albopictus* and *Aedes aegypti* to dengue virus infection after feeding on blood of viremic humans: implication for public health. J. Infect. Dis. 212:1182–1190
63. Diagne CT, Diallo D, Faye O et al (2015) Potential of selected Senegalese Aedes spp. mosquitoes [Diptera: Culicidae] to transmit Zika virus. BMC Infect Dis 15:492
64. Wong PS, Li MZ, Chong CS, Ng LC, Tan CH (2013) *Aedes [Stegomyia] albopictus:* a potential vector of Zika virus in Singapore. PLoS Negl Trop Dis 7:e2348
65. Grad G, Caron M, Mombo IM et al (2014) Zika virus in Gabon [Central Africa]—2007: a new threat from *Aedes albopictus*? PLoS Negl Trop Dis 8:e2681
66. Sourisseau M, Schilte C, Casartelli N et al (2007) Characterization of reemerging Chikungunya virus. PLoS Pathog 3:e89
67. Ng LF, Chow A, Sun YJ et al (2009) IL-1 beta, IL-6, and RANTES as biomarkers of Chikungunya severity. PLoS One 4:4261
68. Ozden S, Huerre M, Riviere JP et al (2007) Human muscle satellite cells as targets of Chikungunya virus infection. PLoS One 2:527
69. Gardner J, Anraku I, Le TT et al (2010) Chikungunya virus arthritis in adult wild-type mice. J Virol 84:8021–8032
70. Labadie K, Larcher T, Joibert C et al (2010) Chikungunya disease in nonhuman primates involve long–term viral persistence in macrophages. J Clin Invest 120:894–906
71. Couderc T, Chretien F, Schilte C et al (2008) A mouse model for Chikungunya: young age and inefficient type-1 interferon signaling are risk factors for severe disease. PLoS Pathog 4:e29
72. Das T, Jaffer-Bandjee MC, Hoarau JJ et al (2010) Chikungunya fever: CNS infection and pathologies of a re-emerging arbovirus. Prog Neurobiol 91:121–129
73. Schwartz O, Albert ML (2010) Biology and pathogenesis of Chikungunya virus. Nat Rev Microbiol 8:491–500
74. Schilte C, Couderc T, Chretien F et al (2010) Type 1 IFN controls Chikungunya virus via its action on nonhematopoietic cells. J Exp Med 207:429–442
75. Couderc T, Khandouri N, Grandadam M et al (2009) Prophylaxis and therapy for Chikungunya virus infection. J Infect Dis 200:516–523
76. Lim ML, Mateo L, Gardner J, Suhrbier A (1998) Alphavirus-specific cytotoxic T lymphocytes recognize a cross-reactive epitope from the capsid protein and can eliminate virus from persistently infected macrophages. J Virol 72:5146–5153
77. Maek-A-Nantawat W, Silachamroon U (2009) Presence of autoimmune antibody in Chikungunya infection. Case Rep Med 2009:840183
78. Malvy D, Ezzedine K, Mamani-Matsuda M et al (2009) Destructive arthritis in a patient with persistent specific IgM antibodies. BMC Infect Dis 9:200–207

79. Hoarau JJ, Jaffer-Bandjee MC, Kreijbich TP et al (2010) Persistent chronic inflammation by Chikungunya arthrogenic alphavirus in spite of a robust immune response. J Immunol 184:5914–5927
80. Simmons CP, Farrar JJ, van Vinh CN, Wills B (2012) Dengue. N Engl J Med 366:1423–1432
81. Halstead SB (1989) Antibody, macrophages, dengue virus infection, shock, and hemorrhage: a pathogenetic cascade. Rev Infect Dis 11(Suppl. 4):S830–S839
82. Balmaseda A, Standish K, Mercado JC et al (2010) Trends in patterns of disease transmission over 4 years in a pediatric cohort study in Nicaragua. J. Infect. Dis. 201:5–14
83. Montoya M, Gresh L, Mercado JC et al (2013) Symptomatic versus inapparent outcome in repeat dengue virus infection is influenced by the time interval between infections and study year. PLoS Negl Trop Dis 7:e2357
84. Guzman MG, Alvarez M, Halstead SB (2013) Secondary infection as a risk factor for dengue hemorrhagic fever/dengue shock syndrome: an historical perspective and role of antibody-dependent enhancement of infection. Arch Virol 158:1445–1459
85. Guzman MG, Kouri G, Bravo J, Valdes L, Vazquez S, Halstead SB (2002) Effect of age on outcome of secondary dengue 2 infections. Int J Infect Dis 6:118–124
86. Martina BE, Koraka P, Osterhaus AD (2009) Dengue virus pathogenesis: an integrated view. Clin Microbiol Rev 22:564–581
87. Huy NT, Giang TV, Thuy DHD, Kikuchi M, Hien TT, Zamora J, Hirayama K (2013) Factors associated with dengue shock syndrome: a systematic review and meta-analysis. PLoS Negl Trop Dis 7:e2412
88. Martina BEE (2014) Dengue pathogenesis: a disease driven by host response. Sci Prog 97:197–214
89. Lim DS, Yawata N, Selva KJ et al (2014) The combination of type 1 IFN, TNF-α, and cell surface receptor engagement with dendritic cells enables NK cells to overcome immune evasion by dengue. J Immunol 193:5065–5075
90. Hottz ED, Medeiros-de-Moraes IM, Vieira-de-Abreu A et al (2014) Platelet activation and apoptosis modulate monocyte inflammatory response in dengue. J Immunol 193:1864–1872
91. Halstead SB (2003) Neutralization and antibody-dependent enhancement of dengue viruses. Adv Virus Res 60:421–467
92. Rothman AL (2011) Immunity to dengue virus: a tale of original antigenic sin and tropical storms. Nat Rev Immunol 11:532–543
93. Hamel R, Dejarnac O, Wichit S et al (2015) Biology of Zika virus infection in human skin cells. J Virol 21:84–86
94. Blazquez AB, Escribano-Romero E, Merino-Ramos T, Saiz JC, Martin-Acebes MA (2014) Stress responses in flavivirus-infected cells: activation of unfolded protein response and autophagy. Front Microbiol 5:266
95. Tappe D, Perez-Giron JV, Zammarchi L et al (24 December 2015) Cytokine kinetics of Zika virus-infected patients from acute to reconvalescent phase. Med Microbiol Immunol. doi:10.1007/s00430-015-0445-7
96. Grant A, Ponia SS, Tripathi S et al (2016) Zika virus targets human STAT2 to inhibit type 1 interferon signaling. Cell Host Microbe 19:882–890
97. Miner JJ, Cao B, Govero J et al (2016) Zika virus infection during pregnancy in mice causes placental damage and fetal demise. Cell 165:1081–1091
98. Cugola FR, Fernandes IR, Russo FB et al (2016) The Brazilian Zika virus strain causes birth defects in experimental models. Nature 534:267–271
99. Li C, Xu D, Ye Q et al (2016) Zika virus disrupts neural progenitor development and leads to microcephaly in mice. Cell Stem Cell 19:120–126
100. Wu KY, Zuo GL, Li XF et al (2016) Vertical transmission of Zika virus targeting the radial cells affects cortex development of offspring mice. Cell Res 26:645–654
101. Mysorekar IU, Diamond MS (2016) Modeling Zika virus in pregnancy. N Engl J Med 375:481–484
102. Suy A et al (2016) Prolonged Zika virus Viremia during pregnancy. N Engl J Med 375:2611–2613

103. Musso D, Nhan T, Robin E et al (2014) Potential for Zika virus transmission through blood transfusion demonstrated during an outbreak in French Polynesia. November 2013 to February 2014. Eur Surveill 19:20761
104. Centers for Disease Control and Prevention (2016) Interim guidance for interpretation of Zika virus antibody test results. MMWR 65:475
105. Staikowsky F, Talarmin F, Grivard P et al (2009) Prospective study of Chikungunya virus acute infection in the Island La Reunion during the 2005-2006 outbreak. PLoS One 4:e7603
106. Gerardin P, Barau G, Michault A et al (2008) Multidisciplinary prospective study of mother-to-child Chikungunya virus infections on the island of La Reunion. PLoS Med 5:e60
107. Sissoko D, Malvy D, Ezzedine K et al (2009) Post-epidemic Chikungunya disease on Reunion Island: course of rheumatic manifestations and associated factors over a 15 month period. PLoS Negl Trop Dis 3:e389
108. Brighton SW, Prozesky OW, de la Harpe AL (1983) Chikungunya virus infection. A retrospective study of 107 cases. S Afr Med J 63:313–315
109. Schilte C, Staikowsky F, Couderc T et al (2013) Chikungunya virus associated long-term arthralgia: a 36-month prospective longitudinal study. PLoS Negl Trop Dis 7:e2137
110. Manimunda SP, Vijayachari P, Uppoor R et al (2010) Clinical progression of Chikungunya fever during acute and chronic arthritic stages and the changes in joint morphology as revealed on imaging. Trans R Soc Trop Med Hyg 104:392–399
111. Ganu MA, Ganu AS (2011) Post-chikungunya chronic arthritis our experience with DMARDs over two year follow up. J Assoc Physicians India 59:83–86
112. Sun J, Luo S, Lin J et al (2012) Inapparent infection during an outbreak of dengue fever in Southeastern China. Viral Immunol 25:456–460
113. Malhotra HS, Garg RK (2014) Dengue-associated hypokalemic paralysis: causal or incident? J Neurol Sci 344:238
114. Huang SY, Lee IK, Liu JW, Kung CT, Wang L (2015) Clinical features and risk factors for rhabdomyolysis among adult patients with dengue virus infection. AmJTrop Med Hyg 92:75–81
115. Lizarraga KJ, Nayer A (2014) Dengue-associated kidney disease. J Nephropathol 3:57–62
116. Mena Lora AJ, Fernandez J, Morales A, Soto Y, Feris-Iglesias J, Brito MO (2014) Disease severity and mortality caused by dengue in a Dominican pediatric population. AmJTrop Med Hyg 90:169–172
117. Nascimento E, Hottz ED, Garcia-Bates TM, Bozza F, Marques ET Jr, Barratt-Boyes SM (2014) Emerging concepts in dengue pathogenesis: interplay between plasmablasts, platelets, and complement in triggering vasculopathy. Crit Rev Immunol 34:227–240
118. Hostick O, Martiez E, Guzman MG, Martin JL, Ranzinger SR (2015) WHO dengue case classification in 2009 and its usefulness in practice: an expert consensus in the Americas. Pathog Glob Health 109:19–25
119. Verma R, Sahu R, Holla V (2014) Neurological manifestations of dengue: a review. J Neurol Sci 346:26–34
120. Sahu R, Verma R, Jain A et al (2014) Neurological complications in dengue virus infection: a prospective cohort study. Neurology 83:1601–1609
121. Dussart P, Petit L, Labeau B et al (2008) Evaluation of two new commercial tests for the diagnosis of acute dengue fever virus infection using NS1 antigen detection in human serum. PLoS Negl Trop Dis 2:e280
122. Guzman MG, Jaenisch T, Gaczkowski R et al (2010) Multi-country evaluation of the sensitivity and specificity of two commercially-available NS1 ELISA assays for dengue diagnosis. PLoS Negl Trop Dis 4:e811
123. Fry SR, Meyer M, Semple MG et al (2011) The diagnostic sensitivity of dengue rapid test is significantly enhanced by using a combined antigen and antibody testing approach. PLoS Negl Trop Dis 5:e1199
124. Kaur P, Kaur G (2014) Transfusion support in patients with dengue fever. Int J Appl Basic Med Res 4(Suppl. 1):S8–S12
125. Duffy MR, Chen TH, Hancock WT et al (2009) Zika virus outbreak on Yap Island, Federated States of Micronesia. N Engl J Med 360:2536–2542

126. Cao-Lormeau VM, Musso D (2014) Emerging arbovirus in the Pacific. Lancet 384:1571–1572
127. Roth A, Mercier A, Lepers C et al (2014) Concurrent outbreaks of dengue, Chikungunya and Zika virus infections—an unprecedented epidemic wave of mosquito-borne viruses in the Pacific 2012-2014. Eur Surveill 19(41). pii: 20929
128. European Centre for Disease Prevention and Control (2015) Rapid risk assessment: Zika virus epidemic in the Americas: potential association with microcephaly and Guillian-Barre syndrome. 10 December 2015. ECDC, Stockholm
129. Vogel G (2016) A race to explain Brazil's spike in birth defects. Evidence points toward the fast-spreading Zika virus as the cause of microcephaly. Science 351:110–111
130. Fleming-Dutra KE, Nelson JM, Fischer M, Erin Staples J, Karwowski MP, Mead P, Villanueva J, Renquist CM, Minta AA, Jamieson DJ, Honein MA, Moore CA, Rasmussen SA (2016) Update interim guidelines for health care providers caring for infants and children with possible Zika virus infection United States, February 2016. MMWR Morb Mortal Wkly Rep 65:182–187
131. Moore CA, Weaver DD, Bull MJ (1990) Fetal brain disruption sequence. J Pediatr 116:383–386
132. Petersen LR, Jamieson D, Powers AM, Honein MA (2016) Zika virus. N Engl J Med 373:1552–1563
133. Brasil P, Pereira JP Jr, Raja Gabaglia C et al (2016) Zika virus infection in pregnant women in Rio de Janeiro preliminary report. N Engl J Med. doi:10.1056/NEJMoa1602412
134. Cauchemez S, Besnard M, Bompard P, et al. Association between Zika virus and microcephaly in French Polynesia, 2013-15: a retrospective study. Lancet 2016; Online. http//dx.doi.org/10.1016/S0140-6736[16]00651-6.
135. Carteaux G, Maquart M, Bedet A et al (2016) Zika virus associated with meningoencephalitis. N Engl J Med 374:1595–1596
136. Mecharles S, Herrmann C, Poullain P et al (2016) Acute myelitis due to Zika virus infection. Lancet 387:1481
137. Centers for Disease Control and Prevention (2016) Interim guidelines for the evaluation and testing of infants with possible congenital Zika virus infection United States. MMWR 65(3):63–67
138. Faye O, Faye O, Diallo M, Weidmann M, Sall AA (2013) Quantitative real-time PCR detection of Zika virus and evaluation with field-caught mosquitoes. Virol J 10:311
139. Gourinat AC, O'Connor O, Calvert E, Goarant C, Dupont-Rouzeyrol M (2015) Detection of Zika virus in urine. Emerg Infect Dis 21:84–86
140. Musso D, Roche C, Nhan TX, Teissier A, Cao-Lormeau VM (2015) Detection of Zika virus in saliva. J Clin Virol 68:53–55
141. Oehler E, Watrin L, Larre P, et al. Zika virus infection complicated by Guillian-Barre syndrome case report, French Polynesia, December 2013. Euro. Surveill. 2014; 19: pii 20720
142. Ventura CV, Maia M, Ventura BV et al (2016) Ophthalmological findings in infants with microcephaly and presumable intra-uterine Zika virus infection. Arq Bras Oftalmol 79:1–3
143. Franca GVA, Schuler-Faccini L, Oliveira WK et al (2016) Congenital Zika virus syndrome in Brazil: a case series of the first 1501 livebirths with complete investigation. Lancet 388:891–897
144. Martines RB, Bhatnagar J, de Oliveira Ramos AM et al (2016) Pathology of congenital Zika syndrome in Brazil: a case series. Lancet 388:898–904
145. Briolant S, Garin D, Scaramizzino N, Jouan A, Crance JM (2004) In vitro inhibition of Chikungunya and Semliki Forest viruses replication by antiviral compounds: synergistic effect of interferon-alpha and ribavirin combination. Antivir Res 61:111–117
146. Brighton SW (1984) Chloroquine phosphate treatment of chronic Chikungunya arthritis. An open pilot study. S Afr Med J 66:217–218
147. de Lamballerie X, Boisson V, Reynier JC et al (2008) On Chikungunya acute infection and chloroquine treatment. Vector Borne Zoonotic Dis 8:837–839
148. Chang LJ, Dowd KA, Mendoza FH et al (2014) Safety and tolerability of Chikungunya virus-like particle vaccine in healthy adults: a phase 1 dose-escalation trial. Lancet 384:2046–2052

149. Capeding MR, Tran NH, Hadinegoro SRS et al (2014) Clinical efficacy and safety of a novel tetravalent dengue vaccine in healthy children in Asia: a phase 3, randomized, observer-masked, placebo-controlled trial. Lancet 384:1358–1365
150. Villar L, Dayan GH, Arredondo-Garcia JL et al (2015) Efficacy of a tetravalent dengue vaccine in children in Latin America. N Engl J Med 372:113–123
151. Hadinegoro SR, Arrendondo-Garcia JL, Capeding MR et al (2015) Efficacy and long-term safety of a dengue vaccine in regions of endemic disease. N Engl J Med 373:1195–1206
152. Jelitha R, Nirmalatiban P, Nyanamalar S, Cabriz MG (2015) Descriptive review of safety, reactivity and immunogenicity of dengue vaccine clinical trials, 2003-2013. Med J Malays 70:67–75
153. Tricou V, Minh NN, Van TP et al (2010) A randomized trial of chloroquine for the treatment of dengue in Vietnamese adults. PLoS Negl Trop Dis 4:e785
154. Larocca RA, Abbink P, Peron JP et al (2016) Vaccine protection against Zika virus from Brazil. Nature. doi:10.1038/nature18952
155. Abbink P, Larocca RA, De La Barrera RA et al (2016) Protective efficacy of multiple vaccine platforms against Zika virus challenge in rhesus monkeys. Science. doi:10.1126/science.aah6157
156. Wise de Valdez MR, Nimmo D, Betz J et al (2011) Genetic elimination of dengue vector mosquitoes. Proc Natl Acad Sci U S A 108:4772–4775
157. Turner JD, Langley KL, Johnston K et al (2009) Wolbachia lipoprotein stimulates innate and adaptive immunity through Toll-like receptors 2 and 6 induce disease manifestations of filariasis. J Biol Chem 284:22364–22368
158. Moreira LA, Iturbe-Ormaetxe I, Jeffery JA et al (2009) A Wolbachia symbiont in *Aedes aegypti* limits infection with dengue, Chikungunya, and Plasmodium. Cell 139:1268–1278
159. Walker T, Johnson PH, Moreira LA et al (2011) The Mel Wolbachia strain blocks dengue and invades caged *Aedes aegypti* populations. Nature 476:450–453
160. Hoffmann AA, Montgomery BL, Popovici J et al (2011) Successful establishment of Wolbachia in Aedes populations to suppress dengue transmission. Nature 476:454–457
161. Callaway E (2016) Rio fights Zika with biggest release yet of bacteria-infected mosquitoes. Nature 539:17–18
162. Blaney JE Jr, Durbin AP, Murphy BR, Whitehead SS (2006) Development of a live attenuated vaccine using reverse genetics. Viral Immunol 19:19–32
163. Kirkpatrick BD, Durban AP, Pierce KK et al (2015) Robust and balance immune responses to all 4 dengue virus serotypes following administration of a single dose of a live attenuated tetravalent dengue vaccine to healthy, flavivirus-naïve adults. J. Infect. Dis. 212:702–710
164. Roy C, Adams AP, Wang E et al (2014) Chikungunya vaccine candidate is highly attenuated and protects nonhuman primates against telemetrically monitored disease following a single dose. J. Infect. Dis. 209:1891–1899
165. Tretyakova I, Hearn J, Wang E, Weaver S, Pushko P (2014) DNA vaccine initiates replication of live attenuated Chikungunya virus in vitro and elicits protective immune response in mice. J. Infect. Dis. 209:1882–1890
166. Simanjuntak Y, Liang JJ, Lee YL, Lin YL (2015) Repurposing of prochlorperazine for use against dengue virus infection. J. Infect. Dis. 211:394–404
167. Vogel G (2013) Lab dishes up mini-brains. Science 341:946–947
168. Lorenzi OD, Major C, Acevedo V et al (2016) Reduced incidence of Chikungunya virus infection in communities with ongoing *Aedes aegypti* Mosquito Trap Intervention Studies—Salinas and Guayama, Puerto Rico, November 2015-februaty 2016. MMWR Morb Mortal Wkly Rep 65. doi:10.15585/mmwr.mm6518e3
169. Reikind MH, Wund MA (2009) Experimental assessment of the impact of northern long-eared bats on ovipositing *Culex* [Diptera: Culicidae] mosquitoes. J Med Entomol 46:1037–1044
170. Honein MA et al (2017) Birth defects among fetuses and infants of US women with evidence of possible Zika virus infection during pregnancy. JAMA 317(1):59–68. doi:10.1001/jama.2016.19006
171. Brasil P et al (2016) Zika virus infection in pregnant women in Rio de Janeiro. N Engl J Med 355:2321–2334

Chapter 7
Ebola and Marburg: Out of Africa

7.1 Introduction

Lassa fever was the first major hemorrhagic fever disease to emerge in Africa from a town [Lassa] in Nigeria in 1959 [1]. Lassa fever virus, an arenavirus, is endemic throughout West Africa and spread by contact with the multimammatc mouse, which sheds the virus in the urine. Most cases are mild or subclinical, but severe disease can result in fever, vomiting, diarrhea, shock, and death [1]. However, Ebola virus [EBV] and Marburg virus [MARV], members of the *FIloviridae* family, are more contagious and deadly than Lassa fever. Fruit bats are suspected to be the natural reservoir of filoviruses, including EBV and MARV, and wild animals in the rain forests of Africa may become infected through contact with bat feces and body secretions through consumption of contaminated water or plants and become intermediate hosts [2]. Humans are incidental hosts that are believed to become infected from contact with infected wild animals, chimpanzees, gorillas, monkeys, forest antelopes, dunkers, and porcupines from hunting for bushmeat. Prior to 2014, these conditions were neglected tropical diseases that caused rare, obscure infections affecting limited populations in Africa without any impact on global health. However, this dramatically changed after March 23, 2014, when the World Health Organization [WHO] was notified of an outbreak of Ebola virus disease [EVD] in Guinea, and declared in August 8, the epidemic to be a "public health emergency of international concern" [3]. By September 13, 2014, five countries in West Africa—Guinea, Liberia, Nigeria, Senegal, and Sierra Leone—reported a total of 4507 cases with 2296 related deaths [3]. Moreover, cases were exported to Europe and the USA [mainly healthcare personnel] which raised public alarm. By the end of 2015, when the epidemic was simmering, there were more than 28,000 cases of EVD with more than 11,000 attributed fatalities [4].

I.W. Fong, *Emerging Zoonoses*, Emerging Infectious Diseases of the 21st Century,
DOI 10.1007/978-3-319-50890-0_7

7.2 Historical Aspects

7.2.1 Marburg Virus [MARV] Infection

MARV was first discovered in 1967, when laboratory workers in Marburg and Frankfurt [Germany] and Belgrade [former Yugoslavia] became infected with an unknown agent [5]. Thirty-two subjects developed severe flu-like illness that resulted in seven fatalities. The source of the infection was traced to African green monkeys [*Chlorocebus aethiops*] that were imported from Uganda and transported to these cities. Infection occurred during dissection of the monkeys to obtain kidney cells to culture poliomyelitis vaccine strains [5]. After 3 months of intensive investigation by scientists, the etiologic agent was cultured and identified and named after the city Marburg, with the most cases, and represented the first isolation of a filovirus. The first report of naturally acquired MARV occurred in South Africa, when a traveler and his companion and a nurse became infected but was quickly contained by infection control practice for suspected Lassa fever [6]. From 1975 through 1985, sporadic small outbreaks of MARV occurred in Africa, but public health concern were diminished when EBV emerged in 1976 and was associated with mortality of up to 90%, whereas case-fatality rate of MARV disease [MVD] was much lower [5].

MARV reemerged in the Democratic Republic of the Congo [DRC] with two large outbreaks in 1998–2000 [7] and subsequently an outbreak for the first time in West Africa, in Angola, in 2004–2005 [8]. Unlike previous outbreaks, these more recent epidemics were associated with high fatality rates, 83% In DRC and 90% in Angola, with a total of 406 cases. The high mortality in Africa compared with a much lower mortality [21.8%] in the initial outbreak in Europe in 1967 may be the result of multiple factors such as access to medical care, level of the healthcare from available facilities, and concomitant illnesses in the local African population such as malaria and acquired immunodeficiency syndrome [AIDS]. Although as of 2012, there had been only a total of 452 cases and 368 deaths due to MVD, the disease is likely underdiagnosed and reported [5].

7.2.2 Ebola Virus [EBV] Infection

EBV first appeared in Zaire, the former name for the DRC, in 1976 when it caused a major outbreak of hemorrhagic fever in the DRC and Sudan [9, 10]. The etiology agent was soon isolated and discovered to be a new virus belonging to the same family of *Filoviridae* viruses as MARV [11]. Investigation at that time indicated cross infection of medical staff with high fatalities, and medical facilities were closed to eliminate dissemination of the virus through the use of unsterilized needles and syringes and contact with infected secretions due to lack of barrier-nursing techniques [12]. The outbreaks rapidly diminished except for a small outbreak in Sudan in 1979 and an isolated case in the DRC in 1977. There was a period of quiescence with respect to Ebola activity in Africa, until 15 years later. In 1994–1996, at least three separate areas of active EBV transmission were recognized: Cote d'Ivoire in 1994 [13], DRC in 1995 [14], and Gabon in 1994–1996 [15]. Outbreaks of EBV have occurred primarily in remote villages in Central and West Africa, close to tropical rain forests. The disease was mainly found in the DRC, Uganda, South Sudan, and Gabon until the recent epidemic in 2014. In 1989–1992, an outbreak of EBV occurred in cynomolgus monkeys [*Macaca fascicularis*] imported from the Philippines and recurred in 1996 again in primate facilities, and the strain was labeled EBV-subtype Reston based on the initial outbreak in the facility in Reston, Virginia [16, 17].

Recent molecular data suggest that EBV have estimated lineage millions of years and evolutionary biologists have detected viral DNA of EBV within genomes of animals, such as remnants in different species of rodents, including the mouse and Norway rats [18]. It is speculated that EBV infected rodents and other mammals, such as bats, at least 20 million years before infecting humans. Historians have also theorized that EBV may have caused an outbreak in Athens in 430 BCE [19]. The ancient plaque of Athens has been a subject of much speculation as to the etiology, including plague, typhoid, typhus, smallpox, and measles [20]. However, the clinical description of the outbreak and manifestations could be compatible with an EBV epidemic, and the geographic origin of the ancient outbreak is fitting. The Athens plague originated in a region south of Egypt called by ancient Greeks "Aethiopia" which refers to a region in sub-Saharan Africa, Sudan, and the DRC—where EBV outbreaks have occurred [19, 21].

7.3 Virology

MARV and EBV are closely related members of the *Filoviridae* family, which are filamentous, single strand, negative-sense, enveloped RNA viruses [22]. *Filoviridae* family consists of three genera: *Ebolavirus*, *Marburgvirus*, and *Cuevavirus*. The viral genome contains seven genes encoding seven structural proteins and several nonstructural proteins for all EBVs and the only representative of *Cuevavirus*, Lloviu virus [isolated from a European bat], while MARV is not capable of encoding nonstructural proteins due to lack of editing site within its glycoprotein GP gene [23]. The surface glycoprotein GP mediates viral attachment and entry into host cells and is an important target for vaccine development [24]. The genus *Marburgvirus* includes only a single species, *Marburg marburgvirus*, with five different lineages, and genomic divergence between isolates is less than 30%. The MARV strains reveal two distinct genetic lineages: Ravn virus [MARV-Rav] isolated in 1987 from a case in southeastern Kenya and which caused the Marburg hemorrhagic fever [MHF] DRC outbreak in 1998–2000 and the Angola strains [MARV-Ang] [5]. The DRC epidemic appears to have been caused by nine different virus variants in tested patients, representing separate crossover events from the natural reservoir to the human population [7]. The Marv-Ang was responsible for the largest documented outbreak of MHF and originated in Angola [8].

Ebola virus family has five species: *Zaire ebolavirus*, *Bundibugyo ebolavirus*, *Taï Forest ebolavirus*, *Sudan ebolavirus*, and *Reston ebolavirus* [2]. Zaire EBV [EBOV] is the causative agent of the largest ever epidemic of viral hemorrhagic fever, recent West Africa outbreak of 2014–2015, and appears to be the most virulent strain. The Sudan EBV [SUDV] shares about 70% amino acid sequence with EBOV and has caused five outbreaks, including the second largest outbreak in Uganda in 2000–2001, which affected 425 people [25]. EBOV and SUDV were responsible for simultaneous outbreak in 1976, when Ebola viruses were first described [9, 10]. Bundibugyo EBV [BDBV] was the causative agent of an outbreak in 2007 in Bundibugyo, Uganda, and reappeared in a 2012 outbreak in the DRC with mortality between 25 and 51% [26, 27]. Taï Forest EBV [TAFV] was isolated from a single, nonlethal human infection, and the Reston EBV [RESTV], which is pathogenic for nonhuman primates, has not been associated with human disease but can cause seroconversion in handlers of infected animals [13, 28].

7.3.1 Ecology

MARV and EBVs are zoonotic pathogens that persist and flourish in natural reservoir hosts in Africa. Humans and nonhuman primates are intermediate spillover hosts that develop severe disease and high mortality. Ecological niche modeling of outbreaks and epidemiological patterns have suggested that EBV is endemic in the rain forests of central and western Africa, while MARV is more prevalent in open, dry areas of eastern and south-central Africa [29, 30]. Most or nearly all the primary natural infections with MARV have been linked to human entry into caves inhabited by bats [mine worker or cave visitors], and bats were suspected to be natural reservoirs or important in the transmissions of the disease [5]. MARV was first detected in the common Egyptian fruit bat [*Rousettus aegyptiacus*] in 2007 and from bats in caves linked to human disease in Uganda [31]. Genomic analysis of MARVs isolated from bats of caves in Uganda and mines in DRC showed the sequences matched very closely to human isolates that caused outbreaks [32]. *R. aegyptiacus* may not be the exclusive reservoir of MARV as other species of bats have been found to have antibodies and RNA of the virus. Recent experiments in Egyptian fruit bats showed that MARV inoculated subcutaneously produced asymptomatic infection with development of strong protective immunity, but direct physical contact and the airborne route failed to produce infection between bats [33]. See Fig. 7.1 for an outline of the biological cycle and ecology of MARV.

EBVs have never been isolated from any fruit bat species, but viral RNA fragments of EBOV and virus-specific antibodies were detected in a few fruit bat species captured around the endemic areas during the 2001–2003 EVD outbreaks in Gabon and the DRC [34]. In a recent seroepidemiological study of migrating fruit bats [*Eidolon helvum*] from Zambia during 2006–2013, filovirus-specific antibodies were detected continuously to various species with overall 8.6% to EBVs and 0.9% to MARV, but filovirus RNA genomes were not detected in the liver and spleen of the bats [35].

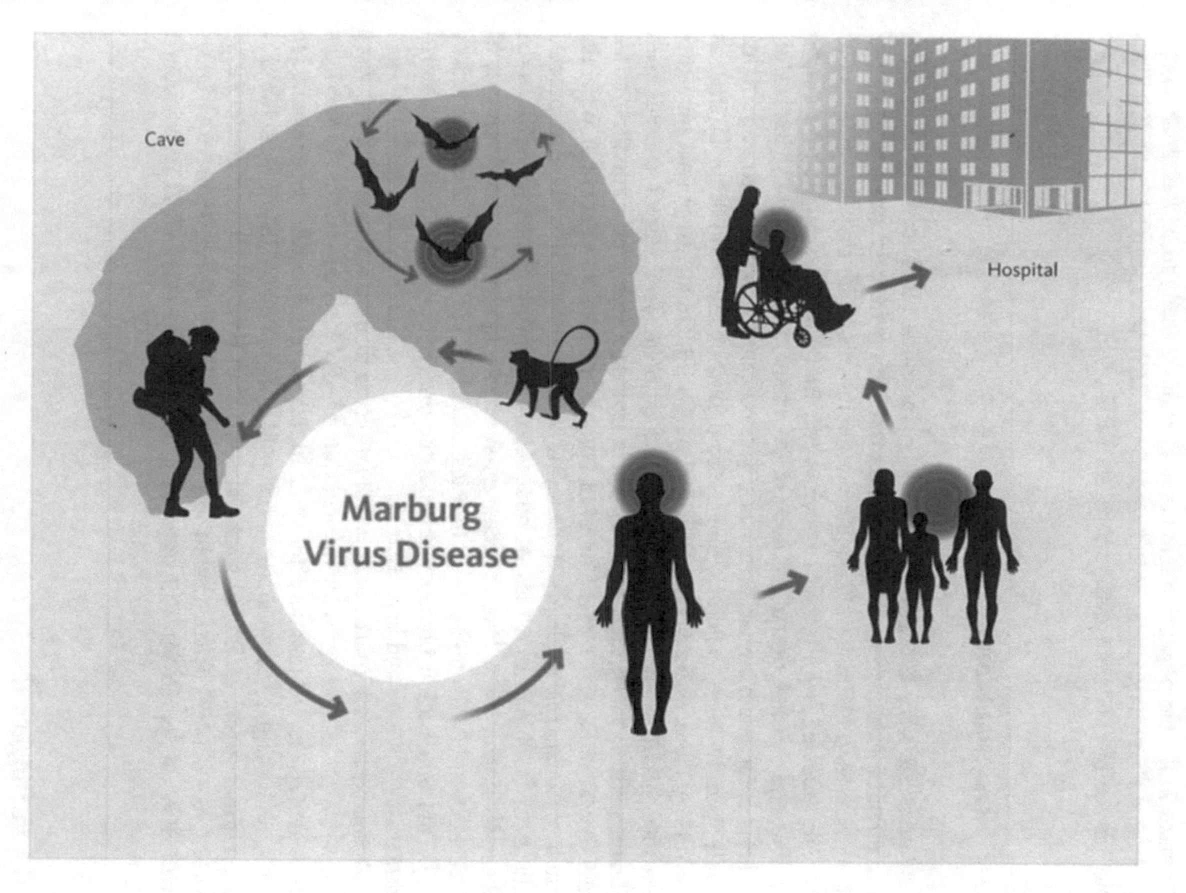

Fig. 7.1 Biological cycle and ecology of Marburg virus

The change of the serological dominant virus species in these bats were interestingly correlated with the viral species causing outbreaks in Central and West Africa during 2005–2014. This study suggests that these migrating bats, which are prevalent across Africa, may be used as surrogate sentinels for determining the species of circulating filoviruses in the bat population. However, they are unlikely to be the reservoirs of MARV and EBVs as no detectable fragments of RNA were found. Figure 7.2 illustrates the ecology and transmission hosts for EBV.

7.3.2 *Transmission*

Filoviruses are very contagious and easily transmitted to humans with contact from infected animals or other humans, even with extremely small amounts of virus particles. Initial naturally acquired infection of MARV may be through contact with infected bat's [several species] secretions in caves or mines or from handling and consumption of bat meat or from handling or consuming intermediate hosts such as nonhuman primates [5]. EVD outbreaks appear to be initiated mainly from hunting, handling, and consumption of infected bushmeat such as monkeys, gorillas, duikers, fruit bats, and others [2]. Human-to-human transmission then becomes the main driver of the outbreak from direct contact with infected patients' secretions or body fluids [saliva, sweat, stool, vomitus, urine, tears, breast milk, and semen] or even body contact of a deceased at funerals. Families, friends, and medical personnel are at particular risk of caring for infected patients or handling of corpses without use of proper protection [36–38]. In the recent EBV epidemic of West Africa, non-survivors and patients developing rapid critical illness had the highest risk of transmitting the virus in the course of illness [39]. This observation probably reflects the presence of very high vial load in these patients' blood and body fluids.

Persistent latent filovirus in survivors of acute disease may transmit the virus weeks to months after recovery on occasion. MARV has been cultured from tears and semen tissue 1–3 months after acute illness [5, 40], and EBV was detected in semen and vaginal secretion for months [41, 42]. Viable EBOV was recently found from aqueous humor 14 weeks after onset of EVD and 9 weeks after clearance of viremia [43]. Also sexual transmission of EBV and persistence of infective EBOV in semen for 179 days or more after onset of EVD was recently reported [44].

7.4 Pathogenesis

MARV and EBVs enter the body through small skin abrasions or mucosal membranes after direct contact with body fluids, secretions, or personal contact with infected person or animal. Macrophages and dendritic cells are the primary

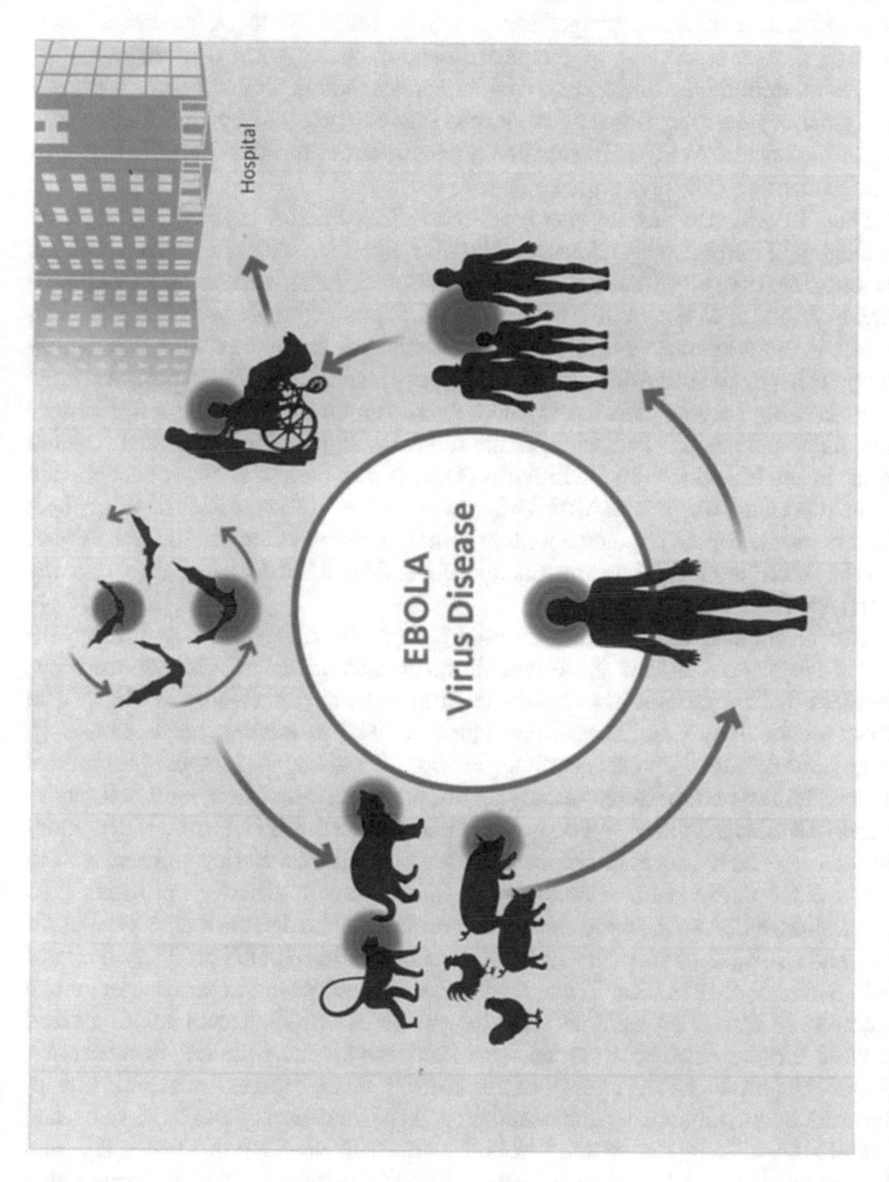

Fig. 7.2 Biological cycle and ecology of Ebola virus

targets of these filoviruses, and spread of the viruses rapidly occurs to invade and replicate in the endothelial cells, lymph nodes, liver, and spleen [5, 45]. Depletion of lymphocytes with lymphopenia is more consistent and greater with Ebola than MARV and is due to enhanced apoptosis possibly related to aberrant T-cell activation signals from macrophages and dendritic cells [45, 46]. Characteristically infected tissues especially of the reticuloendothelial system demonstrate high viral load, minimal inflammation, tissue necrosis, and depletion of macrophages and lymphocytes, reflecting a dysregulated immune response and ineffective control of the invading virus. The surface glycoprotein of the filoviruses is involved in attachment to cells and virus entry [47, 48].

There is evidence that the filoviruses can block interferon [IFN] signaling by employing different mechanisms to abort or interfere with the innate immune response. MARV VP40 inhibits phosphorylation of Janus kinases and STAT proteins in response to type 1 and II IFNs and IL6, preventing downstream signaling [5]. EBV VP24 interacts with STAT1 and member of the nuclear import in family and prevents nuclear translocation of phosphorylated STAT1 [49]. Massive release of proinflammatory cytokines and chemokines [tumor necrosis factor-a, interleukins, inflammatory proteins, nitric oxide radicals, etc.] occurs in the final stage of disseminated infection with the filoviruses [5, 50, 51], but this is better documented in humans with EBV than MARV. The cytokine storm and secondary capillary leak are responsible for the hypotension, septic shock syndrome, and multiorgan failure, and the secondary thrombocytopenia and disseminated intravascular coagulopathy [DIC] result in bleeding.

Various animal models have been used to study the mechanisms and pathogenesis of the filoviruses, and these include nonhuman primates, mice, guinea pigs, hamsters, and marmosets. The clinical features and uniform lethality of MARV and EBV in cynomolgus and rhesus macaques [as well as African green monkeys], closely mimic human disease, and they are considered the gold standard for animal models. Wild-type filoviruses usually do not produce disease in rodents and mouse or inbred strains of guinea pigs do not develop DIC. Recently improved rodent models such as recently developed hamster model and outbred Harley strain of guinea pig do show coagulopathy which is more consistent with disease in primates [52]. In macaques MARV-Ang appears to produce more severe disease than other MARV [53], and among the five strains of EBV the Zaire strain [EBOV] produces the most rapid and fatal disease [52]. These models seem to reflect the experience in human outbreaks of MARV an EBV and is quite consistent with the recent EBV outbreak in West Africa, which was caused by a Zaire strain. For instance, in macaques, MARV-Ang produced rapid onset of disease with fever, depression, anorexia, petechial rash, and lymphopenia and death in 6.7 days; depletion of T-cells, B-cells, and natural killer cells by day 6; and cytokine storm with elevation of TFN-α, Il-6, and chemokine CCL2 early during infection [53]. The pathogenesis of filovirus infection is illustrated in Fig. 7.3 [2].

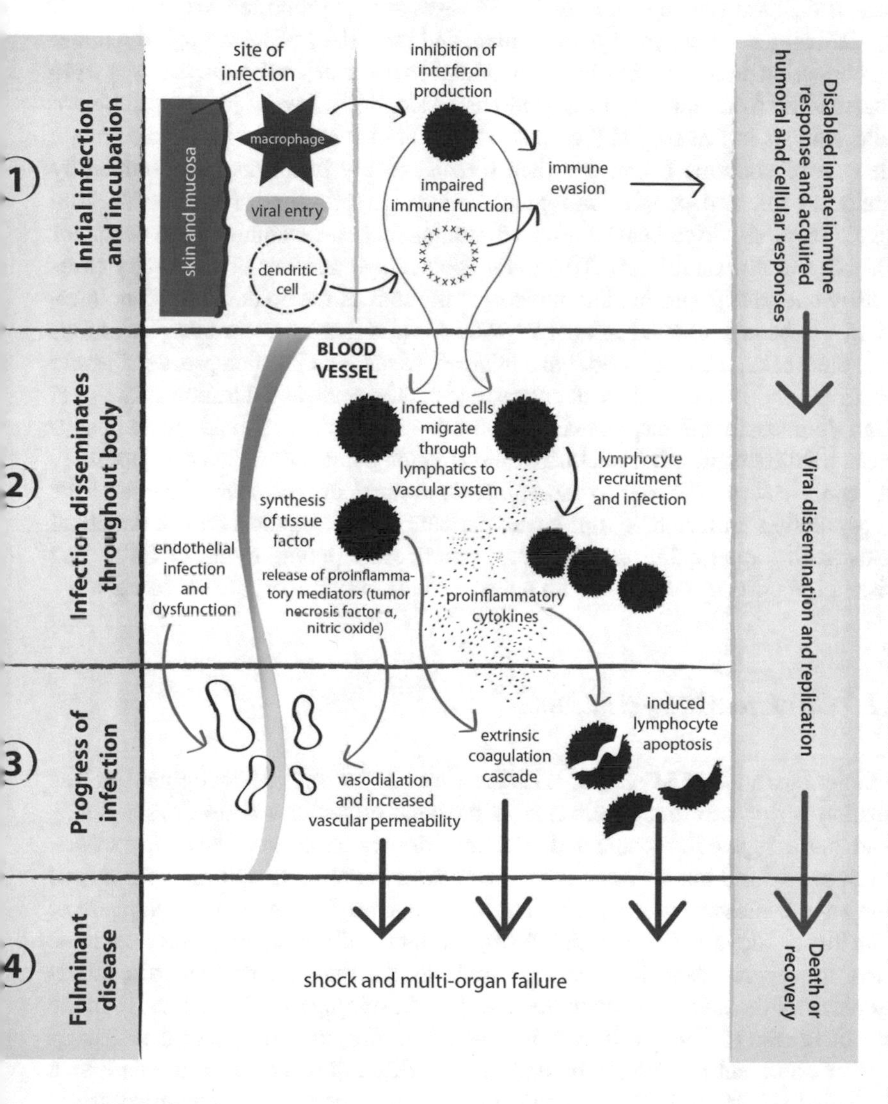

Fig. 7.3 Outline of the pathogenesis of filovirus infection [EVD]. Schema of diagram modelled and modified from Kanpathipallai R. N. Engl. J. Med. 2014; 371:e18

7.5 Scale of Ebola West African Epidemic

Between December 2013 and April 2016, the largest epidemic of EVD has occurred with 28,616 cases and at least 11,310 deaths affecting the populations of Guinea, Liberia, and Sierra Leone primarily, but 36 cases are also reported [exported] from Italy, Mali, Nigeria, Senegal, Spain, Britain, and the USA [54]. Although the initial human infection was probably from an animal [bushmeat], subsequent cases were estimated to be from human-to-human transmission [55]. Investigation with phylogenetic analysis indicated that the West African epidemic started in Guinea from a single case of zoonotic origin and then spread quickly from human to human by migration of infected people throughout Guinea, Liberia, and Sierra Leone. The epidemic in West Africa peaked after 10 months, but cases continued to occur for another 18 months until April 2016 [54]. The average number of secondary cases caused by one case [reproduction number] at the start of the epidemic [before intervention] has been estimated to be 1.71–2.02 [56]. However, as in other infectious disease outbreaks, it is estimated that a minority of patients ["superspreaders"] were responsible for 80% of the subsequent cases [54]. The scale and duration of the West African Ebola outbreak have been attributed to several factors: inadequate healthcare and public health infrastructure of the affected countries, delayed response by WHO and Western countries to the outbreak, cultural practices and beliefs of the local population, failure of symptomatic patients to seek medical care, contacts of patients fleeing quarantine, unsafe burial practices of people dying of EVD, and persistent infection of survivors with delayed transmission from sexual contact.

7.5.1 Clinical Manifestations

The clinical features of MVD and EVD are very similar and are dealt together. The incubation period can range from 2 to 21 days but typically is 3–14 days and may depend on the type of exposure and infectious dose at exposure. The course of disease can be divided into three phases: generalization phase, early organ phase, and late organ/convalescent phase [1, 5]. The initial nonspecific generalization phase of flu-like illness [days 1–4], with high fever, myalgia, arthralgia, profound weakness, prostration, severe headaches, nausea, vomiting, diarrhea, abdominal pain, sometimes pharyngitis, and maculopapular rash [5]. Development of a rash early in the course of illness [5–7 days] occurs in 25–52% in filovirus infection; it is usually diffuse or confluent and non-itchy and may be difficult to detect in the dark skin individuals [57]. Myalgia is common but significant myositis or rhabdomyolysis is rarely reported and may be overlooked. Rhabdomyolysis with creatine kinase >5000 U/L was recently reported in 36% of 38 patients with EVD in Guinea [58]. In the recent EBV outbreak in West Africa, gastrointestinal symptoms with severe diarrhea and profound fatigue were the most common symptoms [59]. Dehydration and electrolyte disturbances [hyponatremia, hypokalemia, and hypocalcemia], lymphadenopathy, leukopenia, and thrombocytopenia were commonly found [2]. The early organ phase occurs from day 5 to 13, with persistence of high fever and initial symptoms and evidence of leaky vasculature. Patients may develop dyspnea, conjunctival injection, and edema; confusion, delirium, encephalopathy, or encephalitis; and later bleeding tendencies. In previous outbreaks of EBV and MARV, hemorrhagic manifestations were seen in more than 75% in the latter part of this phase, consisting of petechiae, mucosal bleeding, bloody diarrhea, hematemesis, and ecchymoses [5]. However, in the 2014–2015 EBV outbreak, overt bleeding was uncommon [19% or less], and fever could be absent [about 18%], even in patients with high mortality [59–61]. Multiple organ dysfunction commonly occur at this stage, elevated liver enzymes in most patients, renal impairment or failure secondary to dehydration, rhabdomyolysis and shock, and pancreatic disturbance with hyperamylasemia [3, 5, 62]. The late phase, usually after day 12–13, may result in death or a prolonged recovery period. Fatalities usually occur 8–16 days following onset of symptoms, and in the EVD outbreak in Sierra Leone, most patients died within 2 days of hospital admission [59], reflecting delay in seeking medical treatment or possibly rapid progressive disease. Dying patients usually have obtundation or coma, severe metabolic abnormalities, coagulopathy, shock, and multiple organ failure [5, 62]. Fatality rates of filovirus outbreaks varied from 20% to 90%, which may depend on accessibility to modern medical care, timing of patients seeking medical treatment, and awareness of the diagnosis or differential diagnosis. In the recent EBOV epidemic in West Africa, early during the outbreak, the mortality was very high 70–73% [59, 62], but with education of the local population on seeking early treatment in specialized centers, the mortality fell with a final tally of 28,601 cases and 11,300 deaths [39.5% mortality] [4]. Clinical prognostic factors associated with higher mortality include age over 45 years, pregnancy, multiorgan failure, vascular leak syndrome, severe metabolic disturbance, and coagulopathy [59–62]. Early markers of high mortality has been found with initial high level of viremia [63], elevation of several chemokines and cytokines, and elevated thrombomodulin and ferritin, but high levels of sCD40L were associated with survival [64].

Survivors of filovirus infection have had a prolonged, extended, convalescence period with weakness and exhaustion, myalgia, excessive sweating, reactive depression, skin desquamation, partial amnesia, and ocular findings. Early and long-term sequelae following EVD outbreaks, called post-EVD syndrome, have recently been analyzed and reported in formal studies. In a cross-sectional study of 603 survivors of EVD from the recent outbreak in Sierra Leone, 82–151 days after hospital discharge, residual sequelae were common [65]. Symptoms included arthralgia [76%], new ocular symptoms [60%], uveitis [18%], and auditory symptoms [26%]. Higher viral load at presentation was associated with uveitis and ocular diagnoses. A smaller but longer study from the EBV outbreak of 2007 in Uganda was also reported, with a follow-up about 29 months after the outbreak, and symptoms were compared in survivors [$n = 49$] to uninfected controls [$n = 157$]. Common clinical ailments were greater in survivors of EVD than controls for all symptoms: headache, 88 vs. 75% [$p = 0.007$]; retro-orbital pain, 29 vs. 5% [$p < 0.0001$]; muscle weakness, 12 vs. 3% [$p = 0.038$]; joint pain, 35 vs. 11% [$p = 0.020$]; joint stiffness,

22 vs. 7% [$p = 0.022$]; blurred vision 39 vs. 17% [$p = 0.018$]; hearing loss, 27 vs. 10% [$p = 0.010$]; fatigue, 57 vs. 25% [$p = 0.001$]; difficulty swallowing, 27 vs. 13% [$p = 0.017$]; and insomnia, 57 vs. 26% [$p = 0.001$] [66]. The high incidence of ocular symptoms had been described by others and includes vision loss, uveitis, conjunctivitis, and blurred vision [67]. The pathogenesis of late sequelae is not well understood but appears to be related to delayed clearance of the virus in immunological protected body compartments, such as chambers of the eye [43], male gonads, placenta, and possibly joints and the brain. There is no evidence that filoviruses causes chronic infection, but there is some evidence that persistent antigen can result in sustained immune activation [WHO Meeting Report on survivors of Ebola virus disease: clinical care of EVD survivors, Freetown, 3–4 August 2015]. This concept of the pathogenesis of the late sequelae of filovirus infection, also described after MVD [6, 57], is supported by infection with EBOV in a macaque with apparent recovery and delayed death. During recovery, the primate developed clinical ocular findings with conjunctivitis, and autopsy revealed strong antigen positivity associated with necrotizing scleritis, perioptic neuritis, and steatitis of the eye with neutrophil and macrophage infiltration, and the brain demonstrated neutrophilic and lymphoplasmacytic necrotizing choriomeningoencephalitis [68].

Of concern is recent report of a life-threatening complication [meningoencephalitis] 9 months after recovery from acute EBV infection, in a Scottish nurse who became infected in Sierra Leone in 2014. EBV was detected by RT-PCR with high levels in the cerebrospinal fluid and low level in plasma [69]. Thus, it is important for recovered patients of EBD to be followed for long term in Africa to assess for late relapse for an estimate of the frequency. Moreover, cases of late relapse could rekindle the outbreak.

7.5.2 Diagnosis

The clinical manifestations of MVD or EVD are nonspecific and can be confused with several diseases endemic in Central and West Africa, and the epidemiology is very important in considering the diagnosis, especially in the presence of an outbreak. The WHO has provided pocket guide for frontline health worker in Africa for case identification of hemorrhagic fever. One of the most important aids in the diagnosis is the history of exposure within 2–21 days prior to the patient's onset of symptoms. These include exposure to blood, excreta, vomit, or sweat of a sick person or deceased body by family, friends, traditional healers, and funeral attendants. Other risky exposures include contact with live or dead bats and monkeys/gorillas, consumption or preparation of bushmeat, visiting or exploring mines or caves, and sexual exposure to a recovered male patient within 3 months of suspected or proven diagnosis and recovery. However, this may need to be extended to 6–12 months after recovery based on a recent report [44]. The differential diagnosis includes malaria, enteric infections, respiratory viral infections including influenza, typhoid fever, chikungunya fever, dengue fever, and bacterial sepsis [70, 71]. Confirmation of the diagnosis of filovirus infection can be accomplished by detection of the virus by molecular methods or antigen by enzyme immunoassay or by serological methods. These tests, however, are performed in a biosafety level 4 facility as the viruses are very virulent and potentially could be spread by aerosol with deadly consequences. During outbreaks of filovirus infections, mobile laboratories [provided by developed countries through WHO] commonly use polymerase chain reaction [PCR] or enzyme-linked immunoassay [ELISA] for rapid screening and diagnosis. Since the MARV outbreak in Angola in 2005, filovirus diagnostic testing with real-time quantitative reverse transcriptase PCR [qRT-PCR] has been considered the method of choice for both field and reference laboratory diagnostic testing [72, 73]. Simpler, lower-cost, rapid diagnostic tests for detecting filovirus antigen include antigen-capture ELISA and lateral-flow immunodiagnostic [LFI] assay that can detect EBV or MARV matrix VP40 and glycoprotein [GP] antigens in blood or body fluids [74]. ELISA for detection of virus-specific IgG or IgM can also be used for confirmation of the diagnosis [74, 75].

7.5.3 Management

Management of filovirus infection is best accomplished in a specialized center geared for level 4 biohazard infectious pathogens, but this is unavailable in endemic Africa; thus, local hospitals or healthcare facilities have to set up temporary isolation units. Treatment of suspected cases should be initiated early in the patient's home or village even before admission to a healthcare facility, as profound dehydration from diarrhea and vomiting is commonly present. Oral rehydration with electrolyte solution can be provided by the first responders or antiemetic drugs such as metoclopramide or chlorpromazine for those with vomiting to allow for oral rehydration. Education of the local population on the manifestation of hemorrhagic fevers, protective measures before transfer to a health facility, and preferable local home-bedside care provided by a recovered subject of the illness may be of value in reducing the high mortality of filovirus disease. Home remedy for fluid loss from diarrhea should be encouraged even before medical contact, such as consumption of homemade soup and coconut water, which is rich in potassium depleted in severe diarrhea. This could ameliorate the profound dehydration of patients that are admitted to hospital.

7.5.3.1 Initial Response

Medical personnel from the local hospital or the District Health Officer will usually keep the patient in a holding room/area until assessed by a rapid response team. The patient and the family are counseled on the disease, transmission risk, and need for isolation and protective measures. The patient is isolated in a designated room or area, away from crowded areas, well ventilated, with good sunlight and adjoining toilet and restricted access. Strict adherence to standard contact and droplet precautions for healthcare, environmental, and laboratory workers is essential to prevent transmission. Strict infection control measures are instituted, including wearing of protective clothing or personal protective equipment [PPE] for all healthcare workers, support staff who clean the room, handle supplies or equipment or waste for disposals, laboratory staff who handle specimens from suspected patients, burial teams who remove bodies, or family members who care for infected patients [see WHO guidelines [76]]. Despite these protective measures, transmission of EBV has occurred in healthcare workers in West Africa, Europe, and the USA. Thus, CDC choice of PPE reflected a goal of no skin exposure, including N95 mask, head coverage with goggles, total body coverage, or a one piece suite with complete coverage of the body and a respirator [77]. Although filoviruses have not been shown or believed to be transmitted by aerosols, certain procedures can generate aerosols of body secretions and should be avoided. Moreover, in experimental animal model, it was possible to infect marmosets with EBV by aerosol route [78]. Donning and doffing of PPE is a key feature of the guidelines, and break in technique may have been responsible for transmission of EBV to a nurse in the US. Thus, a buddy system was instituted for an observer to ensure no break in the method of infection control, especially putting on or removing the PPE.

7.5.3.2 Treatment

There is no proven effective specific therapy for MVD or EVD. Supportive therapy with intravenous crystalloids to correct severe dehydration, correction of electrolyte disturbances, blood transfusion to correct severe anemia from bleeding tendencies, antimalarial therapy for presumed malaria [after appropriate blood smears], and antibiotics for possible bacterial infection after obtaining blood and other cultures are usually instituted. Malaria treatment for suspected or proven EVD in the recent West Africa outbreak was given to patients systematically or based on confirmed malaria diagnosis. At a treatment center in Lofa County, Liberia, there was a temporary shortage of the first-line antimalarial combination [artemether-lumefantrine]; thus, an alternative combination [artesunate-amodiaquine] was used for a period of time. Interestingly, the mortality rate in this observational study revealed a 31% lower risk of death with artesunate-amodiaquine-treated patients [36 of 71] than the artemether-lumefantrine-treated patients [125 of 194], even after adjusted analysis and with a stronger effect observed among patients without malaria [79]. Amodiaquine in vitro has inhibitory activity against EBV, and the improved survival possibly could be related to its antiviral effect [80]. Thus, randomized studies are needed, but this may be difficult now that the epidemic has ended, but this could be further explored in animal models.

Convalescent blood or plasma from recovered patients was recommended by the WHO for EVD in the recent outbreak, as possible lifesaving therapy [81]. This form of treatment has been used in Lassa fever with conflicting results [82–84]. However, in randomized controlled trial, convalescent plasma against Argentinian hemorrhagic fever reduced the mortality compared to normal plasma, 1.1 vs. 16.5% [85].

Clinical evaluation of convalescent plasma to treat EVD was evaluated in the recent outbreak affecting Guinea, Sierra Leone, and Liberia in a nonrandomized comparative trial. The trial enrolled 102 patients but could only provide acceptable safety data but no efficacy result [86]. Case reports of two physicians with severe EVD transported from Africa and treated in the USA with convalescent plasma and investigational antiviral [lipid-bound small-interfering RNA [siRNA], despite multiorgan failure in one case, resulted in survival in both patients [87]. This encouraging report provides further stimulus for appropriate controlled studies.

Although shock and multiorgan failure can result from vascular leakage in viral hemorrhagic fevers, associated with a high mortality, it was more common in the 2014–2015 EVD outbreak to be due to massive volume depletion, requiring 3–5 L of intravenous fluid replacement per day, and frequent potassium replacement and electrolyte correction [88]. Oral antiemetic and antidiarrheal agents should be used to reduce gastrointestinal fluid loss, improve symptoms, and lessen the risk of environmental contamination [89]. Loperamide is a suitable antidiarrheal agent with antiperistaltic and antisecretory effects but inadequately studied in filovirus diseases. Although the mechanism of diarrhea in EVD is unknown, the large-volume diarrhea suggests a secretory mechanism [90]. Multiple randomized controlled studies of loperamide in infectious diarrheas have proven its efficacy and safety [91]. Thus, these readily available, inexpensive treatments may reduce intestinal fluid losses and shock and improve survival. During the early response to the West Africa EBV outbreak, isolation centers often lacked point-of-care electrolyte testing laboratory and shortages of oral potassium supplements [92]. Severe potassium depletion in EVD was often significant and may be associated with generalized weakness, impaired respiratory function, muscle necrosis, and cardiac arrhythmias. Large-volume infusion of normal saline may precipitate development of hyperchloremic acidosis, and this may favor the use of Ringer's solution for severe volume depletion [93]. Even the presence of vascular leakage and mutiorgan failure in EVD is not uniformly fatal, and success has been reported with a combination of ventilation support, antibiotic treatment, renal replacement therapy, and an investigational fibrin-derived peptide [FX06] for vascular leak syndrome in Germany [94].

7.5.4 Experimental Treatment

Preclinical studies on promising agents are usually performed on small animal models such as guinea pigs and mice, and the filoviruses can be adapted for these animals. In the guinea pig MARV model, various treatments [IFN, cytokine inhibition, antibody transfer] had shown promising results with improved or prolonged survival but were unsuccessful in nonhuman primates [5, 95]. Novel antisense therapies to block viral protein expression with phosphorodiamidate morpholino oligomers [PMO] can prevent postexposure lethal MARV infection [100%] and EBOV infection [60%] in primates [96]. However, it is unclear whether or not it would be effective for established infection. Small-molecule inhibitor showed complete protection against MARV and EBOV in the mouse model soon after infection [97], and a lipid nanoparticle of a small-interfering RNA [siRNA] directed against MARV nucleoprotein protected against a lethal challenge in nonhuman primates [98, 99].

The lipid-bound siRNA [TKM-100802], which targets the L polymerase, viral proteins 24 and 35, has been shown to be protective against lethal EBOV challenge in guinea pig models [100], and a combination of modified siRNA protected against lethal challenge in nonhuman primates [101]. TKM-100802 was used in two patients with severe EVD in the US along with convalescent plasma with improvement, but the effect of the siRNA cannot be determined [87]. One of the most promising treatments for the highly fatal EBOV infection is the administration of specific antibodies targeting the virus surface glycoprotein [GP]. It was first shown that immunoglobulin from a surviving EBOV-infected macaque conferred protection against EBOV infection 2 days after challenge in other macaques [102]. Several studies have demonstrated the effectiveness of monoclonal antibodies against EBOV and enhanced as a combination cocktail and even as delayed treatment in nonhuman primates [103–105]. Two monoclonal antibody combinations, ZMapp and ZMab, have been used under emergency compassionate protocols in the recent EBOV outbreak in West Africa, because of their efficacy in nonhuman primates. ZMab is a combination of six monoclonal antibodies, and ZMapp contain two antibodies from ZMab and another [13C6] to create a more potent cocktail that reversed clinical signs in all six macaques even 5 days after inoculation with EBOV [106, 107]. At least seven patients with EVD have been treated with ZMapp with five surviving, and six patients were treated with ZMab with all surviving [108]. However, it will be difficult to prove efficacy by a randomized controlled study as the epidemic has ended. In a small randomized, controlled trial of 71 patients, fatality rate was 37% [13/35] in patients receiving standard care and 22% [8/356] in cases treated with ZMapp, but did not meet the efficacy target [109].

Other experimental agents in the pipeline include favipiravir, a broad antiviral oral pyrazine compound that inhibits RNA-dependent RNA polymerase; TKM-Ebola, an injectable combination of modified small-interfering RNAs targeting EBOV polymerase, viral protein 24, and gene VP35; BCX4430, a broad oral antiviral compound that inhibits RNA-dependent RNA polymerase; and AV1-7537, an injectable PMO that binds to one of EBV seven genes and prevents replication [2]. Favipiravir was recently assessed in 126 EVD patients in a non-comparator proof-of-concept trial [JIKI trial] in West Africa between December 2014 and April 2015 [108]. There was no serious adverse events reported but efficacy could not be adequately assessed. The baseline viral load was an accurate predictor of mortality; pretrial mortality in patients with lower viral load [77 copies/ml] was >30.5% but with favipiravir it was 20%, but with higher viral load, the mortality was 91%. In another retrospective smaller case–control study, the survival rate of favipiravir treated patients was 64.8% [11/17] versus 27.8% [5/18] in the untreated control group, and the viral load was reduced >100-fold in the 52.9% of the treated cases [110].

7.5.5 *Vaccines in Development*

Several filovirus vaccines are in development, but the two most promising types are the virus-like particle vaccines and the virus-vector-based vaccines. Virus-like particle vaccines have been shown to be protective in guinea pigs and nonhuman primates for MARV and EBV, and there is a potential for a pan-filovirus vaccine [111–113]. Viral vectors to express filoviral antigens such as GP-antigen are in vaccines being tested. The two systems being assessed are based on replication-defective adenoviral vectors and the recombinant vesicular stomatitis virus [VSV]-based vaccines for both MARV and EBV [114–116]. The recombinant[r] VSV-based vaccine appears to be more effective in preventing filovirus disease in nonhuman primates [up to 100%] than the adenovirus-vectored vaccines which gives partial protection [116, 117]. Phase 1 study to assess safety and immunogenicity of a recombinant adenovirus type 5 vector-based EBV vaccine in healthy adults was recently reported from China [118]. No serious adverse event was reported and only mild fever and mild pain at the injection site were noted. A single injection of the high-dose vaccine produced robust immunogenic response, with glycoprotein-specific humoral and T-cell response against EBV in 14 days [118].

Phase 1 trial of a rVSV vaccine, rVSV-ZEBOX with EBOV DNA, was performed in healthy adults in Africa and Europe and showed good immunogenic response in all participants, and mild to moderate adverse events were common but transient [119]. Antibodies to the EBV glycoprotein were detected in all participants with all doses of vaccine, but higher levels of neutralizing antibodies were present with the higher dose of vaccine. Fever occurred in 35% of vaccines but lasted a median of 1 day, and joint pains occurred in 33% of European participants and lasted a median of 8 days. EBV and MARV glycoprotein DNA vaccines, singly and in combination, have also undergone phase 1 trial in healthy Ugandan adults [120]. Although the vaccines elicited antibody and T-cell responses specific to the glycoprotein with single or combined vaccines, the immunogenic response was not as robust as to the rVSV-ZEBOX vaccine. Maximum antibody responses were seen only in 57% of subjects receiving the EBV vaccine, and 63% also had a T-cell response to the EBOV glycoprotein [120]. Cross-protection with one filovirus vaccine does not occur in animals when challenged with another species of filovirus, but combined vaccines can produce protection against both viruses. Preliminary study in outbred Harley strain of guinea pigs showed that a single injection of a single-vector [rVSV] trivalent filovirus vaccine could produce 100% protective efficacy against homologous filovirus challenge against MARV, EBOV, SEBOV [Sudan EBV] [121].

The only phase 2–3 clinical efficacy trial of a filovirus vaccine was recently reported, and the results are "extremely promising" according to the WHO. In April 2015, new cases of EVD were still occurring in Basse Guinee region of Guinea, and a cluster-randomized trial with rVSV-ZEBOX vaccine was initiated, using a ring vaccination strategy to target all contacts of newly infected patients for vaccination [122]. Adult contacts were randomized to receive the vaccine immediately or after 3 weeks and were followed at home on six visits for 12 weeks. After almost 4 months, no new cases of EVD occurred in 4123 adults who received immediate vaccine, but 16 cases occurred among 3528 subjects who received delayed vaccine, with vaccine efficacy of 74.7–100% with 95% confidence interval estimate [122]. Secondary analysis of the ring vaccination indicated that the vaccine efficacy in eligible adults was 75.1% and for everyone [eligible or not for vaccination] was 76.3%. Only one serious adverse event [febrile episode that resolved] was attributed to the vaccine.

7.6 Future Directions

There are calls for major changes to the global health responses to crises as a result of the recent West African EBV outbreak. Criticisms have been directed at the WHO and government leaders of developed nations for the tardy response to the public health emergency in Africa. The International Monetary Fund has also been blamed for the root cause of the epidemic, by not providing adequate funding for poor countries to overhaul and transform their [African] healthcare systems to meet the challenges of infectious diseases outbreaks. It would take many years and probably hundreds of billions of dollars [donated by rich countries] to upgrade the healthcare systems of all the African countries to a level on par with developed nations. In the meantime, it has been proposed to establish an African Center of Disease Control, similar to the CDC in the USA, as referral base center to investigate disease outbreaks and supervise national public health centers and to perform technically difficult and expensive tests. Reforms to the existing global health system structure to a more organized system with an empowered WHO at the head, to allow for more rapid international responses, with defined responsibilities to coordinate donor countries provisions and resource mobilization have been suggested [123].

Now that the epidemic in West Africa has ended, it is unlikely that sufficient data will be collected to definitely prove the efficacy of rVSV-ZEBOV for commercial license. Should further studies be done if another EBV outbreak occurs with similar design and with the same vaccine? Doing further efficacy trials in future outbreaks may be considered unethical in view of the high mortality rate. Thus, it would be reasonable to give rVSV-ZBOV in a ring vaccination strategy early to all contacts of EVD in future outbreaks and monitor for outcome. Further studies in nonhuman primates should be performed with the trivalent filovirus vaccine with rVSV-vector. If the results of future studies confirm the efficacy shown in guinea pigs, human studies could be done as a comparator to rVSV-ZEBOV vaccine in future outbreaks with early ring vaccination protocol.

ZMapp is a promising therapy for EVD and a controlled randomized study in West Africa, sponsored by National Institute of Allergy and Infectious Diseases [NIAID], at affixed dose of 50 mg/kg with other candidate treatments as comparator was planned before the end of outbreak [124]. Future drug trials for severe filovirus infections could assess combinations of ZMapp with IFN-α, or combination with favipiarvir, or another promising antiviral, GS-5734, a prodrug of the adenine nucleotide analog which demonstrated 100% efficacy in monkeys with treatment initiated 3 days after infection [125].

Modeling and empirical studies during the West African Ebola outbreak have indicated that large epidemics of EVD are preventable and rapid response can interrupt transmission and limit the size of outbreaks [54]. Future filovirus or EVD outbreaks should be expected and planned for in advance with recognized means of control: rapid detection and diagnosis, extensive surveillance, prompt patient isolation, comprehensive contact tracing, appropriate clinical supportive care, early use of promising vaccines and experimental treatments, rigorous enforcement of effective infection control measures, and education and engagement of the communities, such as safe burial practices.

Addendum Recent studies indicate that EBOV has evolved to become more infectious and deadly to account for the massive West African outbreak. The recent epidemic affected 100-fold more people than previous outbreaks. This appears to be related to mutation of the gene encoding the virus enveloped glycoprotein [A82V], with increased ability to infect primate cells, including human dendritic cells, with associated increased mortality [126]. Another disturbing report outlines the resurgence of EVD linked to a survivor with virus persistence in the seminal fluid for more than 500 days [127].

References

1. Anderson DM, Keith J, Novak PD (2000) Lassa fever. In: Dorland's illustrated medical dictionary, 29th edn. Saunders, Philadelphia, p 663
2. Kanapathipillai R (2014) Ebola virus disease: current knowledge. N Engl J Med 371:e18
3. WHO Ebola Response Team (2014) Ebola virus disease in West Africa the first 9 months of the epidemic and forward projections. N Engl J Med 371:1481–1495
4. World Health Organization. Ebola situation report. 2 December 2015 [http://apps.who.int/ebola/current-situation/ebola-situation-reort-2-december-2015]
5. Brauburger K, Hume AJ, Muhlberger E, Olejnik J (2012) Forty-five years of Marburg virus research. Viruses 4:1878–1927
6. Gear JS, Cassel GA, Gear AJ et al (1975) Outbreak of Marburg virus disease in Johannesburg. Br Med J 4:489–493
7. Bausch DG, Nichol ST, Muyembe-Tamfum JJ et al (2006) Marburg hemorrhagic fever associated with multiple lineages of virus. N Engl J Med 355:909–919
8. Towner JS, Khristova ML, Sealy TK et al (2006) Marburgvirus genomics and association with a large hemorrhagic fever outbreak in Angola. J Virol 80:6497–6516
9. World Health Organization (1978) Ebola hemorrhagic fever in Zaire, 1976. Report of an International Commission. Bull WHO 56:271–293
10. World Health Organization (1978) Ebola hemorrhagic fever in Sudan, 1976. Report of a World Health Organization International Study Team. Bull WHO 56:247–270
11. Johnson KM, Webb PA, Lange JV, Murphy FA (1977) Isolation and characterization of a new virus [Ebola virus] causing acute hemorrhagic fever in Zaire. Lancet 1:569–571
12. Peters CJ, LeDue JW (1999) An introduction to Ebola: the virus and the disease. J Infect Dis 179(Suppl. 1):9–16
13. Le Guenno B, Formenty P, Wyers M, Gounon P, Walker F, Boesch C (1995) Isolation and partial characterizarion of a new strain of Ebola virus. Lancet 345:1271–1274
14. Muyembe-Tamfum JJ, Kipassa M (1995) Ebola hemorrhagic fever in Kilwit, Zaire. International Scientific and Technical Committee and World Health Organization Collaborating Center for Hemorrhagic Fevers [letter]. Lancet 345:1448
15. Georges AJ, Leroy EM, Renaut AA et al (1999) Ebola hemorrhagic fever outbreaks in Gabon, 199–1997: epidemiologic and health control issues. J Infect Dis 179(Suppl. 1):S65–S75

16. Jahrling PB, Geisbert TW, Dalgard DW et al (1990) Preliminary report: isolation of Ebola virus from monkeys imported to USA. Lancet 335:502–505
17. Rollin PE, Williams RJ, Bressler DS et al (1999) Ebola [subtype Reston] virus among quarantined nonhuman primates recently imported from the Philippines to the United States. J Infect Dis 179(Suppl. 1):S108–S114
18. Taylor DJ, Ballinger MJ, Zhan JJ, Hanzly LE, Bruenn JA (2014) Evidence that ebolaviruses and cuevaviruses have been diverging from marburgviruses since the Miocene. Peer J 2:e556
19. Kazanjian P (2015) Ebola in antiquity? J Infect Dis 61:963–968
20. Cunha BA (2004) The cause of the plague of Athens: plague, typhoid, typhus, smallpox or measles? Infect Dis Clin N Am 18:29–43
21. Olson PE, Hames CS, Benenson AS, Genovese EN (1996) The Thucydides syndrome: Ebola déjà vu? or Ebola reemergent? Emerg Infect Dis 2:155–156
22. Feldman H, Sanhez A, Geisbert TW (2013) Filoviridae. In: Knipe DM, Howley PM (eds) Field's virology, 6th Edition. Wolters Kluwer, Alphen aan den Rijn, Philadelphia
23. Reynard O, Volchkov VE (2015) Characterization of a novel neutralizing monoclonal antibody against Ebola virus GP. J Infect Dis 212(Suppl. 2):S372–S378
24. Chan SY, Speck RF, Ma MC, Goldsmith MA (2000) Distinct mechanisms of entry by envelope glycoproteins of Marburg and Ebola [Zaire] viruses. J Virol 74:4933–4937
25. Centers for Disease Control and Prevention (2001) Outbreak of Ebola hemorrhagic fever Uganda, August 2000–January 2001. MMWR Morb Mortal Wkly Rep 50:73–77
26. Towner JS, Sealy TK, Khristova ML et al (2008) Newly discovered Ebola virus associated with hemorrhagic fever outbreak in Uganda. PLoS Pathog 4:e1000212
27. Kuhn JH, Becker S, Ebihara H et al (2010) Proposal for a revised taxonomy of the family Filoviridae: classification, names of taxa and viruses, and virus abbreviations. Arch Virol 155:2083–2103
28. Miranda ME, Ksiazek TG, Retuya TJ et al (1999) Epidemiology of Ebola [subtype Reton] virus in the Philippines, 1996. J Infect Dis 179(Suppl. 1):S115–S119
29. Peterson AT, Carrol DS, Mills JN, Johnson KM (2004) Potential mammalian filovirus reservoirs. Emerg Infect Dis 10:2073–2081
30. Peterson AT, Lash RR, Carrol DS, Johnson KM (2006) Geographic potential for outbreaks of Marburg hemorrhagic fever. Am J Trop Med Hyg 75:9–15
31. Towner JS, Amman BR, Sealy TK et al (2009) Isolation of genetically diverse Marburg viruses from Egyptian fruit bats. PLoS Pathog 5:e1000536
32. Swanepoel R, Smit SB, Rollins PE et al (2007) Studies of reservoir hosts for Marburg virus. Emerg Infect Dis 13:1847–1851
33. Paweska JT, van Vuren PJ, Fenton KA et al (2015) Lack of Marburg virus transmission from experimentally infected to susceptible in-contact Egyptian fruit bats. J Infect Dis 212(Suppl. 2):S109–S118
34. Em L, Kumulungui B, Pourrut X et al (2005) Fruit bats as reservoirs of Ebola virus. Nature 438:575–576
35. Ogawa H, Miyamoto H, Nakayama E et al (2015) Seroepidemiological prevalence of multiple species of filoviruses in fruit bats [*Eidolon helvum*] migrating in Africa. J Infect Dis 212(Suppl. 2):S101–S108
36. Bausch DG, Borchert M, Grein T et al (2003) Risk factors for Marburg hemorrhagic fever, Democratic Republic of the Congo. Emerg Infect Dis 9:1531–1537
37. Dowell SF, Mukunu R, Ksiazek TG, Khan AS, Rollin PE, The Commission de Lutte contre les Epidemies a Kliwit (1999) Transmission of Ebola hemorrhagic fever: a study of risk factors in family members, Kikwit, Democratic Republic of the Congo. J Infect Dis 179(Suppl. 1):S87–S91
38. Roels TH, Bloom AS, Buffington J et al (1999) Ebola hemorrhagic fever, Kikwit, Democratic Republic of the Congo, 1995: risk factors for patients without a reported exposure. J Infect Dis 179(Suppl. 1):S92–S97
39. Yamin D, Gertler S, Ndeffo-Mbah ML, Skrip LA, Fallah M, Nyenswah TG, Altice FL, Galvani AP (2015) Effect of Ebola progression on transmission and control in Liberia. Ann Intern Med 162:11–17

40. Slenczka WG (1999) The Marburg virus outbreak of 1967 and subsequent episodes. Curr Top Microbiol Immunol 235:49–75
41. Rowe AK, Bertolli J, Khan AS et al (1999) Clinical, virologic and immunologic follow-up of convalescent Ebola hemorrhagic fever patients and their contacts, Kikwit, Democratic Republic of the Congo. J Infect Dis 179(Suppl. 1):S28–S35
42. Rodriguez LL, De Roo A, Guimard Y et al (1999) Persistence and genetic stability of Ebola virus during the outbreak in Kikwit, Democratic Republic of the Congo, 1995. J Infect Dis 179(Suppl. 1):S170–S176
43. Varkey JB, Shantha JG, Crozier I et al (2015) Persistence of Ebola virus in ocular fluid during convalescence. N Engl J Med 372:16–18
44. Mate SE, Kugelman JR, Nyenswah JT et al (2015) Molecular evidence of sexual transmission of Ebola virus. N Engl J Med 373:2448–2454
45. Stroher U, West E, Bugany H, Klenk HD, Schnittler HJ, Feldmann H (2001) Infection and activation of monocytes by Marburg and Ebola viruses. J Virol 2001(75):11025–11033
46. Geisbert TW, Hensley LE, Gibb TR, Steele KE, Jaax NK, Jahrling PB (2000) Apoptosis induced in vitro and in vivo during infection by Ebola and Marburg viruses. Lab Invest 80:171–186
47. Will C, Muhlberger E, Linder D, Slenczka W, Klenk HD, Feldmann H (1993) Marburg virus 4 encodes the virion membrane protein, a type 1 transmembrane glycoprotein. J Virol 67:1203–1210
48. Agopian A, Castrano S (2014) Structure and orientation of Ebola fusion peptide inserted in lipid membrane models. Biochim Biophys Acta 1838:117–126
49. Zhang AP, Bornholdt ZA, Liu T et al (2012) The Ebola virus interferon antagonist VP24 directly binds STAT1 and has a novel, pyramidal fold. PLoS Pathog 8:e1002550
50. Baize S, Leroy EM, Georges AJ et al (2002) Inflammatory responses in Ebola virus-infected patients. Clin Exp Immunol 128:163–168
51. Huitchinson KL, Rollin PE (2007) Cytokine and chemokine expression in humans infected with Sudan ebolavirus. J Infect Dis 196(Suppl. 2):S357–S363
52. Geisbert TW, Strong JE, Feldmann H (2015) Consideration in the use of nonhuman primate model of Ebola virus and Marburg virus infection. J Infect Dis 212(Suppl. 2):S91–S97
53. Fernando L, Qui X, Melito PL et al (2015) Immune response to Marburg virus Angola infection in nonhuman primates. J Infect Dis 212(Suppl. 2):S234–S241
54. Ebola Response Team WHO (2016) After Ebola in West Africa—unpredictable risks, preventable epidemics. N Engl J Med 375:587–596
55. Gire SK, Goba A, Andersen KG et al (2014) Genomic surveillance elucidates Ebola virus origin and transmission during the 2014 outbreak. Science 345:1369–1372
56. Meltzer MI, Atkins CY, Santibanez S et al (2014) Estimating the future number of cases in the Ebola epidemic—Liberia and Sierra Leone, 2014–2015. MMWR Suppl 63:1–14
57. Kortepeter MG, Bausch DG, Bray M (2011) Basic clinical and laboratory feature of filoviral hemorrhagic fever. J Infect Dis 204(Suppl. 3):S810–S816
58. Courmac JM, Karkowski L, Bordes J et al (2016) Rhabdomyolysis in Ebola virus disease. Results of an observational study in a treatment center in Guinea. Clin Infect Dis 62:19–23
59. Qin E, Bi J, Zhao M et al (2015) Clinical features of patients with Ebola virus disease in Sierra Leone. Clin Infect Dis 61:491–495
60. Ebola Response Team WHO (2014) Ebola virus disease in West Africa: the first 9 months of the epidemic and forward projections. N Engl J Med 371:1481–1495
61. Schieffelin JS, Shaffer JG, Goba A et al (2014) Clinical illness and outcomes in patients with Ebola in Sierra Leone. N Engl J Med 371:2092–2100
62. Sharma N, Cappell MS (2015) Gastrointestinal and hepatic manifestations of Ebola virus infection. Dig Dis Sci 60:2590–2603
63. de La Vega MA, Caleo G, Audet J et al (2015) Ebola viral load at diagnosis associates with patient outcome and outbreak evolution. J Clin Invest 125:4421–4428
64. McElroy AK, Erickson BR, Fliestra TD, Rollin PE, Nichol ST, Towner JS, Spiropoulou CF (2014) Ebola hemorrhagic fever: novel biomarker correlates of clinical outcome. J Infect Dis 210:558–566

65. Mattia JG, Vandy MJ, Chang JC et al (2016) Early clinical sequelae of Ebola virus disease in Sierra Leone: a cross sectional study. Lancet Infect Dis 16:331. doi:10.1016/s1473-3099(15)00489-2
66. Clark DV, Kibuuka H, Millard M et al (2015) Long-term sequelae after Ebola virus disease in Bundibugyo, Uganda: a retrospective cohort study. Lancet Infect Dis 15:905–912
67. Moshirfar M, Fenzyl CR, Li Z (2014) What we know about ocular manifestation of Ebola. Clin Ophthalmol 8:2355–2357
68. Alves DA, Honko AN, Kortepeter MG, Sun M, Johnson JC, Lugo-Roman LA, Hensley LE (2016) Necrotizing scleritis, conjunctivitis, and other pathologic findings in the left eye and brain of an Ebola virus-infected rhesus macaque [*Macaca mulatta*] with apparent recovery and delayed time of death. J Infect Dis 213:57–60
69. Jacobs M, Rodger A, Bell DJ et al (2016) Late Ebola virus relapse causing meningoencephalitis: a case report. Lancet 388:498–503
70. O'Shea MK, Clay KA, Craig DG et al (2015) Diagnosis of febrile illnesses other than Ebola virus disease at an Ebola treatment unit in Serra Leone. Clin Infect Dis 61:795–798
71. Boggild AK, Esposito DH, Kozarsky PE et al (2015) Differential diagnosis of illness in returning travelers arriving from Sierra Leone, Liberia, or Guinea: a cross-sectional study from the GeoSentinel Surveillance Network. Ann Intern Med 162:757–764
72. Grolla A, Jones SM, Fernando L et al (2011) The use of a mobile laboratory unit in support of patient management and epidemiological surveillance during the 2005 Marburg outbreak in Angola. PLoS Negl Trop Dis 5:e1183
73. Spengler JR, McElroy AK, Harmon JR, Stroher U, Nichol ST, Spiropoulou CF (2015) Relationship between Ebola virus real-time quantitative polymerase chain reaction-based threshold cycle value and virus isolation from human plasma. J Infect Dis 212(Suppl. 2):S346–S349
74. Boisen ML, Oottamasathien D, Jones AB et al (2015) Development of prototype filovirus recombinant antigen immunoassays. J Infect Dis 212(Suppl. 2):S359–S367
75. Ksiazek TG, West CP, Rollins PE, Jahrling PB, Peters CJ (1999) ELISA for the detection of antibodies to Ebola viruses. J Infect Dis 179(Suppl. 1):S192–S198
76. WHO (2014) Clinical management of patients with viral hemorrhagic fever: a pocket guide for the front-line health worker. http://www.who.int./csr/resources/publications/clinical-management-partners/en/
77. Centers for Disease Control and Prevention (2014) Guidelines on personal protective equipment to be used by healthcare workers during management of patients with Ebola virus disease in US hospitals, including procedures for putting on [donning] and removing [doffing]. http://www.cdc.gov/vhf/ebola/healthcare-us/ppe/guidance.html
78. Smither SJ, Nelson M, Eastaugh L, Nunez A, Salguera FJ, Lever MS (2015) Experimental respiratory infection of marmosets [*Callithrix jacchus*] with Ebola virus Kikwit. J Infect Dis 212(Suppl. 2):S336–S345
79. Gignoux E, Azman AS, de Smet M et al (2016) Effect of artesunate-amodiaquine on mortality related to Ebolavirus disease. N Engl J Med 374:23–32
80. Madrid PB, Chopra S, Manger ID et al (2013) A systematic screen of FDA-approved drugs for inhibitors of biological threat agents. PLoS One 8:e60579
81. World Health Organization (2014) Use of convalescent blood or plasma collected from patients recovered from Ebola virus disease, as an empirical treatment during outbreaks. In: Interim guidance for national health authorities and blood transfusion services. WHO, Geneva
82. Jahrling PB, Peters CJ (1984) Passive antibody therapy of Lassa fever in cynomolgus monkeys: importance of neutralizing antibody and Lassa virus strain. J Infect Dis 44:528–533
83. McCormick JB, King JB, Webb PA et al (1986) Lassa fever. Effective therapy with ribavirin. N Engl J Med 314:20–26
84. Frame JD, Verbrugge GP, Gill RG, Pinneo L (1984) The use of Lassa fever convalescent plasma in Nigeria. Trans R Soc Trop Med Hyg 78:319–324
85. Maiztegui JI, Fernandez NJ, de Damilano AJ (1979) Efficacy of immune plasma in treatment of Argentine hemorrhagic fever and association between treatment and late neurological syndrome. Lancet 2:1216–1217

86. van Griensven J, De Weiggheleire A, Delamou A et al (2016) The use of Ebola convalescent plasma to treat Ebola virus disease in resource-constrained settings: a perspective from the field. Clin Infect Dis 62:69–74
87. Kraft CS, Hewlett AL, Koepsell S et al (2015) The use of TKM-100802 and convalescent plasma in 2 patients with Ebola virus disease in the United States. Clin Infect Dis 61:496–502
88. Lyon GM, Mehta AK, Varkey JB et al (2014) Clinical care of two patients with Ebola virus disease in the United States. N Engl J Med 371:2402–2409
89. Chertow DS, Kleine C, Edwards JK (2014) Ebola virus disease in West Africa clinical manifestations and management. N Engl J Med 371:2054–2057
90. Chertow DS, Uyeki TM, Dupont HL (2015) Loperamide therapy for voluminous diarrhea in Ebola virus disease. J Infect Dis 211:1036–1037
91. Riddle MS, Arnold S, Tribble DR (2008) Effect of adjunctive loperamide in combination with antibiotics on treatment outcomes in traveler's diarrhea: a systematic review and meta-analysis. Clin Infect Dis 47:1007–1014
92. Clay KA, Johnston AM, Moore A, O'Shea MK (2015) Targeted electrolyte replacement in patients with Ebola virus disease. Clin Infect Dis 61:1030–1031
93. West TE, von Saint Andre-von Arnim A (2014) Clinical presentation and management of severe Ebola virus disease. Ann ATS 11:1341–1350
94. Wolf T, Kann G, Becker S et al (2015) Severe Ebola virus disease with vascular leakage and multiorgan failure: treatment of a patient in intensive care. Lancet 385:1428–1435
95. Mehedi M, Groseth A, Feldmann H, Ebihara H (2011) Clinical aspects of Marburg hemorrhagic fever. Futur Virol 6:1091–1006
96. Warren TK, Warfield KL, Wells J et al (2010) Advanced antisense therapies for postexposure protection against lethal filovirus infections. Nat Med 16:991–994
97. Warren TK, Warfield KL, Wells J et al (2010) Antiviral activity of a small-molecule inhibitor of filovirus infection. Antimicrob Agents Chemother 54:2152–2159
98. Thi EP, Mire CE, Ursic-Bedoya R et al (2014) Marburg virus infection in nonhuman primates: therapeutic treatment by lipid-encapsulated siRNA. Sci Transl Med 6:250ra116
99. Ursic-Bedoya R, Mire CE, Robbins M et al (2014) Protection against lethal Marburg virus infection mediated by lipid encapsulated small interfering RNA. J Infect Dis 209:562–570
100. Geisbert TW, Hensley LE, Kagan E et al (2006) Postexposure protection of guinea pigs against a lethal Ebola virus challenge is conferred by RNA interference. J Infect Dis 193:1650–1657
101. Geisbert TW, Lee AC, Robbins M et al (2010) Postexposure protection of nonhuman primates against a lethal Ebola virus challenge with RNA interference: a proof-of-concept study. Lancet 375:1896–1905
102. Dye JM, Herbert AS, Kuehne AI et al (2012) Postexposure antibody prophylaxis protects nonhuman primates from filovirus disease. Proc Natl Acad Sci U S A 109:5034–5039
103. Pettitt J, Zeitlin L, Kim do H et al (2013) Therapeutic intervention of Ebola virus infection in rhesus macaques with the MB-003 monoclonal antibody cocktail. Sci Transl Med 5:199ra113
104. Olinger GG Jr, Pettitt J, Kim D et al (2012) Delayed treatment of Ebola virus infection with plant-derived monoclonal antibodies provides protection in rhesus macaques. Proc Natl Acad Sci U S A 109:18030–18035
105. Marzi A, Yoshida R, Miyamoto H et al (2012) Protective efficacy of neutralizing monoclonal antibodies in a nonhuman primate model of Ebola hemorrhagic fever. PLoS One 7:e36192
106. Qui X, Wong G, Audet J et al (2014) Reversion of advanced Ebola virus disease in nonhuman primates with ZMapp. Nature 514:47–53
107. Davidson E, Bryan C, Fong RH, Barnes T, Pfaff JM, Mabila M, Rucker JB, Doranz BJ (2015) Mechanism of binding to Ebola virus glycoprotein by the ZMapp, ZMab, and MB-003 cocktail antibodies. J Virol 89:10982–10992
108. Sissoko D, Laouenan C, Folkessons E et al (2016) Favipiravir for treatment of Ebola virus disease [the JIKI Trial]: a historically-controlled, single arm proof-of-concept trial in Guinea. PLoS Med 13:e1001967

109. The PRERVAIL II Writing Group (2016) A randomized, controlled trial of ZMapp for Ebola virus infection. N Engl J Med 375:1448–1456
110. Bai CQ et al (2016) Clinical and virological characteristics of Ebola virus disease patients treated with Favipiravir (T-705)-Sierra Leone, 2014. Clin Infect Dis 63:1288–1294
111. Swenson DL, Warfield KL, Negley DL, Schmaljohn A, Aman MJ, Bavari S (2005) Virus-like particles vaccine exhibit potential as a pan-filovirus vaccine for both Ebola and Marburg viral infections. Vaccine 23:3033–3042
112. Warfield KL, Aman MJ (2011) Advances in virus-like particle vaccines for filoviruses. J Infect Dis 204:S1053–S1059
113. Swenson DL, Warfield KL, Larsen T, Alves DA, Coberly SS, Babvari S (2008) Monovalent virus-like particle vaccine protects guinea pigs and nonhuman primates against multiple Marburg viruses. Expert Rev Vaccines 7:417–429
114. Wang D, Schmaljohn AL, Raja NU et al (2006) De novo synthesis of Marburg virus antigens from adenovirus vectors induce potent humoral and cellular immune responses. Vaccine 24:2975–2986
115. Geisbert TW, Daddario-Dicaprio KM, Geisbert JB et al (2008) Vesicular stomatitis virus based vaccines protect nonhuman primates against aerosol challenge with Ebola and Marburg viruses. Vaccine 26:6894–6900
116. Geisbert TW, Feldmann H (2011) Recombinant vesicular stomatitis virus-based vaccines against Ebola and Marburg virus infections. J Infect Dis 204:S1075–S1081
117. Marzi A, Feldmann H (2014) Ebola virus vaccines: an overview of current approaches. Expert Rev Vaccines 13:521–531
118. Zhu FC, Hou LH, Li JX et al (2015) Safety and immunogenicity of a novel recombinant adenovirus type-5 vector-based Ebola vaccine in healthy adults in China: preliminary report of a randomized, double-blind, placebo-controlled, phase 1 trial. Lancet 385:2272–2279
119. Agnandji ST, Huttner A, Zinser ME et al (2016) Phase 1 trial of rVSV Ebola vaccine in Africa and Europe: preliminary report. N Engl J Med 374:1647. doi:10.1056/nejmoa1502924
120. Kibuuka H et al (2015) Safety and immunogenicity of Ebola and Marburg virus glycoprotein DNA vaccines assessed separately and concomitantly in healthy Ugandan adults: a phase 1b, randomized, double-blind, placebo-controlled clinical trial. Lancet 385:1545–1554
121. Mire CE, Geisbert JB, Versteeg KM, Mamaeva N, Agans KN, Geisbert TW, Connor JH (2015) A single-vector, single-injection trivalent filovirus vaccine: proof of concept study in outbred guinea pigs. J Infect Dis 212(Suppl. 2):S384–S388
122. Henao-Retreppo AM, Longini IM, Egger M et al (2015) Efficacy and effectiveness of an rVSV-vectored vaccine expressing Ebola surface glycoprotein: interim results from the Guinea ring vaccination cluster-randomized trial. Lancet 386:857–866
123. Gostibn LO, Friedman EA (2015) A retrospective and prospective analysis of the West African Ebola virus disease epidemic: robust national systems at the foundation and an empowered WHO at the apex. Lancet 385:1902–1909
124. National Institute of Allergy and Infectious Diseases. Putative investigational therapeutics in the treatment of patients with known Ebola infection. Full text view. ClinicalTrials.gov. https://clinicaltrials.gov/ct2/show/NCT02363322?term=zmapp+ebola&rank=1
125. Madelain V, Tram Nguyen TH, Olivo A, de Lamballerie X, Guedj J, Taburet AM, Mentre F (2016) Ebola virus infection: review of the pharmacokinetic and pharmcodynamic properties of drugs considered for testing in human efficacy trials. Clin Pharmacokinet 55:907. doi:10.1007/s40262-015-0364-1
126. Diehl WE et al (2016) Ebola virus glycoprotein with increased infectivity dominated the 2013–2016 epidemic. Cell 167:1088–1098
127. Diallo B et al (2016) Resurgence of Ebola virus disease in Guinea linked to a survivor with virus persistence in seminal fluid for more than 500 days. Clin Infect Dis 63:1353–1356

Chapter 8
Hepatitis E: A Zoonosis

8.1 Introduction

Hepatitis E is of major global public health importance but has received little attention in industrialized countries until recently. It is endemic in many developing countries where it is responsible for annual sporadic disease and intermittent large epidemics and may be responsible for 50% or more of acute hepatitis in these populations. It was first suspected in a large waterborne epidemic of acute hepatitis in Kashmir, India, in 1980; then labeled as non-A, non-B hepatitis [1]; and subsequently named hepatitis E virus [HEV]. The World Health Organization [WHO] estimates that about two billion people or one-third of the world's population live in endemic areas of HEV and are at risk for infection. It has been estimated that HEV affects 20 million people globally, most asymptomatic, and causes 3.5 million acute hepatitis, with about 70,000 deaths each year [2]. It was initially thought that HEV was limited to developing countries and that cases in industrialized countries were confined to returning travelers from endemic regions. However, over the past decade or more, there is evidence that locally transmitted cases of autochthonous HEV are increasingly reported from most developed countries, more common than previously recognized, and might be more common than hepatitis A [3]. In industrialized countries HEV disease was frequently misdiagnosed as drug-induced hepatitis, and cumulative evidence now indicate that HEV is primarily a zoonosis in developed countries.

8.2 Virology

8.2.1 *Evolutionary History*

Bayesian analysis suggests that the most recent common ancestor for modern HEV existed between 536 and 1344 years ago [4]. The progenitor of HEV appeared to have given rise to anthropotropic forms, which evolved to genotypes 1 and 2, and to

I.W. Fong, *Emerging Zoonoses*, Emerging Infectious Diseases of the 21st Century,
DOI 10.1007/978-3-319-50890-0_8

enzootic forms which evolved to genotypes 3 and 4 [4]. Review of previous published monographs suggests that HEV caused outbreaks in Western Europe in the eighteenth century, such as what occurred in Ludenscheid, Palatinate [Germany], in 1794 [5], and caused large outbreaks of jaundice in the 1950s in India [1]. Population dynamics indicate that HEV genotypes 3 and 4 experienced population expansion starting in the late nineteenth century until around 1940–1945 and then declined around 1990; and genotype 1 increased in infected population size about 30–35 years ago [4]. Genotype 4 exhibited different population dynamics with different evolutionary history in China and Japan [4]. Although identification of HEV particles could be identified in stools of infected volunteers by immune electron microscopy and experimental infection could be achieved by fecal-oral route in 1983 [6], the agent was not fully identified until 1990 [7] and named in 1991.

8.2.2 *Virology and Classification*

HEV is a single-stranded, positive sense, nonenveloped, small RNA virus in the family *Hepeviridae* [8]. The genome of the virus contains three open reading frames [ORFs] surrounded by two short noncoding regions. ORF1 encodes the nonstructural proteins which play a major role in the adaptation of the virus to its host and viral replication; ORF2 encodes the capsid protein, important in virion assembly and immunogenicity; and ORF3 encodes a small multifunctional protein, involved in morphogenesis and release of new virions [8, 9]. HEV is icosahedral in symmetry with a diameter of about 28–35 nm, with 180 capsomers formed by the ORF2-encoded capsid protein, which contains the viral genome [10]. HEV was recognized to have a single strain or serotype, but four genotypes [1–4] were detected in humans, with different geographical distribution and mode of transmission. Genotypes 1 and 2 are strictly found in humans and are responsible for endemic and epidemic acute hepatitis in developing countries. Genotypes 3 and 4 are responsible for sporadic disease worldwide through zoonotic transmission and are present in numerous animals. Recently, there has been a consensus proposal for new classification of the family *Hepeviridae*. The family is divided into two genera: *Orthohepevirus*, all mammalian and avian HEVs, and *Piscihepevirus,* cutthroat trout virus present in fish [11]. Species within the genus *Orthohepevirus* are designated *Orthohepevirus A* [isolates from humans, pig, wild boar, deer, mongoose, rabbit, and camel], *Orthohepevirus B* [isolates from chicken and bird], *Orthohepevirus C* [isolates from rat, greater bandicoot, Asian musk shrew, ferret, and mink], and *Orthohepevirus D* [isolates from bat].

8.3 Distribution of Human Genotypes

HEV genotypes 1 [Asia and Africa] and 2 [Mexico and Africa] are endemic in developing countries and transmitted by the fecal-oral route mainly through contaminated water. Large epidemics are common in these regions during the rainy season and with flooding and natural disasters, and sporadic cases may occur throughout the year from poor sanitary conditions. Genotype 1 comprises the human Burma strain [prototype], originating from a large outbreak in Rangoon in 1982, and several strains from Asia and Africa [12]. Genotype [Gt] 1 has been divided into five subtypes [a–e]. Gt1a is the most frequent subtype responsible for outbreaks in different areas of Asia, for instance, India, Pakistan, Vietnam, Burma, and Nepal, and was found in the sewage of Barcelona; subtype 1b was mainly identified in outbreaks in China, Bangladesh, Pakistan, Haiti, and Cuba; subtype 1c was mainly found with sporadic hepatitis in India, China, and Japan and caused outbreak in Kyrgyzstan; subtype 1d has been exclusively found in northern Africa, Algeria, and Morocco; and subtype 1e was found in different areas of Central, northern Africa, including Nigeria, Egypt, Sudan, Djibouti, and Algeria [13]. Genotype 2 was first detected in a hepatitis outbreak in Mexico in 1986–1987and includes several strains isolated in outbreaks in Chad and Nigeria [12]. Subtype 2a comprises the Mexican prototype strain and subtype 2b strains from patients in African countries, such as Nigeria, Chad, Central African Republic, Democratic Republic of Congo, Egypt and Namibia [13].

Genotype 3 includes human and animal strains isolated from the United States [USA], Canada, Argentina, Spain, France, the UK, Austria, the Netherlands, New Zealand, and others. Genotype 3 comprises 10 subtypes [a–j] and frequently is present in many mammals of endemic countries, such as pig, boar, deer, and others. HEV genotype 3 was discovered in 1997 from domestic pigs in the USA and soon after in a hepatitis patient with closely related genomic sequence [14, 15]. In the subsequent years, HEV-3 has been detected in humans and several animal species across the world. Several studies reported detection of HEV RNA in commercial food products such as pig livers or sausages [16, 17]. Workers and farmers in the pig industry appear to become frequently infected with animal HEV, as reflected by anti-HEV antibodies [18]. HEV-3 has a propensity to cause cross-species infection, as demonstrated by several case series in Japan [from eating raw deer meat] [19, 20] and in Southern France where identical strains of HEV-3 were found in pig liver sausages and patients with acute hepatitis [17]. HEV-4 has been detected from humans and pigs mainly from Eastern Asia, China, Taiwan, Japan, India, and Vietnam and is subdivided into seven subtypes [a–g] [13]. In China genotype 4 is transmitted mainly from consumption of pig's and rabbit's undercooked meats.

8.4 Epidemiology

HEV-1 and HEV-2 are hyperendemic [prevalence of hepatitis of 25% or episodic major waterborne outbreaks of acute hepatitis] in many developing countries such as India, Bangladesh, China, Egypt, and Mexico [21]. Seroprevalence in developing countries such as India and in Southeast Asia ranges from 27% to 80% in the general population [22]. Contamination of drinking water by human feces is the major mode of transmission by these two strictly human pathogens, although person-to-person transmission and food contamination is possible. Contamination of drinking and irrigation water from inadequate disposal and management of sewage lead to many epidemics in developing countries. Utilization of untreated river water for daily use such as drinking, bathing, and waste disposal is a problem in many developing countries of Southeast Asia and Africa. Epidemics and increased rates of infection are prevalent with annual flooding and natural disasters in regions where river, pond, or well water are the only or main source of drinking water. The use of untreated wastewater for agriculture irrigation is another source of HEV-1/HEV-2 outbreak or increased incidence, as demonstrated in Turkey [23]. Although large outbreaks were mainly associated with contaminated drinking water, a large outbreak of over 3218 residents in Northern Uganda was attributed to person-to-person contact and poor hygienic practices [24].

HEV is considered endemic when the prevalence rate of non-A, non-B hepatitis is less than 25% [21], which is mainly due to infection with genotypes 3 and 4 in industrialized countries. Endemic countries include most of Western Europe, the USA, New Zealand, many countries in South America and Asia, and the Middle East [21, 25]. Prevalence rates vary considerably from country to country and from study to study in the same regions, which may be related to lack of standardized serological assays. A recent review of the epidemiological data of 107 publications summarized the HEV seroprevalence: 2–7.8% in Europe, Japan, and South America and 18.2% to 20.6% in the USA, Russia, the UK, Southern France, and Asia [26]. The level of HEV IgG seroprevalence was related to occupational exposure to swine and type of HEV serological kits used.

HEV-3 is widely distributed in many countries of the world and has been detected in sporadic cases of hepatitis and domestic pig, boar, and deer except from African countries until recently. A recent study from Nigeria found the prevalence rate was 76.7% by PCR for HEV RNA genotype 3 overall in domestic pigs [27]. HEV-4 is isolated or detected by molecular methods in humans and domestic pigs almost exclusively in Asian countries [28]. Although genotype 4 has been detected in swine in Italy and France, and two cases of autochthonous infections had been reported in France from eating pork liver sausages [29, 30]. In parts of rural China and Hong Kong, genotype 4 is more common than genotype 1 and may be predominantly of zoonotic transmission. Infection of pigs, naturally or experimentally, is subclinical, but pathology can show mild microscopic changes in the liver and lymph nodes [31]. HEV infection in swine is common in many developed countries and age dependent and by 18 weeks of age up to 86% become infected [32]. Viremia occurs in 1–2 weeks with fecal shedding lasting for 3–7 weeks, greatest from 1–4 months of age [21]. Transmission between pigs and herds is common by fecal-oral route, by close contact and contamination of water and feed by feces.

8.5 Pathogenesis

The cellular biology of HEV is not well understood due to a lack of robust in vitro cell culture system. It has been suggested that heparin sulfate proteoglycan likely act as receptors for attachment of the viral protein capsid and that heat shock cognate protein 70 may be involved in cell entry [8]. However, a recent report indicates that HEV enters liver cells through a dynamin-2, clathrin, and membrane cholesterol pathway [33]. Type 1 interferons [IFNs] are important components of the innate immunity for combating invading viruses. In an in vitro hepatoma cell culture system, it was demonstrated that HEV inhibited IFN-β expression by ORF1 products [34]. It has been surmised that the immune response to the HEV is responsible for the clinical symptoms and liver injury rather than on the virus itself, based on human and animal studies [35]. This is suggested by the observation that the onset of icteric symptoms typically coincides with a rise in antibodies and a decline with rise in viral load [36]. However, the results in different groups of patients have not been consistent. In one study in India, severe, fulminant liver disease was correlated with greater stimulation of the Th-1 and Th-2 immunity, with higher antibodies, IFN-¥, and inflammatory cytokines but undetected virus compared to patients with acute self-limited hepatitis with detectable HEV RNA [37]. In contrast other studies showed that pregnant female with fulminant liver failure had higher viral load compared to those with mild hepatitis [38].

Animal models are indispensable for research on infectious agents to elucidate the pathogenic mechanisms for producing disease. HEVs have a wide natural host range of animal species that would be suitable for studies on pathogenesis, such as swine, rat, rabbit, chicken, and nonhuman primates. However, an animal model is lacking to study the underlying mechanisms in fulminant hepatitis and chronic infection until recently. In a swine model, quantitative proteomic analysis was used to identify cellular factors and pathways affected during acute infection [39]. Several proteins [known to be involved in other virus life cycles] were upregulated in HEV-infected livers, such as nuclear ribonucleoprotein K, apolipoprotein E, and prohibitin. HEV impairs several cellular processes which could account for various types of disease manifestation. However, in this model, there was no significant liver inflammation or necrosis. Some differences were observed between three subtypes of HEV, indicating that genetic variability may induce variations in disease pathogenesis [39]. Inoculation of cynomolgus monkeys with HEV genotype 3 results in subclinical hepatitis, similar to infection in most healthy young adults. Virus RNA was detected in the serum and feces between 5 and 53 days after inoculation, and mild inflammation of the liver with elevated liver enzymes was observed [40]. All of the monkeys showed severe lymphocytopenia with raised monocytes at the time of elevated transaminases. Recently, experimental infection has been induced in pregnant rabbits with HEV demonstrating high mortality and fulminant hepatitis with liver necrosis, secondary to proliferating virus in the liver and absence of antibodies [41].

The largest human study on the pathogenesis of HEV was performed in India, with 46 acute and 78 recovered patients and 71 healthy controls [42]. High specific and robust nonspecific IFN-¥-producing T-cell response in the acute infection suggests a role in the clearance of the HEV infection. Significantly high levels of interleukin [IL]-1α and sIL-2Rα during the acute phase suggest a role in the pathogenesis of acute HEV infection.

8.6 Clinical Features

The manifestation of HEV infection varies greatly from asymptomatic or subclinical infection, to acute self-limited hepatitis [resembling acute hepatitis A], to severe fulminant hepatitis, and occasionally to chronic infection. Factors determining clinical presentation include age, immune status, presence of underlying liver disease, and pregnancy. The majority of infected subjects from either hyperendemic regions or industrialized countries with HEV infection are asymptomatic. Infected children with HEV rarely develop clinical disease [43], and during a vaccine trial in China, it was observed that <5% of natural seroconverters develop clinical hepatitis [44]. In an outbreak of zoonotic foodborne transmitted HEV from consumption of roasted pig in France, 70.6% of the infected subjects were asymptomatic [45]. However, jaundice has been noted for about 75% of patients infected with HEV-3 or HEV-4 in previous reports [46].

Acute hepatitis E has an incubation period of 3–8 weeks [mean of 40 days] with a short prodromal period of flu-like illness, myalgia, arthralgia, weakness, nausea, and vomiting [47, 48]. Peak incidence of the epidemic form [genotypes 1 and 2] in developing countries occurs in 15–35-year-old subjects, with a predominance of male to female, 3:1 [1, 47]. However, zoonotic infection with genotypes 3 and 4 affects mainly middle-aged and elderly men [3, 13]. Most cases of acute hepatitis with symptoms and jaundice lasts for days to several weeks and have an uncomplicated course, with mortality of about 1% in the normal general population [47, 49]. Pregnant women, however, are prone to more severe fulminant hepatitis with mortality of 5–25% and increased risk of abortions and stillbirths [50, 51]. Acute infection is especially severe during the second and third trimester of pregnancy. Pregnancy is well known to be associated with moderate immune suppression in order for tolerance of the fetus, secondary to increased levels of progesterone, estrogen, and human chorionic gonadotropin. Pregnancy is associated with downregulation of nuclear factor-kappa-B with shift in the Th-1 cell/Th-2 cell balance toward a Th-2 bias, with decreased cellular immunity [52, 53]. HEV infection of human epithelial cells results in upregulation of interleukin [IL]-6, IL-8, and tumor necrosis factor [TNF]-α [54]. In healthy pregnant women, Th1-type responses prevail until the mid-second trimester, with IFN-γ decreasing and IL-6 increasing from the tenth to 40th weeks [55]. Impaired cellular immunity in pregnant women with HEV has been demonstrated by decreased lymphocyte response to phytohemagglutinin compared to non-pregnant women with HEV [52] and decreased natural killer [NK] activity in the third trimester [56]. It has also been suggested that micronutrient deficiencies [which are common in pregnant women of Southeast Asia] contribute to the impaired immune response to HEV [57].

The more severe disease in pregnancy may be due to decreased clearance and ineffective control of replicating virus, as some studies have found high viral load in blood of infected pregnant patients. It has been suggested that the increased viral replication in pregnancy may be due to both impaired cellular immunity [as indicated by decreased CD4 cell count] and higher than normal [compared to non-HEV-infected pregnant women] of estrogen, progesterone, and β-HCG [58]. Genetic variation in the gene that regulates progesterone receptor expression may be a risk factor for fulminant hepatitis in HEV infection in pregnancy. Mortality in infected pregnant women has been related to reduce expression of the progesterone receptor [38]. There is also evidence that HEV can replicate in the placenta, and this may play a role in high fetal loss, vertical transmission, and maternal mortality [59]. This has been reduplicated in experimental infection of pregnant rabbits with evidence of placenta infection, stillbirth, vertical transmission, and high mortality [41]. Miscarriages in pregnant women with HEV acute hepatitis has been reported in about 8–16% [60, 61].

Patients with underlying liver disease that become acutely infected with HEV will usually develop worsening of their condition and risk of decompensation of chronic stable liver disease. This is particularly common in patients with chronic alcoholism and underlying hepatic steatosis, liver fibrosis, or cirrhosis and can result in hepatic failure with ascites and encephalopathy with a high mortality [62, 63]. In patients with liver cirrhosis and acute superimposed HEV, the liver histopathology can be nonspecific and be mistaken for acute alcoholic hepatitis [3]. It has been reported in industrialized countries that patients with zoonotic HEV and underlying chronic liver disease may respond to ribavirin and avoid liver transplantation [64].

Blood exposure from infected animals is another potential means of zoonotic transmission of HEV. A case of acute hepatitis E in a researcher following a scalpel injury while working on a pig has been described [65]. Blood transfusion from infected asymptomatic blood donors is also another means of transmission. In southwestern France 52.5% of blood donors had been infected at some time to HEV [66]. Moreover, since 2014, there have been reported cases of transfusion transmission of HEV infection and detection of viremic, asymptomatic blood donors [67].

8.6.1 Extrahepatic Manifestations of HEV

Several non-hepatic manifestations of HEV have been described albeit on rare occasions. Mild pancreatitis that occurs between the 2nd and 3rd week after onset of illness has been described in hyperendemic countries [68]. Hematological complications such as immune-mediated thrombocytopenia and hemolytic anemia have been reported in both endemic and non-endemic countries [69]; and Henoch-Schonlein purpura has been noted in a child with acute HEV infection [70]. The most well-recognized extrahepatic manifestations of HEV, however, are neurological complications. At least 91 cases of HEV-associated neurological manifestations have been reported in both developing countries [genotype 1] and developed countries [genotype 3] [71]. Patients usually present with primarily neurological symptoms and are generally anicteric or with mild hepatitis from sporadic infection, mostly in middle-aged men. HEV-associated neurological complications include Guillain-Barre syndrome [n = 36], neuralgic amyotrophy [n = 30], encephalitis/

myelitis [12], and miscellaneous conditions [$n = 14$], such as Bell's palsy, vestibular neuronitis, myositis, and mononeuritis multiplex [71]. The pathogenesis of neurological disease associated with HEV is unclear but may be immune related in some entities and secondary to direct invasion of the central nervous system [CNS] in others. Evidence that may support HEV as a neurotropic pathogen includes ability of the virus to be cultured on a range of neurological cell lines, presence of HEV RNA in the cerebrospinal fluid [CSF] in some cases of neurological injury, compartmentalization of HEV quasispecies observed between serum and CSF, and improvement of painful peripheral neuropathy in a patient with clearance of the virus from serum and CSF after antiviral therapy [71, 72].

8.6.2 Chronic HEV Infection

Chronic HEV infections occur primarily in immunosuppressed patients with organ transplantation, solid or hematological malignancies on chemotherapy, and advanced human immunodeficiency virus [HIV] infection with low CD4 cell count. Chronic HEV infection was first described in Europe in liver and renal transplant recipients in 2008. Kamar et al. initially reported acute HEV infection in 14 transplant recipients from Southwest France, 8 of who developed chronic infection with persistent elevated transaminases, and development of significant liver fibrosis within 18 months [73]. Chronic HEV infection is mainly diagnosed in industrialized countries, such as Europe, the USA, and Japan, and almost exclusively from zoonotic transmission from genotype 3 [74], except for a recent report of a child with persistent genotype 4 infection [75]. The source of infection in most cases were unknown, but undercooked or raw pork and deer meat were most commonly implicated, and recently chronic infection in a liver transplant recipient was attributed to consumption of camel meat and milk [76]. Phylogenetic analysis showed that the patient's HEV sequence belonged to the camelid HEV genotype 7 that can infect humans [76]. Camel species of HEV are now grouped in *Orthohepevirus A* which includes human and pig species.

In a retrospective study of 85 transplant recipients who were infected with HEV, 32% were symptomatic at the time of diagnosis and 56 [66%] patients became chronically infected [77]. Chronic infection was associated with the use of tacrolimus and low platelet count. In symptomatic patients jaundice was rare and symptoms were nonspecific, fatigue, mild diarrhea, and arthralgia. Liver transaminases were modestly elevated and alanine transaminase was usually less than 300 IU/L. The prevalence of posttransplantation infection with HEV is estimated to be 1–2% in non-endemic areas [78], and 10% of those with chronic infection develop cirrhosis within 2 years [79]. While some patients can clear the virus spontaneously, especially with reduction of immunosuppressive drugs, others may progress to liver failure and death.

Isolated cases of chronic HEV been described in hematological malignancies such as lymphoma treated with rituximab, or reactivation in acute lymphoblastic leukemia after allogenic stem cell transplant [75]. A case series of 6 patients with hematological malignancies, recognized to have zoonotic HEV genotype 3 infections after developing significant increase in transaminases, was reported from a single center in France [80]. Only one of the patients was symptomatic with jaundice, and five of the patients cleared their HEV viremia, but three patients had prolonged infection over 6 months. In HIV-infected patients, the prevalence of chronic HEV is low and varies from 0% to 0.5%, and all cases with chronic HEV infection had CD4 counts below 200 cells [74]. In patients with advanced HIV infection, persistent HEV can rapidly develop into cirrhosis within 18 months [80–83]. In general chronic HEV is associated with impaired HEV-specific T-cell immunity, with weak CD4+ and CD8+ T-cell responses that become detectable after viral clearance [84]. Thus, the prognosis and long-term course depend on the degree of immunosuppression and the reversibility. There are also isolated reports of chronic HEV infection in immunocompetent patients [85, 86].

Extrahepatic manifestations can also be present in patients with chronic HEV and most notable are the neurological complications. Renal disease such as membranous or membranoproliferative glomerulonephritis has been reported in patients with organ transplantation and chronic HEV, with resolution after clearance of the virus [76]. Rheumatologic complications with arthralgia, skin rashes, and cryoglobulinemia in the presence of chronic HEV had also been reported [87].

8.6.3 Diagnosis

Acute diagnosis of HEV can be made by serological tests to detect IgM antibody which appears early with onset of illness and may become undetectable during recovery [47], but can persist for 3–6 months [74]. IgG antibody can appear a few days to weeks after IgM but persists for years and gradually decrease with time and may eventually disappear [74]. Thus, for acute diagnosis in normal hosts, the presence of IgM anti-HEV antibody or fourfold rise of IgG antibody with convalescence is required. However, although commercial serology HEV diagnostic kits are available, their reliability has been variable and as of recently none of the assays has received FDA approval [47]. Serological tests are usually carried out by enzyme-linked immunoassays [ELISA] using antigens corresponding to immunodominant epitopes from ORF2 and/or ORF3 belonging to Mexico or Burma strains [88]. The sensitivity and specificity of commercially available kits in immunocompetent subjects for detection of IgG are very good [>93.6%] and for IgM are excellent [>99.5%] [89]. HEV RNA can be detected in the serum and feces at the time of presentation and during the incubation period, and up to 4–6 weeks after onset of illness [35]. Thus, PCR can be used for acute diagnosis and is more sensitive than serology but is more frequently used in research setting or for the diagnosis of chronic HEV infection.

Chronic HEV infection is diagnosed by molecular methods to detect HEV RNA in serum or stools for a minimum of 3–6 months from the time of first diagnosis [74]. The reverse transcriptase [RT]-PCR is the most commonly used test, but there is also variation in the performance of different assays [90]. A multiplex real-time quantitative [q] RT-PCR that detects less than 50 copies of HEV genotypes 3 and 4 has been developed [91], but does not detect genotypes 1 and 2 which do not cause chronic disease. However, a popularly used HEV RT-qPCR was recently found to be robust to detect HEV RNA from seven genotypes within the species *Orthohepevirus A* [92]. Loop-mediated isothermal amplification [LAMP] has also been used to detect HEV RNA. The LAMP assay is quicker than RT-PCR and does not require any special equipment such as a PCR machine and is therefore more suitable for resource-limited countries and in field use. The only study to date on this assay on HEV RNA detection found it to be 100-fold more sensitive than a conventional RT-PCR with good specificity [93]. However, it is surprising that no other studies have been reported since 2009 to confirm these findings.

Although detection of HEV by PCR is the gold standard for diagnosis of chronic HEV infection, a new commercially available ELISA directed at the HEV capsid appears to be promising to discriminate chronic from acute infection. In a recently published study, the anti-HEV Ag-specific ELISA had a sensitivity of 65% and a specificity of 92% in detecting ongoing chronic HEV infection but was less sensitive than real-time PCR [94]. However, HEV Ag remained detectable for >100 days after HEV RNA clearance in chronic HEV patients treated with ribavirin, and higher HEV Ag was detected in chronically infected subjects compared to acutely infected patients.

8.7 Treatment

In most healthy subjects, acute HEV is a benign self-limited disease, and only supportive treatment is required. For severe fulminant hepatitis that occurs in pregnancy and occasionally in non-pregnant older adults, even in zoonotic foodborne cases [95], no specific treatment is available, but patients should be hospitalized and monitored regularly with supportive treatment of vomiting, bleeding episodes, and dehydration.

Chronic HEV in organ transplant recipients should have reduction of the immunosuppression, including doses of corticosteroids and tacrolimus, and viral clearance can occur in 30% of cases [96]. In patients that have persistent viremia, the standard treatment is 3 months of ribavirin [600 mg median dose] that can produce sustained virological response or cure in 78% [96]. Risk factors for relapse include initial lymphopenia, HEV RNA in the serum after 1 month of treatment or decrease in viral load after 7 days of treatment of <0.5 $\log_{10}$ copies/ml, and presence of HEV RNA in the stools at end of treatment [96]. Relapse can be treated for 6 months with ribavirin and 40% may develop sustained virological response. Ribavirin can also be used for treatment of chronic HEV infection in hematological malignancies and HIV disease [72, 97], but the experience is smaller than in organ transplantation. Some patients have been treated with pegylated IFN-α alone or in combination with ribavirin. Ribavirin treatment has been associated with anemia in 54% of patients and the need for blood transfusion in 12%.

A systematic review of antiviral treatment in chronic HEV has been recently published. Ribavirin was used in 105 patients and pegylated IFN in 8 patients with chronic HEV [98]. Ribavirin treatment resulted in sustained virological response 6 months after treatment in 64% of patients, and pegylated IFN-α resulted in sustained response in only 2 of 8 [28%] patients. While ribavirin treatment produced anemia which required erythropoietin or blood transfusion in 45% of patients, IFN treatment led to acute rejection in two transplant recipients. Ribavirin has been used on a few occasions for acute severe or fulminant HEV hepatitis with apparent improvement in HIV-infected subjects [99], in immunocompetent patient with severe acute HEV-3 infection [100], and in patients with acute HEV with underlying chronic liver disease [64, 101]. Since these cases represent uncontrolled experiences, it is difficult to determine any therapeutic benefit of ribavirin. Figure 8.1 summarizes the transmission, course, treatment, and outcome of the various genotypes of HEV.

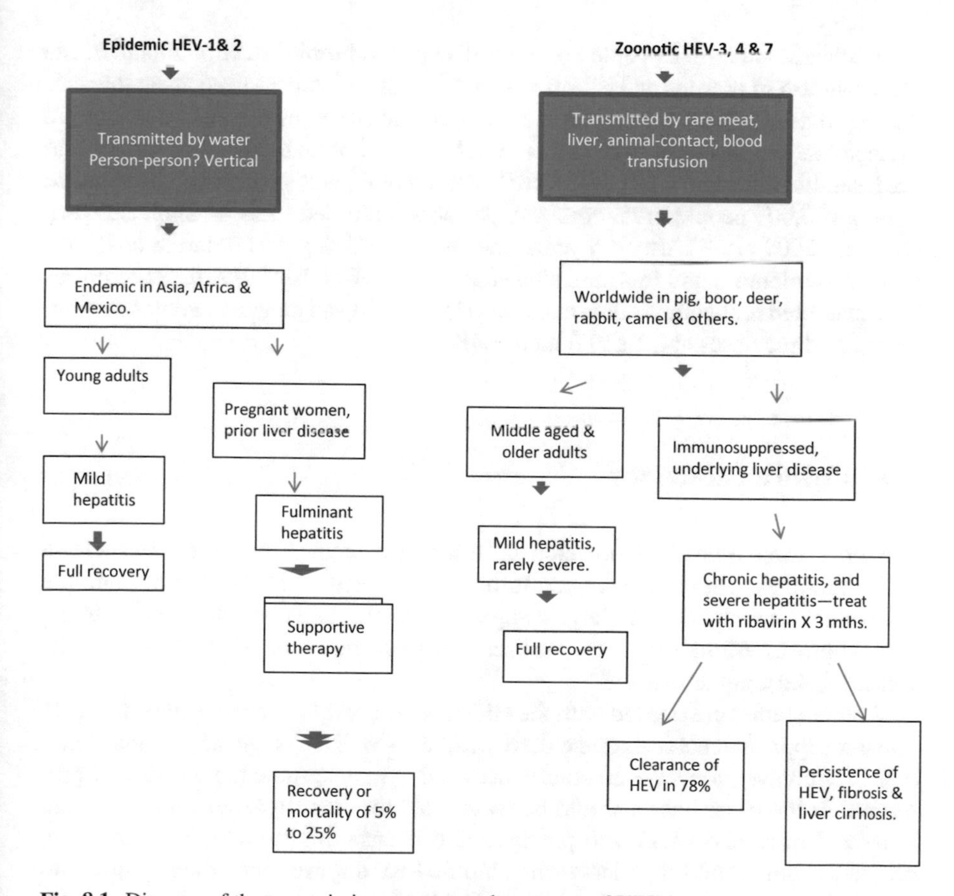

Fig. 8.1 Diagram of the transmission, course, and treatment of HEV by genotypes

8.8 Prevention

In hyperendemic regions where year-round sporadic cases and annual epidemics are a constant threat during the rainy season, the ideal solution for prevention is to provide a source of clean water supply, improve sanitation including proper disposal of waste and sewage, and maintain good hygienic practice [such as proper hand washing] by educating the local population. However, unfortunately in many developing poor countries, these facilities are not available in remote rural areas and sometimes in larger towns.

In industrialized countries where zoonotic transmission is the main source of HEV infection, eating raw or undercooked pork or sausages is preventable through education to avoid such infection. Cooking the meat to temperatures greater than 70°C should decrease the viral load and reduce or prevent infection [102]. Public health departments may need to publicize the risks to immunosuppressed people and those with underlying liver disease and provide preventive guidelines.

Vaccination of at-risk people for severe disease or chronic infection should be the ideal solution in preventing HEV. It may not be cost-effective to vaccinate the general population against HEV, even in hyperendemic countries as most infections are asymptomatic or result in mild illness. An effective vaccine has been developed and marketed in China since 2012. The HEV 239 recombinant vaccine had been tested in over 100,000 people in randomized, placebo-controlled trials in China between 2007 and 2009 [103]. After 4.5 years, the vaccine efficacy was found to be 86.8% with good tolerance and few mild side effects. Since October 2012, the vaccine has been marketed for healthy adults aged 16–65 years old and pregnant women, administered in three doses at 0,1 and 6 months [89].

8.9 Future Directions

Further research is needed to develop and test more reliable assays for HEV IgG and IgM antibodies. These assays need to be standardized and validated by multiple independent investigators, in large samples of patients and controls, in order to get FDA approval. Similarly, a standardized approved real-time RT-PCR need to be validated and commercialized.

Further studies are needed with the HEV 239 vaccine to determine its efficacy in immunosuppressed subjects and elderly people, as well as its efficacy against genotype 3. Moreover, longer-term studies are needed to determine the duration of protection. In the meantime, it would be reasonable for other hyperendemic countries in Asia, Africa, and Mexico to purchase and license this vaccine for their use in pregnant women and for patients with chronic liver diseases, including people with chronic hepatitis B and chronic hepatitis C infection.

References

1. Wong DC, Purcell RH, Sreenivasan MA, Prasad SR, Pavri KM (1980) Epidemic and endemic hepatitis in India: evidence for a non-A, non-B hepatitis virus etiology. Lancet 2:876–879
2. Haffar S, Bazerbachi F, Lake JR (2015) Making the case for the development of a vaccination against hepatitis E virus. Liver Int 35:311–316
3. Dalton HR, Bendall R, Ljaz S, Banks M (2008) Hepatitis E: an emerging infection in developed countries. Lancet Infect Dis 8:698–709
4. Purdy MA, Khudyakov YE (2010) Evolutionary history and population dynamics of hepatitis E virus. PLoS One 5:e14376
5. Teo CC (2012) Fatal outbreaks of jaundice in pregnancy and the epidemic history of hepatitis E. Epidemiol Infect 140:767–787
6. Balayan MS, Andjaparidize AG, Savinskaya SS, Ketiladze ES, Braginsky DM, Savinov AP, Poleschuk VF (1983) Evidence for a virus in non-A, non-B hepatitis transmitted via the fecal-oral route. Intervirology 20:23–31
7. Reyes GR, Purdy MA, Kim JP, Luk KC, Young LM, Fry KE, Bradley DW (1990) Isolation of a cDNA from the virus responsible for enterically transmitted non-A, non-B hepatitis. Science 247:1335–1339
8. Cao D, Mengn XJ (2012) Molecular biology and replication of hepatitis E virus. Emerg Microbes Infect 1:e17
9. Kamar N, Bendall R, Legrand-Abravanel F et al (2012) Hepatitis E. Lancet 379:2477–2488
10. Xing L, Kato K, Li T, Takeda N, Miyamura T, Hammar L, Cheng R (1999) Recombinant hepatitis E capsid protein self-assembles into a dual-domain T=1 particle presenting native virus epitopes. Virology 265:35–45
11. Smith DB, Simmonds P, Members of the International Committee on the Taxonomy of Viruses Hepeviridae Study Group et al (2014) Consensus proposals for classification of the family *Hepeviridae*. J Gen Virol 95:2223–2232
12. Johne R, Dremsek P, Reetz J, Heckel G, Hess M, Ulrich RG (2014) *Hepeviridae*: an expanding family of vertebrate viruses. Infect Genet Evol 27:212–229
13. Perez-Gracia MT, Suay B, Mateos-Lindemann ML (2014) Hepatitis E: an emerging disease. Infect Genet Evol 22:40–59
14. Kwo PY, Schlauder GG, Carpenter HA et al (1997) Acute hepatitis E by a new isolate acquired in the United States. Mayo Clin Proc 72:1133–1136
15. Schlauder GG, Gj D, Erker JC et al (1998) The sequence and phylogenetic analysis of a novel hepatitis E virus isolated from a patient with acute hepatitis reported in the United States. J Gen Virol 79:447–456
16. Wenzel JJ, Preiss J, Schemmerer M et al (2011) Detection of hepatitis E virus [HEV] from porcine livers in Southeastern Germany and high sequence homology to human HEV isolates. J Clin Virol 52:50–54
17. Colson P, Borentain P, Queyriaux B et al (2010) Pig liver sausages as a source of hepatitis E virus transmission to humans. J Infect Dis 202:825–834
18. Christensen PB, Engle RE, Hjort C et al (2008) Time trend of the prevalence of hepatitis E antibodies among farmers and blood donors: a potential zoonosis in Denmark. Clin Infect Dis 47:1026–1031
19. Tei S, Kitajima N, Takahashi K et al (2003) Zoonotic transmission of hepatitis E virus from deer to human beings. Lancet 362:371–373
20. Tomiyama D, Inoue E, Osawa Y et al (2009) Serological evidence of infection with hepatitis E virus among wild Yero-deer, *Cervus Nippon yesoensis*, in Hokkaido. Jpn J Viral Hepat 16:524–528
21. Yugo DM, Meng X-J (2013) Hepatitis E virus: foodborne, waterborne and zoonotic transmission. Int J Environ Res Public Health 10:4507–4533
22. Abe K, Li TC, Ding X et al (2006) International collaborative survey on epidemiology of hepatitis E virus in 11 countries. Southeast Asian J Trop Med Public Health 37:90–95

23. Ceylan A, Ertem M, Ilcin E, Ozekinci T (2003) A special risk group for hepatitis E infection: Turkish agriculture workers who use untreated waste water for irrigation. Epidemiol Infect 131:753–756
24. Teshale EH, Grytdal SP, Howard C et al (2010) Evidence of person-to-person transmission of hepatitis E virus during a large outbreak in Northern Uganda. Clin Infect Dis 50:1006–1010
25. Teo CG (2010) Much meat, much malady: changing perceptions of the epidemiology of hepatitis E. Clin Microbiol Infect 16:24–32
26. De Schryver A, De Schrijver K, Francois G et al (2015) Hepatitis E virus infection: an emerging occupational risk? Occup Med [Oxford] 65:667–672
27. Owolodun OA, Gerber PF, Gimenez-Lirola LG, Kwaga JK, Opriessnig T (2014) First report of hepatitis E virus circulating in domestic pigs in Nigeria. AmJTrop Med Hyg 91:699–704
28. Okamoto H (2007) Genetic variability and evolution of hepatitis E virus. Virus Res 127:216–228
29. Monne I, Ceglie L, Dim G et al (2015) Hepatitis E virus genotype 4 in a pig farm, Italy, 2013. Epidemiol Infect 143:529–533
30. Colson P, Romanet P, Moal V et al (2012) Autochthonous infections with hepatitis E virus genotype 4, France. Emerg Infect Dis 18:1361–1364
31. Halibur P, Kasorndorkbua C, Gilbert C et al (2001) Comparative pathogenesis of infection of pigs with hepatitis E virus as recovered from a pig and a human. J Clin Microbiol 39:918–923
32. Leblanc D, Ward P, Gagne MJ et al (2007) Presence of hepatitis E virus in a naturally infected swine herd from nursery to slaughter. Int J Food Microbiol 117:160–166
33. Holla P, Ahmad I, Ahmed Z, Jameel S (2015) Hepatitis E virus enters liver cells through a dynamin-2, clathrin and membrane cholesterol-dependent pathway. Traffic 16:398–416
34. Nan Y, Yu Y, Ma Z, Khattar SK, Fredericksen B, Zhang YJ (2014) Hepatitis E virus inhibits type 1 interferon induction by ORF1 products. J Virol 88(20):11924–11932
35. Krain LJ, Nelson KE, Labrique AB (2014) Host immune status and response to hepatitis E virus infection. Clin Microbiol Rev 27:139–165
36. Zhang JZ, Im SW, Lau SH et al (2002) Occurrence of hepatitis E antbodies, and viremia in sporadic cases of non-A, –B, and –C acute hepatitis. J Med Virol 66:40–48
37. Saravanabalaji S, Tripathy AS, Dhoot RR, Chapha MS, Kakrani AL, Arankalle VA (2009) Viral load, antibody titers and recombinant open reading frame 2 protein-induced TH1/TH2 cytokines and cellular immune responses in self-limiting and fulminant hepatitis E. Intervirology 52:78–85
38. Bose PD, Das BC, Kumar A, Gondal R, Kumar D, Kar P (2011) High viral load and deregulation of the progesterone receptor signaling pathway: association with hepatitis E-related pregnancy outcome. J Hepatol 54:1107–1113
39. Rogee S, Le Gall M, Chafey P et al (2015) Quantitative-proteomics identifies host factors modulated during acute hepatitis E virus infection in the swine model. J Virol 889:129–143
40. De Carvalho LG, Marchevsky RS, dos Santos DR et al (2013) Infection by Brazilian and Dutch swine hepatitis E virus strains induces hematological changes in *Macaca fascicularis*. BMC Infect 13:495
41. Xia J, Liu L, Wang L et al (2015) Experimental infection of pregnant rabbits with hepatitis E virus demonstrating high mortality and vertical transmission. J Viral Hepat 22:850–857
42. Tripathy AS, Das R, Rathod S, Arankalle VA (2012) Cytokine profiles, CTL response and T cell frequencies in the peripheral blood of acute patients and individuals recovered from hepatitis E infection. PLoS One 7:e31822
43. Buti M, Plans P, Dominguez A et al (2008) Prevalence of hepatitis E virus infection in children in northeast of Spain. Clin Vaccine Immunol 15:732–734
44. Zhu FC, Zhang J, Zhang XF et al (2010) Efficacy and safety of a recombinant hepatitis E vaccine in healthy adults: a large scale, randomized, double-blind placebo-controlled, phase 3 trial. Lancet 376:895–902
45. Guillois Y, Abravanel F, Miura T et al (2016) High proportion of asymptomatic infections in an outbreak of hepatitis E associated with a spit-roasted piglet, France, 2013. Clin Infect Dis 62:351–357

46. Kamar N, Dalton HR, Abravanel F, Izopet J (2014) Hepatitis E virus infection. Clin Microbiol Rev 27:116–138
47. Hoofnagle JH, Nelson KE, Purcell RH (2012) Hepatitis E. N Engl J Med 367:1236–1244
48. Wedemeyer H, Pischke S, Manns MP (2012) Pathogenesis and treatment of hepatitis E virus infection. Gastroenterology 142:1388–1397
49. Emerson SU, Purcell RH (2003) Hepatitis E virus. Rev Med Virol 13:145–154
50. Khuroo MS, Teli MR, Skidmore S, Sofi MA, Khuroo MI (1981) Incidence and severity of viral hepatitis in pregnancy. Am J Med 70:252–255
51. Khuroo MS, Kamil S (2003) Etiology, clinical course of sporadic acute viral hepatitis in pregnancy. J Viral Hepat 10:61–69
52. Pal R, Aggarwal R, Naik SR, Das V, Naik S (2005) Immunological alterations in pregnant women with acute hepatitis E. J Gastroenterol Hepatol 20:1094–1101
53. Navaneethan U, Al Mohajer M, Shata MT (2008) Hepatitis E and pregnancy: understanding the pathogenesis. Liver Int 28(9):1190–1199
54. Devhare PB, Chatterjee SN, Arankalle VA, Lole KS (2013) Analysis of antiviral response in human epithelial cells infected with hepatitis E virus. PLoS One 8:e63793
55. Aris A, Lambert F, Bessette P, Moutquin JM (2008) Maternal circulating interferon-gamma and interleukin-6 as biomarkers of Th1/Th2 immunity status throughout pregnancy. J Obstet Gynecol Res 34:7–11
56. Kraus TA, Engel SM, Sperling RS et al (2012) Characterizing the pregnancy immune phenotype results of the Viral Immunity and Pregnancy [VIP] study. J Clin Immunol 32:300–311
57. Labrique AB, Klein S, Kmush B et al (2012) Immunologic dysregulation and micronutrient deficiencies associated with risk of intrapartum hepatitis E infection in pregnant Bangladeshi women. FASEB J 26:127.4
58. Jilani N, Das BC, Hussain SA et al (2007) Hepatitis E virus infection and fulminant hepatic failure during pregnancy. J Gastroenterol Hepatol 22:676–682
59. Bose PD, Das BC, Hazam RK, Kumar A, Medhi S, Kar P (2014) Evidence of extrahepatic replication of hepatitis E virus in human placenta. J Gen Virol 95(95):1266–1271
60. Petra S, Kumar A, Trivedi SS, Puri M, Sarin SK (2007) Maternal and fetal outcomesd in pregnant women with acute hepatitis E virus infection. Ann Intern Med 147:28–33
61. Krain LJ, Atwell JE, Nelson KE, Labrique AB (2014) Fetal and neonatal health consequences of vertically transmitted hepatitis E virus infection. AmJTrop Med Hyg 90:365–370
62. Dalton HR, Bendall RP, Rashid M et al (2011) Host risk factors and autochthonous hepatitis E infection. Eur J Gastroenterol Hepatol 23:1200–1205
63. Kumar Acharya S, Sharma K, Singh P, Mohanty K, Madan K, Kumar Jha J, Kumar PS (2007) Hepatitis E [HEV] infection in patients with cirrhosis is associated with rapid decompensation and death. J Hepatol 46:387–394
64. Peron JM, Dalton H, Izopet J, Kamar N (2011) Acute autochthonous hepatitis E in western patients with underlying chronic liver disease: a role for ribavirin? J Hepatol 54:1323–1324
65. Sarker S, Rivera EM, Engle RE et al (2015) An epidemiological investigation of a case of acute hepatitis E. J Clin Microbiol 53:3547–3552
66. Mansuy JM, Bendall R, Legrand-Abravanel F et al (2011) Hepatitis E virus antibodies in blood donors, France. Emerg Infect Dis 17:2309–2312
67. Dreier J, Juhl D (2014) Autochthonous hepatitis E: a new transfusion-associated risk? Transfus Med Hemother 41:29–39
68. Thapa R, Biswas B, Mallick D, Ghosh A (2009) Acute pancreatitis complicating Hepatitis E virus infection in a 7-year old boy with glucose 6 phosphate dehydrogenase deficiency. Clin Pediatr 48:199–201
69. Singh NK, Gangappa M (2007) Acute immune thrombocytopenia associated with hepatitis E in an adult. Am J Hematol 82:942–943
70. Thapa R, Mallick D, Biswas B (2010) Henoch Schonlein pupura triggered by acute hepatitis E virus infection. J Emerg Med 39:218–219

71. Dalton HR, Kamar N, van Eijk JJ, Mclean BN, Cintas P, Bendall RP, Jacobs BC (2016) Hepatitis E virus and neurological injury. Nat Rev Neurol 12:77–85
72. Dalton H, Keane F, Bendall R, Mathew J, Ijas S (2011) Treatment of chronic hepatitis E in a HIV positive patient. Ann Intern Med 155:479–480
73. Kamar N, Selves J, Mansuy JM et al (2008) Hepatitis E virus and chronic hepatitis in organ-transplant recipients. N Engl J Med 358:811–817
74. Murali AR, Kotwal V, Chawla S (2015) Chronic hepatitis E: a brief review. World J Hepatol 7:2194–2201
75. Geng Y, Zhang H, Huang W, Harrison T, Geng K, Li Z, Wang Y (2014) Persistent hepatitis E virus genotype 4 infection in a child with acute lymphoblastic leukemia. Hepat Mon 14:e15618
76. Lee G-H, Tan B-H, Teo EC-Y et al (2016) Chronic infection with camelid hepatitis E virus in a liver transplant recipient who regularly consumes camel meat and milk. Gastroenterology 150:355–357
77. Kamar N, Garrouste C, Haagsma EB et al (2011) Factors associated with chronic hepatitis in patients with hepatitis E infection who have received solid organ transplants. Gastroenterology 140:1481–1489
78. Haagsma EB, Niesters HG, van den Berg AP et al (2009) Prevalence of hepatitis E virus infection in liver transplantation recipients. Liver Transpl 15:1225–1228
79. Lhjomme S, Abravanel F, Dubois M, Sandres-Saune K, Rostaing L, Kamar N, Izopet J (2012) Hepatitis E virus quasispecies and the outcome of acute hepatitis E in solid-organ transplant patients. J Virol 86:10006–10014
80. Tavitian S, Peron JM, Huynh A et al (2010) Hepatitis E virus excretion can be prolonged in patients with hematological malignancies. J Clin Virol 49:141–144
81. Dalton HR, Bendall RP, Keane FE, Tedder RS, Ijaz S (2009) Persistent carriage of hepatitis E virus in a patient with HIV infection. N Engl J Med 361:1025–1027
82. Colson P, Dhiver C, Poizot-Martin I, Tmalet C, Gerolami R (2011) Acute and chronic hepatitis E in patients infected with human immunodeficiency virus. J Viral Hepat 18:227–228
83. Jagjit Singh GK, Ijas S, Rockwood N et al (2013) Chronic hepatitis E as a cause for cryptogenic cirrhosis in HIV. J Infect 66:103–106
84. Suneetha PV, Pischke S, Schlaphoff V et al (2012) HEV-specific T-cell responses are associated with control of HEV infection. Hepatology 55:695–708
85. Gonzalez Tallon AI, Moreira Vicente V, Mateos Lindemann ML, Achecar Justo LM (2011) Chronic hepatitis E in an immunocompetent patient. Gastroenterol Hepatol 34:398–400
86. Grewal P, Kamili S, Motamed D (2014) Chronic hepatitis E in an immunocompetent patient: a case report. Hepatology 59:347–348
87. Pischke S, Behrendit P, Manns MP, Wedemeyer H (2014) HEV-associated cryoglobulinemia and extrahepatic manifestation of hepatitis E. Lancet Infect Dis 14:678–679
88. Perez-garcia MT, Garcia M, Suay B, Mateos-Lindemann ML (2015) Current knowledge on hepatitis E. J Clin Transl Hepatol 3:117–126
89. Blasco-Perrin H, Abravanel F, Blasco-Baque V, Peron JM (2016) Hepatitis E, The neglected one. Liver Int 36(Suppl S1):130–134
90. Baylis SA, Hanschmann KM, Blumel J, Nubling CM (2011) Standardization of hepatitis E virus [HEV] nucleic acid amplification technique-based assays: an initial study to evaluate a panel of HEV strains and investigate laboratory performance. J Clin Microbiol 49:1234–1239
91. Zhang X, Li A, Shuai J et al (2013) Validation of an internally controlled multiplex real time RT-PCR for detection and typing of HEV genotypes 3 and 4. J Virol Methods 193:432–438
92. Giron-Callejas A, Clark G, Irving WL, McClure CP (2015) In silico and in vitro interrogation of a widely used HEV RT-qPCR assay for detection of the species Orthohepevirus A. J Virol Methods 214:25–28
93. Lan X, Yang B, Li BY, Yin XP, Li XR, Liu JX (2009) Reverse transcription-loop-mediated isothermal amplification assay for rapid detection of hepatitis E virus. J Clin Mirobiol 47:2304–2306

94. Behrendt P, Bremer B, Todt D et al (2016) Hepatitis E virus [HEV] ORF2 antigen levels differentiate between acute and chronic HEV infection. J Infect Dis 214:361–368
95. Miyashita K, Kang JH, Saga A et al (2012) Three case of acute or fulminant hepatitis E caused by ingestion of pork meat entrails in Hokkaido, Japan: zoonotic foodborne transmission of hepatitis E and public health concerns. Hepatol Res 42:870–878
96. Kamar N, Izopet J, Tripon S et al (2014) Ribavirin for chronic hepatitis E virus infection in transplant recipients. N Engl J Med 370:1111–1120
97. Haiji H, Geroami R, Solas C, Moreau J, Colson P (2013) Chronic hepatitis E resolution in a human immunodeficiency virus [HIV]-infected patient treated with ribavirin. Int J Antimicrob Agents 41:595–597
98. Peters van Ton AM, Gevers TJ, Drenth JP (2015) Antiviral therapy in chronic hepatitis E: a systematic review. J Viral Hepat 22:965–973
99. Robbins A, Lambert D, Ehrhard F et al (2014) Severe acute hepatitis E in an HIV infected patient: successful treatment with ribavirin. J Clin Virol 60:422–423
100. Gerolami R, Borentain P, Raissouni F, Motte A, Solas C, Colson P (2011) Treatment of severe acute hepatitis E by ribavirin. J Clin Virol 52:60–62
101. Goyal R, Kumar A, Panda S, Acharya S (2012) Ribavirin therapy for hepatitis E virus-induced acute on chronic liver failure: a preliminary report. Antivir Ther 17:1091–1096
102. Schielke A, Filter M, Appel B, Johne R (2011) Thermal stability of hepatitis E assessed by a molecular biological approach. Virol J 8:487
103. Zhang J, Zhang X-F, Huang S-J et al (2015) Long-term efficacy of a hepatitis E vaccine. N Engl J Med 372:914–922

Chapter 9
Zoonotic Malaria: *Plasmodium knowlesi*

9.1 Introduction

Malaria is an ancient disease of humans that continues to cause high morbidity and substantial mortality of the world's population, with hundreds of thousands of deaths per year in children. Over the past decade, the global mortality rate from malaria has declined with improved diagnosis and treatment. In 2015 the World Health Organization [WHO] estimated that the global malaria-attributed deaths had declined to no more than 635,000, but the prevalence of malaria for 2014 was still over 134 million cases worldwide [1]. It has caused the greatest health affliction of humans more than any disease in the history of mankind cumulatively. Prior to the start of the twenty-first century, human malaria was known to be caused by four species of *Plasmodium* parasites [*P. falciparum*, *P. vivax*, *P. ovale*, *P. malariae*] adapted to humans as the natural intermediate hosts and with *Anopheles* mosquitoes as the primary hosts, where sexual multiplication occurred. In 2004 a large concentration of simian malaria was found in the human population of Sarawak, Malaysian Borneo [2]. Since then this zoonosis has spread through the Malaysian Peninsula and throughout Southeast Asia, and it is the leading cause of malaria in Malaysian Borneo. Transmission of *P. knowlesi* to humans has occurred in all countries of Southeast Asia except for Laos [3]. The natural hosts of this parasite are Asian monkeys, long-tailed macaques [*Macaca fascicularis*], and pig-tailed macaques [*Macaca nemestrina*], which are distributed throughout Southeast Asia [4, 5]. The vectors and primary hosts of *P. knowlesi* are forest-dwelling mosquitoes of the *Anopheles leucosphyrus* group, which have an overlapping distribution in Southeast Asia as the Asian macaques [3].

I.W. Fong, *Emerging Zoonoses*, Emerging Infectious Diseases of the 21st Century,
DOI 10.1007/978-3-319-50890-0_9

9.1.1 Evolution of Malaria

Plasmodium parasites, a genus of the phylum Apicomplexa, consist of nearly 200 species that infect reptiles, birds, and mammals [6]. All *Plasmodium* species are digenetic with two host species, with a definitive host where the sexual reproduction of parasite occurs and an intermediate host. Vertebrates are the typical intermediate hosts and mosquitoes are the definitive hosts or vectors. There are over 20 plasmodia parasites of apes and monkeys, some of which have been implicated in human infection to a limited degree [7]. The four human parasites, *P. falciparum*, *P. ovale*, *P. malariae*, and *P. vivax*, are remotely related to each other and probably evolved from simian plasmodia parasites independently of each other at various estimated times from thousands to millions of years [8]. Based on phylogenetic analysis of mitochondrial cytochrome b gene sequences, *P. falciparum* is closely related to the chimpanzee parasite`, *Plasmodium reichenowi*, and divergence from the ancestral plasmodia may have occurred 8–11 million years ago. The time of divergence of host species from chimpanzee to humans may have occurred from the time of evolution of the hominids, but others believe that *P. falciparum* has only become a disease burden to humans more recently, about 3330 years ago with an upper limit of 6640 years [9]. *P. malariae* and a New World monkey parasite, *Plasmodium brasilianum*, are genetically indistinguishable, and lateral transfer between hosts from monkeys to humans is believed to occur within the last 15,000 years [8]. *P. vivax* is genetically indistinguishable from another parasite of New World monkeys, *Plasmodium simium*, and may have undergone a host switch to humans about the same period as *P. malariae* [8].

Transmission of plasmodia species from monkeys to humans has been detected for several simian species, including *P. simium* [10], *P. brasilianum* [11], *Plasmodium cynomolgi* [12], and probably *Plasmodium semiovale* [13]. Experimentally malaria can be transmitted from humans to monkeys and this may occur naturally [14]. Through analysis of *P. knowlesi* mitochondrial [mt] DNA sequences, it was estimated that the most recent ancestor was 98,000–478,000 years ago [5]; and the emergence of *P. knowlesi* in Southeast Asia predates the settlement of humans about 70,000 years ago [15]. It is also surmised that *P. knowlesi* migrated with the natural macaque host to Borneo and Southeast Asia during the Pleistocene era [16] and underwent rapid expansion at the time of human population growth in Southeast Asia between 30,000 and 40,000 years ago [17]. Figure 9.1 summarizes the evolution of human malaria.

The ability of animal *Plasmodium* species to transition to become human pathogens may be elucidated by comparison of the genomes of various species. It was known that genes that determine phenotypic differences between plasmodia species are frequently found in only a subset of species, and it was postulated that the genomes may contain species subset-specific genes that determine human pathogenicity, transmissibility, and virulence. In a study of genome comparison of human and nonhuman malaria parasites, 13 genes were found in the human parasites but were absent in simian plasmodia including *P. knowlesi*; some of these genes are specifically upregulated in sporozoites or gametocytes that could be linked to parasite transmission in humans [18].

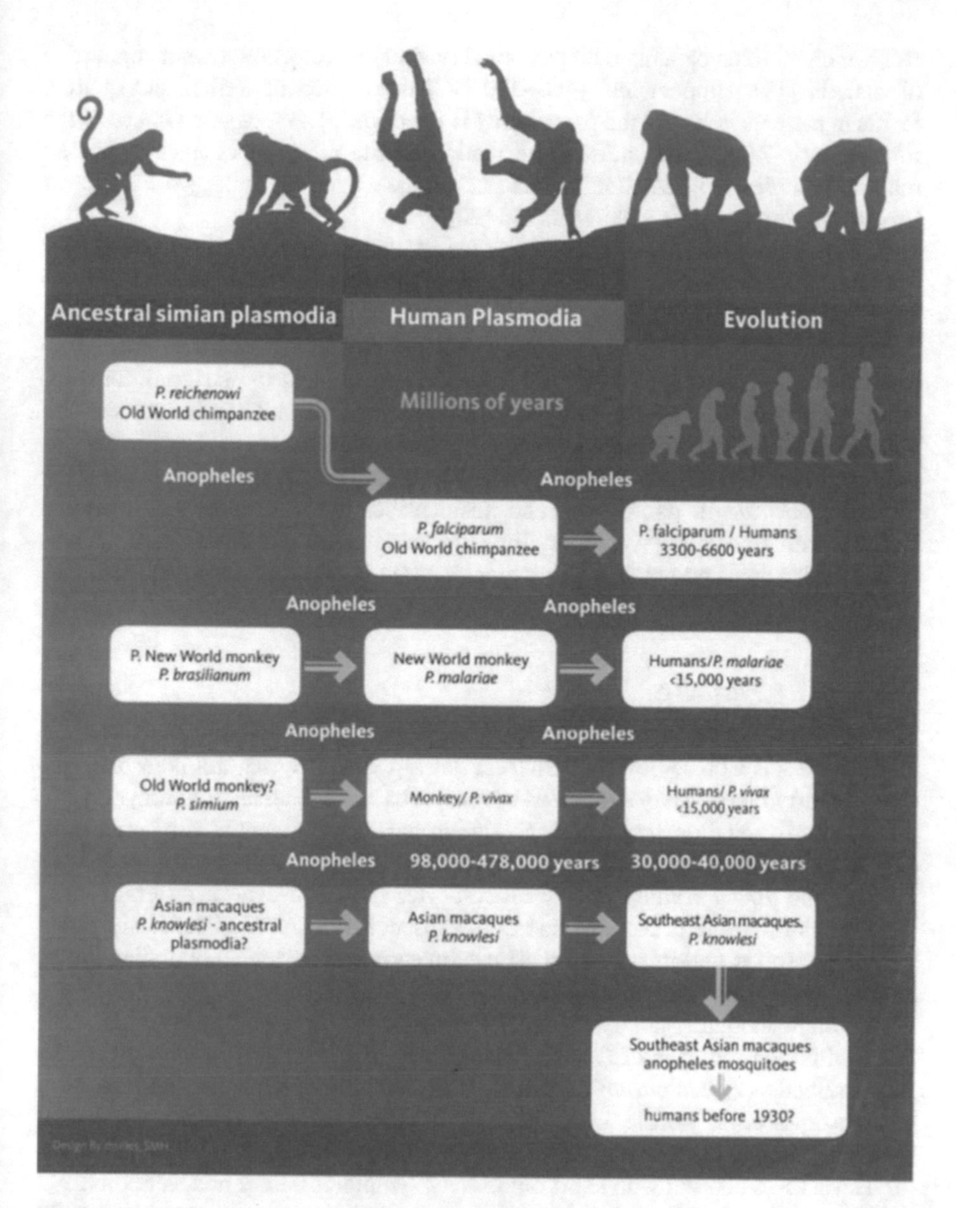

Fig. 9.1 Zoonotic plasmodia and relationship to human malaria parasites

9.1.2 Historical Aspects

Historical records from different cultures suggest that malaria has afflicted humans in the Old World for thousands of years. These include Chinese medical writings from 2700 BC, clay tablet inscriptions from Mesopotamia from about 2000 BC, the Ebere Egyptian Papyrus of 1570 BC, and Vedic period Indian writings of 1500–800

BC, which mention patients with periodic fever and spleen enlargement suggestive of malaria [19]. Hippocrates' [460–370 BC] description of tertian and quartan fevers in patients indicates the presence *of P. vivax* and *P. malariae* in Greece in the fifth century BC [5]. Scientific confirmation of the presence of ancient human malaria includes the detection of malaria antigen from the enlarged spleens of Egyptian mummies over 3000 years old [20].

P. knowlesi was discovered in 1932 in long-tailed macaques and experimentally it could be transmitted to humans [21]. In the mid-thirties *P. knowlesi* was inoculated in humans for treatment of general paresis of the insane, a form of tertiary neurosyphilis [22, 23]. The first naturally acquired human case of *P. knowlesi* infection was described in 1965 [14], and infection was thought to be rare in humans. However, it is very likely that human infection in Malaysia and Southeast Asia with the simian plasmodia was more prevalent for many years before the large focus was recognized in Malaysian Borneo in 2004. Morphologically *P. knowlesi* is very similar to *P. malariae* on microscopy and misidentification was likely very common [24]. Examination of archival blood films further demonstrated that *P. knowlesi* was widely distributed and not newly emergent in 2004 in Malaysian Borneo [25].

9.1.3 The Parasite

P. knowlesi has a close phylogenetic relationship with *P. vivax* and both parasites use the Duffy blood group antigen as a receptor to invade human erythrocytes [26], and the Duffy binding protein of the simian parasite is essential for invasion of human and monkey erythrocytes [27]. Moreover, both plasmodia have a preference for invading young erythrocytes or reticulocytes [28]. However, it differs from *P. vivax* in its life cycle in humans due to the absence of a chronic latent liver phase with hypnozoites, and it is the only primate malaria species with a quotidian [24 h] asexual blood stage development [7].

Although morphologically the mature trophozoites resemble and can be confused with the band form of *P. malariae*, the early trophozoites [ring forms] can be misidentified as *P. falciparum* especially when multiple trophozoites are present in a single erythrocyte [accole form] [7]. Also similar to *P. falciparum* is the potential for high degree of parasitemia, which can vary from <500 parasites/μl blood to very high levels >764,000/μl with short duration of symptoms before presentation [29].

The genome of *P. knowlesi* is made up of 14 chromosomes containing 5188 protein-encoding genes, 80% of which are identified in *P. falciparum* and *P. vivax* [7]. The *P. falciparum* contains about 5300 genes, and the genome is unusually A + T rich, while *P. knowlesi* has lower A + T content [30]. However, *P. knowlesi* was the first malaria parasite species in which antigenic variant families were demonstrated in the genome [31]. The two major variant families [SICAvar and KIR] are dispersed throughout the genome and appear to be important in the pathogenesis. The SICA antigens encoded by the SICAvar gene were identified at the surface of the erythrocytes and are associated with parasite virulence. The KIR proteins match over one-half of the host cell CD99 extracellular domain involved in the T-cell immunoregulatory molecule, which represents an unusual form of molecular mimicry [31]. This may be involved in the adaptation to its macaque host.

9.1.4 Pathogenesis

The life cycle of *P. knowlesi* is similar to *P. malariae* and *P. falciparum* in lacking a chronic latent liver stage. There are no symptoms in any malaria species in the early stages after inoculation of sporozoites by the female anopheles mosquitoes; then sporozoites initially migrate to the liver to undergo asexual multiplication to develop into schizonts. The liver schizonts rupture to release thousands of merozoites to invade the erythrocytes, to initiate the erythrocyte cycle. These merozoites develop into early trophozoites or ring forms which develop into mature trophozoites [band forms] that undergo asexual multiplication to form schizonts, containing numerous merozoites. Rupture of the schizonts release second stage merozoites that infect more erythrocytes to complete the red blood cell [RBC] cycle. Some of the merozoites develop into male and female gametocytes that are taken up by the female mosquitoes for the sexual multiplication stage to complete the life cycle. The duration of the erythrocyte stage depends on the plasmodia species. The duration of the erythrocyte stage is shortest for *P. knowlesi* [about 24 h], while for *P. falciparum*, *P. vivax*, and *P. ovale*, it is 48 h, and for *P. malariae* it is 72 h [3]. Fever peak occurs following rupture of schizonts from erythrocytes to release secondary merozoites to infect more RBC, to produce quotidian, tertian, or quartan fever patterns according to the RBC cycle. Early in all malaria infection and in mixed infection, the fever can be daily until a pattern is established.

The variability in clinical manifestations of *P. knowlesi* malaria from mild to severe disease may be related to host factors and parasite variable expression of virulence factors. It is quite feasible, as with other plasmodia species, that people with mild infection had previous unrecognized infection with development of partial immunity. Genetic factors with variation in innate immunity even in apparently normal hosts are probably important in host factors but are not well defined. Hyperparasitemia is associated with severe disease for both *P. falciparum* and *P. knowlesi* [32], but why some subjects with infection have high parasite load and others low levels remain unclear. Recent investigation suggests that certain genetic variants of *P. knowlesi* are more virulent in humans. Variation in proteins involved in parasite invasion of erythrocytes may be related to invasiveness of a particular genotype and results in high parasite load [33]. Gene clusters of two important members of the invasion gene family in *Plasmodium* species [*P. knowlesi* normocyte-binding protein xa and xb] [34], sequenced in patient isolates, were associated with clinical and laboratory markers of disease severity [33]. In most cases of *P. knowlesi* human malaria, the disease is mild with low parasitemia, as the parasite proliferates poorly in the blood due to a preference for young RBC [reticulocytes]; however, recent investigation demonstrated that *P. knowlesi* can adapt to invade a wider age range of RBCs, resulting in proliferation to produce hyperparasitemia [35].

In human studies patients with uncomplicated *P. knowlesi* infection had lower levels of the proinflammatory cytokines IL-8 and TNF-α than those with complicated disease [36]. While anti-inflammatory cytokines IL-1ra and IL-10 were detected in all patients, they correlated with parasitemia in both *P. knowlesi* and *P. vivax*. Immunoproteonomics in sera of patients with malaria have been used to assess for markers of specific plasmodia parasites and disease severity. While several immunoreactive proteins were identified in malaria-infected subjects, only serotransferrin and hemopexin represent valid markers for *P. knowlesi* infection, and while haptoglobin was observed with *P. vivax*, it was undetectable in *P. knowlesi*-infected subjects [37]. There is also evidence that schizont-infected cell agglutination variant proteins encoded by the SICAvar multigene family in *P. knowlesi* are expressed at the surface of the erythrocytes and are associated with virulence and serve as determinants of naturally acquired immunity, and the spleen influences these processes [38]. The molecular mechanisms for the simian plasmodia to adapt to human erythrocytes to produce zoonotic disease have been unclear. Recent studies suggest that recognition and sharing of human erythrocyte receptor[s] by *P. knowlesi* tryptophan-rich antigens [PKTRags] with human parasite ligands could be a strategy adopted by the monkey malaria parasite to establish the heterologous human host [39].

In severe *P. falciparum* infection with coma, there is sequestration of parasite-infected RBCs in the microvascular vessels of the brain, which is considered to be pathogenic [40]. Sequestration is mediated by the expression of the var-gene family of proteins, PFEMPI, on the surface of the infected erythrocytes [41]. Even though cerebral malaria has not been reported with *P. knowlesi*, sequestration of parasitized RBCs has been found in the microvasculature of the brain of a fatal case [42]. In a cytoadherence study, *P. knowlesi*-infected RBCs bound to intercellular adhesion molecule-1 and/or vascular cell adhesion molecule-1 but not to CD36, a receptor expressed on human endothelial cells [43]. Thus, microvascular sequestration of infected RBCs may play a role in the pathogenesis of severe disease in *P. knowlesi* infection.

9.1.5 Vector and Transmission

Anopheles leucosphyrus group of mosquitoes are the vectors for *P. knowlesi* which are found throughout Southeast Asia and also can transmit *P. falciparum* and *P. vivax* [44]. This diverse group of mosquitoes with many species is primarily jungle vectors that feed on monkeys and humans encroaching on their habitat. The geographic distribution of this group of anopheles mosquitoes has been shown to extend from Southwestern India, eastward to southern China, Taiwan, mainland Southeast Asia, Indonesia, and the Philippines [45]. However, the primary vector for transmission of *P. knowlesi* may vary from region to region in the same country or between countries, probably influenced by density of the species and the ecological niche. Experimental studies carried out in the 1960s to assess susceptibility and capability of different anophelines to transmit *P. knowlesi* to rhesus macaques found that *A. balabacensis* was the most competent vector, followed by *A. stephensi*, *A. maculatus*, and *A. freeborni* [46]. Human-to-human, monkey-to-human, and human-to-monkey transmission of *P. knowlesi* was demonstrated with *A. balabacensis*, with incubation periods in the vector of 12–13 days [47].

However, in Kapit, Malaysian Borneo, where most of the simian malaria has been described in humans, *A. latens* is considered the primary vector [48]. This mosquito species is attracted to both long-tailed macaques and humans and feeds between 7 and 10 pm in the forest, with preference for macaques at higher elevation. A highly zoophilic species, *A. cracens*, which feeds on macaques in the canopy of trees and humans at ground level, appears to be the main vector of *P. knowlesi* in Peninsular Malaysia with peak feeding times at 8–9 pm [49]. In a more recent study from Sabah, Malaysia, *A. balabacensis* was confirmed to be the primary vector of *P. knowlesi*, mostly biting humans in the early evening between 6–8 pm, with infection highest in forest and small-scale farm sites [50]. In southern Vietnam forested areas and peridomestic areas close to villages, *A. dirus* is the only known vector of malaria. This species of anopheles was found to harbor sporozoites of *P. knowlesi* alone or in combination with *P. vivax* and/or with *P. falciparum* [51].

9.1.5.1 Natural Hosts

The main natural hosts for *P. knowlesi* are the long-tailed and pig-tailed macaques which are distributed throughout Southeast Asia, but the parasite can be found as well in the banded leaf monkey [*Presbytis melalophos*] [52]. Infections in the simian hosts usually produce mild transient disease with chronic, low-grade parasitemia [7]. In Malaysia species-specific microsatellite genotyping identified two host-associated populations of *P. knowlesi* from wild macaques and humans [53]. Two-thirds of human *P. knowlesi* infections were of the long-tailed macaque type, and one-third were of the pig-tailed macaque type.

Long-tailed macaques are found in a wide range of habitats from Brunei, Cambodia, Indonesia, Southern Thailand, and Peninsular Malaysia to Sumatra, Java, Borneo, the Philippines, Singapore, and southern Vietnam and may have the second widest distribution of monkeys after rhesus macaques [54]. Pig-tailed macaques are found from Eastern India and Bangladesh to most of mainland Southeast Asia to Sumatra and Borneo.

The geographic distribution of autochthonous human *P. knowlesi* infection ranges from Malaysian Borneo, Peninsular Malaysia, the Indonesian Borneo, the Palawan Island in the Philippines, the border between China and Myanmar, Central and south-central Vietnam, Thailand near the southern Myanmar border and the Island of Ko Phayam, the Pailin Province in Cambodia, and the forested area of Lim Chu Kang in Singapore [7]. Surprisingly, *P. knowlesi* has not emerged in India and Bangladesh although there is potential for the parasite to spread to these countries [55, 56]. However, since many countries in the region have not performed specific studies with sensitive diagnostic techniques to determine the prevalence of *P. knowlesi* in humans and monkeys, the geographic range and extent of the reservoirs are probably underestimated. Based on spatial data on parasite occurrence and the ranges of the identified host and vector species, investigators have assigned evidence score to 475 regions in 19 countries where *P. knowlesi* reservoir may be present [57].

Populations that are at particular risk of acquiring knowlesi malaria are people who live in the forest fringe of endemic areas or venture into the ecological habitats of the macaques and anopheles mosquitoes for work or leisure. The majority of patients are adults who are farmers, hunters, logging camp workers, and children living in forest communities [3]. Travelers from various countries who visit endemic areas and travel to rural/forested regions with contact to wild monkeys are at risk for simian malaria. Between 2005 and 2012, 15 cases of *P. knowlesi* infections have been recognized and published in international travelers [58], but it is likely that this represent the minority and most cases were probably mislabeled as other malaria or were not reported. This is reflected by a report in 2014, which reviewed the literature up to December 2013 and found 103 articles on *P. knowlesi* malaria in travelers [59]. The apparent increase in incidence of *P. knowlesi* malaria across Southeast Asia since the recognition of a large human focus in Malaysian Borneo in 2004 is not well explained. This may be partly explained by increased recognition of the parasite by improved diagnostic methods. However, longitudinal studies in Malaysia have shown that *P. knowlesi* incidence has increased over the past decade [60]. Possible contributing factors to the increased incidence may include a combination of factors such as changing land use, changes in human behavior, changes in the macaque hosts or vectors pattern of behavior, and possibility of human-to-human transmission as playing a role [61].

9.1.6 Clinical Disease

The clinical spectrum of *P. knowlesi* malaria is quite variable from mild nonspecific febrile illness to severe disease that can result in death. Although adults are mainly affected, there is a wide age distribution, and *P. knowlesi* is the most common cause of malaria in areas of Malaysia in all age groups except <5 years of age, where it occurs equally with *P. falciparum* and *P. vivax* [62]. There is a seasonal variation in the incidence of cases and it is correlated with rainfall in the preceding 3–5 months, which is probably related to vector breeding and density at these times. There is limited clinical data on disease in children, as infection rate is low due largely to the exposure risk being mainly confined to proximity to forested areas. However, in Kudat district of Malaysia, *P. knowlesi* is the most common cause of childhood malaria which is generally uncomplicated but commonly associated with anemia and thrombocytopenia [63]. This is in contrast to *P. falciparum* malaria where the highest morbidity and mortality occur in young children under 5 years of age.

In most case series of *P. knowlesi*, adults predominate, and children under 15 years of age represent 10% or less and males account for about 74% of cases [64]. Similar to other types of malaria, the clinical manifestations at presentation most commonly consist of chills and fever [100%], headache [94.4%], rigors [89.7%], malaise [89.7%], myalgia [87.9%], cough [48–56.1%], abdominal pain [31–52.3%], nausea [56.1%], vomiting [33.6%], and diarrhea [18–29%] [3, 65]. Unlike other forms of malaria, however, daily fever persists even after 2 weeks and there is no cyclical pattern of tertian or quartan fever. However, in most cases, a diagnosis is made within 5 days after onset of illness. Clinical signs are usually nonspecific for any febrile illness, but enlarged liver and spleen may be present in 15–26% [3]. Severe malaria varies in different series from 9% to 39%, with higher rates from referral hospitals, but overall averages 9–10% [7, 44]. Clinical severe infection may present with respiratory distress with tachypnea, hypoxemia, hypotension, jaundice, and multiorgan failure, but cerebral malaria or severe anemia has not been reported [3, 7, 44]. Previous studies had reported mortality rates of 1.8–10.7%, but prospective studies indicate a lower mortality of 1–2% [65].

The most common laboratory abnormality is thrombocytopenia which occurs in 98–100% of cases, with a third <50,000 platelets/μl, but bleeding was rare [3]. Anemia with a hemoglobin less than 10 g/dl was uncommon in adults with less than 5% on admission, but more commonly reported in children in about 56% and with 25% <10 g/dl [7, 63]. Mild-moderate hyponatremia was seen in more severe cases in 29% [66], but the best predictors of severe disease were parasite count ≥35,000/μl or a platelet count of ≤45,000/μl [67]. In a prospective study from Sabah, Malaysia, severe malaria was most commonly due to *P. knowlesi* than other plasmodia, and the most common severity criteria included parasitemia >100,000/μl, jaundice, respiratory distress, hypotension, and acute kidney injury [68].

9.1.6.1 Diagnosis

The traditional microscopy examination by skilled personnel of thick and thin blood smears can lead to misdiagnosis of *P. knowlesi* infection as the more benign infection with *P. malariae*, and the early trophozoites can be mistaken for *P. falciparum*. The popular immune-chromatographic card tests used for rapid diagnosis have low sensitivity and can erroneously misidentify *P. knowlesi* as other plasmodia and vice versa [69]. Molecular methods that have been shown to be more sensitive and specific than microscopy have been developed. Nested PCR is the most commonly used method in epidemiological studies and to retrospectively confirm the diagnosis in referral centers and detect low parasitemia [3, 44]. A small proportion of samples without *P. knowlesi* can give false-positive results with *P. vivax* parasites. A single-step, non-nested PCR has been found to be 100% specific [70]. A highly sensitive and specific multiplex quantitative PCR for malaria for detection of all species at low parasitemia of 1–6 parasites/μl, including *P. knowlesi* at 100% sensitivity and specificity, has been developed [71]. However, although the processing is rapid, the cost is prohibitive for malaria diagnosis in endemic areas, and this applies to other PCR assays as well.

Affordable molecular methods that can be easily applied in developing countries where the need is greatest are needed. One such method is the loop-mediated isothermal amplification [LAMP], which does not require specialized equipment [72].

LAMP assays are relatively inexpensive, requiring a simple heating block, primers, and negative and positive controls, and provide a result in <60 min [44]. A LAMP method for diagnosis of *P. knowlesi* infection has been developed and the sensitivity of the test was 100-fold that of a single PCR assay [73]. In a small study of 13 cases of *knowlesi* malaria, the LAMP assay was 100% sensitive and specific [74]. This technology would be suitable for use in most endemic countries with malaria, and a commercialized assay to detect and differentiate all species of human malaria parasites is needed, but verification of its proficiency should be assessed in larger number of malaria cases.

9.1.6.2 Treatment

Experimentally several schizonticidal antimalarial drugs have antiparasitic effect against *P. knowlesi* in nonhuman primates, but recurrent exposure to mefloquine, proguanil, and pyrimethamine could result in development of resistance in macaques [3]. Ex vivo susceptibility of human isolates of *P. knowlesi* from Malaysian Borneo has been performed, using a WHO schizont maturation assay modified for the quotidian life cycle. *P. knowlesi* was found to be highly susceptible to artemisinins, variably and moderately susceptible to chloroquine, and less sensitive to mefloquine [75]. Delayed diagnosis and misleading confusion with *P. malariae* malaria is a real concern, not only for patients with severe malaria on presentation but for those with apparent mild disease. Due to the replication cycle of 24 h of *P. knowlesi*, early aggressive treatment may be necessary to prevent sudden increase in parasitemia and associated complications [3].

For uncomplicated *P. knowlesi* malaria, standard doses of chloroquine appear to be the treatment of choice, although case reports indicate that other agents are also effective *including* artemisinins, quinine, mefloquine, atovaquone/proguanil, and doxycycline [3]. A prospective observational study had shown that chloroquine is quite effective, although one-third of patients experienced increase in parasitemia, twice that for vivax malaria, during the first 6 h of treatment [76]. As recommended by Malaysian Ministry of Health guidelines, two doses of 15 mg primaquine were used 24 h apart for the gametocidal effect. Presumably this was used to try and decrease the risk of human-to-human transmission by mosquitoes, although there is no evidence to support this strategy, but theoretically it may be of some benefit. More recently, a randomized, open-label controlled study was performed to compare artesunate-mefloquine versus chloroquine in uncomplicated *P. knowlesi* malaria in Malaysia [77]. A total of 252 patients were enrolled in the study with almost equal numbers in each arm. Patients receiving artesunate-mefloquine had faster clearance of parasitemia with quicker resolution of fever by about 3 h, lower risk of anemia at 28 days, and less days spent in hospital, but the final cure rates were similar. This study supports a unified treatment with artesunate-based combination for all Plasmodium species in co-endemic areas.

Severe *P. knowlesi* malaria can carry a significant mortality if not treated aggressively and rapidly. Patients are best treated in a hospital and may require intravenous saline to maintain adequate blood pressure and oxygen for respiratory distress. Previous experience in Sabah, Malaysia, suggested that failure to give intravenous artesunate for severe malaria, even with *P. knowlesi*, resulted in significant mortality [78]. Subsequently, all severe malaria irrespective of parasite species was treated early and aggressively with intravenous artesunate in a prospective observational study in Sabah, Malaysia. Severe malaria occurred in 38 of 130 [29%] of *P. knowlesi*, 13 of 122 [11%] of *P. falciparum*, and 7of 43 [16%] *P. vivax* infections based on criteria of high parasitemia >100,000/μl, jaundice, respiratory distress, hypotension, and acute kidney injury [68]. With this treatment there was zero mortality. It would be interesting to compare oral artesunate combination to the intravenous artesunate for severe malaria in non-vomiting patients to determine any difference in outcome. As oral therapy is so much more convenient and can be given immediately in rural areas with limited access to a hospital, my sense is that there likely would be no difference in the outcome in hemodynamic stable patients.

9.1.7 Prevention

There is great concern that emergence and spread of zoonotic malaria will hamper and prevent the global campaign to control and eventually eliminate human malaria. There is a real possibility that *P. knowlesi* could spread further throughout Asia and even Africa, if the parasite evolve to adapt to other nonhuman primates such as rhesus macaques, which are more widespread in many developing countries. In general the three approaches for malaria control include effective treatment of clinical cases, vector control or prevention of mosquito bites, and vaccination. In treatment approach to reduce the human reservoir, there is missing data on the value of large-scale use of a gametocide agent such as primaquine. The widespread use of this agent would be limited because of the risk of hemolytic anemia in the presence of G6PD-deficiency, and the need to do assay for normal presence of this enzyme. A large reservoir of human malaria parasites is present in asymptomatic but infected individuals with low-grade undetected infection with partial immunity to the parasites. No studies have been performed to determine the value in reducing asymptomatic infection and any possible benefit to the prevalence of clinical cases in the endemic community.

Vector control was largely dependent on the use of insecticide sprays which were partially effective but eventually led to widespread insecticide resistance in mosquitoes. Novels methods to reduce mosquito population such as natural larvicidal agents and sterilization of male mosquitoes that are then released in the environment may have limited success. Methods to reduce mosquito bites such as protective clothing and use of insect repellants with DEET have limited long-term compliance in endemic tropical and subtropical climates. In some countries in Africa, insecticide-impregnated bed-nets have had some success in malaria reduction in children. However, this would not be effective for zoonotic malaria where the risk exposure is different.

Development of an effective vaccine for malaria has been the great hope for successful control. To date research and development by vaccine/pharmaceutical companies have largely concentrated on developing an effective *P. falciparum* vaccine. Results of phase 3 trials indicate it is feasible to develop a partially effective vaccine. The RTS,S/AS01 malaria vaccine targets the circumsporozoite protein of *P. falciparum* and has partial efficacy [33.4–50.3%] against clinical and severe malaria in infants and children at 1 year [79]. There have been limited preliminary studies on a *P. knowlesi* malaria vaccine development. A DNA prime, poxvirus [COPAK] boost vaccination with four antigens, led to survival from an otherwise lethal infection in rhesus monkeys and to self-limited parasitemia in 60% of animals [80]. In this study, nine rhesus monkeys were immunized with radiation attenuated *P. knowlesi* sporozoites, and five did not develop infection after challenge with live sporozoites [80]. Further investigation indicated that CD8+ T-cells are the important effector cells protecting against the simian malaria.

9.1.8 Future Direction

Large prospective studies over the next decade in humans and monkeys of multiple species with sensitive PCR or LAMP methods should be performed, to delineate the geographic boundaries of *P. knowlesi* and the ability to naturally infect other species of primates and spread across new territories. Molecular-gene and ex vivo studies need to continue to monitor for development of antimalarial drug resistance. Further investigations are needed to develop an affordable and reliable rapid diagnostic test that can be readily used in rural Asia. The LAMP assay seems most promising, but much larger studies need to confirm its reliability.

Further clinical randomized, controlled trials are needed to define the most effective oral agents for uncomplicated knowlesi malaria. Although the data suggest that artesunate combination could be used to treat all forms of malaria for the blood stage, there is risk of increased artesunate resistance with overuse which is already a concern. Vaccine development needs to continue and expand to human trials. Eventually, an ideal malaria vaccine should not only be effective and safe but should provide a broad coverage for all human plasmodia species. Hence, innovative studies in nonhuman primates should explore the feasibility of developing a multi-plasmodia malaria vaccine.

Most importantly, is the need for more novel, basic scientific investigation to determine the mechanisms by which the simian malaria parasite is able to mutate and adapt to human hosts. This suggests that other simian malaria species may undergo similar genetic mutations to continue to generate new species of zoonotic malaria parasites. If we are able to unravel this basic means, then we could possibly design novel ways for their prevention.

References

1. World Health Organization. World malaria report 2015. WHO: http://www.who.int/malaria/visual-refresh/en/.
2. Singh B, Kim Sung L, Matusop A et al (2004) A large focus of naturally acquired *Plasmodium knowlesi* infections in human beings. Lancet 363:1017–1024
3. Singh B, Daneshvar C (2013) Human infections and detection of *Plasmodium knowlesi*. Clin Microbiol Rev 26:165–184
4. Fooden J (1982) Ecogeographic segregation of macaque species. Primates 23:574–579
5. Lee KS, Divis PC, Zakaria SK et al (2011) *Plasmodium knowlesi*: reservoir hosts and tracking the emergence in humans and macaques. PLoS Pathog 7:e1002015. doi:10.1371/journal.ppat.1002015
6. Garnham PCC (1966) Malaria parasites and other haemosporidia. Blackwell, Oxford
7. Antinori S, Galimberti L, Milazzo L, Corbellino M (2013) *Plasmodium knowlesi*: the emerging zoonotic malaria parasite. Acta Trop 125:191–201
8. Rich SM, Ayala FJ (2006) Evolutionary origins of human malaria parasites. In: Dronamraju KR, Arese P (eds) Malaria: genetic and evolutionary aspects. Springer, New York, pp 125–146
9. Tishkoff SA, Varkonyi R, Cahinhinan N et al (2001) Haplotype diversity and disequilibrium at human G6PD: recent origin of alleles that confer malarial resistance. Science 293:455–462
10. Deane LM, Deane MP, Ferreira NJ (1966) A naturally acquired human infection by *Plasmodium simium* of howler monkeys. Trans R Soc Trop Med Hyg 60:563–564
11. Contacos PG, Lunn JS, Coatney GTR, Kilpatrick JW, Jones FE (1963) Quartan-type malaria parasite of New World monkeys transmissible to man. Science 142:676
12. Eyes DE, Coatney GR, Getz ME (1960) Vivax–type malaria parasite of macaques transmissible to man. Science 131:1812–1813
13. Qari SH, Shi Y-P, Povoa MM et al (1993) Global occurrence of *Plasmodium vivax*-like human malaria parasite. J Infect Dis 168:485–489
14. Chin W, Contacos PG, Coatney GR, Kimball HR (1965) A naturally acquired quotidian-type malaria in man transferable to monkeys. Science 149:865
15. Macaulay V, Hill C, Achilli A, Rengo C et al (2005) Single, rapid coastal settlement of Asia revealed by analysis of complete mitochondrial genomes. Science 308:1034–1036
16. Voris HK (2000) Maps of Pleistocene sea levels in Southeast Asia: shorelines, river systems and time duration. J Biogeogr 27:1153–1167
17. Atkinson QD, Gray RD, Drummond AJ (2008) mtDNA variation predicts population size in humans and reveals a major southern Asian chapter in human prehistory. Mol Biol Evol 25:468–474
18. Frech C, Chen N (2011) Genome comparison of human and non-human malaria parasites reveals species subset-specific genes potentially linked to human disease. PLoS Comput Biol 7:e1002320
19. Sherman IW A brief history of malaria and the discovery of the parasite's life cycle. In: Sherman IW (ed) Malaria: parasite biology, pathogenesis, and protection. American Society of Microbiology, Washington, DC, pp 3–10
20. Miller RL, Ikram S, Armelagos GJ et al (1994) Diagnosis of *Plasmodium falciparum* infections in mummies using the rapid manual ParaSight™– Ftest. Trans R Soc Trop Med Hyg 88:31–32
21. Knowles R, Das Gupta BM (1932) A study of monkey malaria, and its experimental transmission to man. Indian Med Gaz 67:301–320
22. Van Rooyen CE, Pile GR (1935) Observations on infection by *Plasmodium knowlesi* [ape malaria] in the treatment of general paralysis of the insane. Brit Med J 2:662–666
23. Chopra RN, Das Gupta BA (1936) A preliminary note on the treatment of neurosyphilis with monkey malaria. Indian Med Gaz 71:187–188
24. Lee KS, Cox-Smith J, Singh B (2009) Morphological features and differential counts of *Plasmodium knowlesi* parasites in naturally acquired human infections. Malar J 8:73

25. Lee KS, Cox-Smith J, Brooke G, Matusop A, Singh B (2009) *Plasmodium knowlesi* from archival blood films: further evidence that human infections are widely distributed and not newly emergent in Malaysian Borneo. Int J Parasitol 39:1125–1128
26. Singh SK, Singh AP, Pandey S, Yazdani SS, Chitnis CE, Sharma A (2003) Definition of structural elements in *Plasmodium vivax and Plasmodium knowlesi* Duffy binding necessary for erythrocyte invasion. Biochem J 374:193–198
27. Fong MY, Rashdi SA, Yusof R, Lau YL (2015) Distinct genetic differences between the Duffy binding protein [PkDBPalphall] of *Plasmodium knowlesi* clinical isolates from North Borneo and Penninsular Malaysia. Malar J 14:91
28. Kumar AA, Lim C, Moreno Y et al (2015) Enrichment of reticulocytes from whole blood using aqueous multiphase systems of polymers. Am J Hematol 90:31–36
29. Cox-Singh J, Davis TM, Lee KS et al (2008) *Plasmodium knowlesi* malaria in humans is widely distributed and potentially life threatening. Clin Infect Dis 46:165–171
30. Choi JY, Augagneur Y, Ben Mamoun C, Voelker DR (2012) Identification of gene encoding *Plasmodium knowlesi* phosphatidylserine decarboxylase by genetic complementation in yeast and characterization of in vitro maturation of encoded enzyme. J Biol Chem 287:222–232
31. Pain A, Bohme U, Berry AE et al (2008) The genome of the simian and human parasite *Plasmodium knowlesi*. Nature 455:799–803
32. World Health Organization (2013) Management of severe malaria a practical handbook, 3rd edn. WHO, Geneva. Available online at:AM, Pinheiro MM, Divis PC, et al. Disease progression in *Plasmodium knowlesi* malaria is linked to variation in invasion gene family members. PLoS Negl Trop Di 8:e3086
33. Meyer EV, Semenya AA, Okenu DM et al (2009) The reticulocyte binding-like proteins of *P. knowlesi* locate to the micronemes of merozoites and define two new members of the invasion ligand family. Mol Biochem Parasitol 165:111–121
34. Lim C, Hansen E, DeSimone TM et al (2013) Expansion of host cellular niche can drive adaptation of a zoonotic malaria parasite to humans. Nat Commun 4:1638
35. Cox-Singh J, Singh B, Daneshavar C, Planche T, Parker-Williams J, Krishna S (2011) Anti-inflammatory cytokines predominate in acute human *Plasmodium knowlesi* infections. PLoS One 6:e20541
36. Chen Y, Chan CK, Kerishnan JP, Lau YL, Wong YL, Gopinath SC (2015) Identification of circulating biomarkers in sera of *Plasmodium knowlesi*-infected malaria patients comparison against *Plasmodium vivax* infection. BMC Infect Dis 15:49
37. Lapp SA, Korir-Morrison C, Jiang J, Corredor V, Galinski MR (2013) Spleen-dependent regulation of antigenic variation in malaria parasites: *Plasmodium knowlesi* SICAvar expression profiles in splenic and asplenic hosts. PLoS One 8:e78014
38. Tyagi K, Gupta D, Saini E et al (2015) Recognition of human erythrocyte receptors by the tryptophan-rich antigens of monkey malaria parasite *Plasmodium knowlesi*. PLoS One 10:e0138691
39. Ponsford MJ, Medana IM, Prapansilp P et al (2012) Sequestration and microvascular congestion are associated with coma in human cerebral malaria. J Infect Dis 205:663–671
40. Smith JD (2014) The role of PFEMPI adhesion domain classification in *Plasmodium falciparum* pathogenesis research. Mol Biochem Parasitol 195:82–87
41. Cox-Singh J, Hiou J, Lucas SB et al (2010) Severe malaria a case of fatal *Plasmodium knowlesi* infection with post-mortem findings: a case report. Malar J 9:10
42. Fatih FA, Siner A, Ahmed A et al (2012) Cytoadherence and virulence the case of *Plasmodium knowlesi* malaria. Malar J 11:33
43. Millar SB, Cox-Singh J (2015) Human infections with *Plasmodium knowlesi* zoonotic malaria. Clin Microbiol Infect 21:640–648
44. Sallum MAM, Peyton EL, Harrison BA, Wilkerson RC (2005) Revision of the *Leucosphyrus* group of *Anopheles* [Cellia] [*Diptera, Culicidae*]. Revista Brasileira de Entomologia 49(Suppl. 1):1–152
45. Collins WE, Contacos PG, Guinn EG (1967) Studies on transmission of simian malarias. II. Transmission of the H strain of *Plasmodium knowlesi* by *Anopheles balabacensis balabacensis*. J Parasitol 53:841–844

46. Collins WE, Contacos PG, Skinner JC, Guinn EG (1971) Studies on transmission of simian malaria. IV. Further studies on the transmission of *Plasmodium knowlesi* by *Anopheles balabacensis balabacensis* mosquitoes. J Parasitol 57:961–966
47. Vythilingam I, Tan CH, Asmad M, Chan ST, Lee KS, Singh B (2006) Natural transmission of *Plasmodium knowlesi* to humans by *Anopheles latens* in Sarawak, Malaysia. Trans R Soc Trop Med Hyg 100:1087–1088
48. Jiram AI, Vythilingam I, NoorAzian YM, Yusof YM, Azahari AH, Fong MY (2012) Entomologic investigation of *Plasmodium knowlesi* vectors in Kuala Lipis, Pahang, Malaysia. Malar J 11:213
49. Wong ML, Chau TH, Leong CS et al (2015) Seasonal and spatial dynamics of the primary vector of *Plasmodium knowlesi* within a major transmission focus in Sabah, Malaysia. PLoS Neglected Trop Dis 9:e0004135
50. Marchand RP, Culleton R, Maeno Y, Quang NT, Nakazawa S (2011) Co-infection of *Plasmodium knowlesi, P. falciparum, and P. vivax* among humans and *Anopheles dirus* mosquitoes, Southern Vietnam. Emerg Infect Dis 17:1232–1239
51. Eyles DE, Laing ABG, Warren M, Sandoshan AA (1962) Malaria parasites of the Malayan leaf monkeys of the genus *Presbytis*. Med J Malays 17:85–86
52. Divis PC, Singh B, Anderios F et al (2015) Admixture in humans of two divergent *Plasmodium knowlesi* populations associated with different macaque host species. PLoS Pathog 11:e1004888
53. Fooden J (2006) Comparative review of fascicularis-group species of macaques [Primates: *Macaca*]. Fieldana Zool 107:1–43
54. Subbarao SK (2011) Centenary celebration article: *Plasmodium knowlesi:* from macaque monkeys to humans in Southeast Asia and the risk of its spread in India. J Parasit Dis 35:87–93
55. Fuerhrer HP, Swoboda P, Harl J et al (2014) High prevalence and genetic diversity of *Plasmodium malariae* and no evidence of *Plasmodium knowlesi* in Bangladesh. Parasitol Res 113:1537–1543
56. Moyes CL, Henry AJ, Golding N et al (2014) Defining the geographical range of the *Plasmodium knowlesi* reservoir. PLoS Neglected Trop Dis 8:e2780
57. Cramer JP (2015) *Plasmodium knowlesi* malaria: overview focusing on travel-associated infections. Curr Infect Dis Rep 17:469
58. Muller M, Schagenhauf P (2014) *Plasmodium knowlesi* in travelers, update 2014. Int J Infect Dis 22:55–64
59. Williams T, Rahman HA, Jelip J et al (2013) Increasing incidence of *Plasmodium knowlesi* malaria following control of *P. falciparum* and *P. vivax* malaria in Sabah, Malaysia. PLoS Neglected Trop Dis 7:e2026
60. Conlan JV, Sripa B, Attwood S, Newton PN (2011) A review of parasitic zoonoses in a changing Southeast Asia. Vet Parasitol 182:22040
61. Barber BE, William T, Dhararaj P et al (2012) Epidemiology of *Plasmodium knowlesi* malaria in north-east Sabah, Malaysia: family clusters and wide age distribution. Malar J 11:401
62. Barber BE, William T, Jikal M et al (2011) *Plasmodium knowlesi* malaria in children. Emerg Infect Dis 17:814–820
63. Naing DKS, Anderios F, Lin Z (2011) Geographic and ethnic distribution of *P. knowlesi* infection in Sabah, Malaysia. Int J Collab Res Int Med Public Health 3:391–400
64. Daneshvar C, Davis TM, Cox-Singh J et al (2009) Clinical and laboratory features of human *Plasmodium knowlesi* infection. Clin Infect Dis 49:852–860
65. William T, Menon J, Rajahram G et al (2011) Severe *Plasmodium knowlesi* in a tertiary care hospital, Sabah, Malaysia. Emerg Infect Dis 17:1248–1255
66. Willmann M, Ahmed A, Siner A et al (2012) Laboratory markers of disease severity in *Plasmodium knowlesi:* a case control study. Malar J 11:363
67. Barber BE, William T, Mj G et al (2013) A prospective comparative study of knowlesi, falciparum, and vivax malaria in Sabah, Malaysia: high proportion with severe disease from *Plasmodium knowlesi* and *Plasmodium vivax* but no mortality with early referral and artesunate therapy. Clin Infect Dis 56:383–397

68. Jeremiah S, Janagond AB, Parija SC (2014) Challenges in diagnosis of *Plasmodium knowlesi* infections. Trop Parasitol 4:25–30
69. Lucchi NW, Poorak M, Oberstaller J et al (2012) A new single-step PCR assay for the detection of the zoonotic malaria parasite *Plasmodium knowlesi*. PLoS One 7:e31848
70. Reller ME, Chen WH, Dalton J, Lichay MA, Dumler JS (2013) Multiplex 5′ nuclease quantitative real-time PCR for clinical diagnosis of malaria species-level identification and epidemiological evaluation of malaria-causing parasites, including *Plasmodium knowlesi*. J Clin Microbiol 51:2931–2938
71. Tomita N, Mori Y, Kanda H et al (2008) Loop-mediated isothermal amplification [LAMP] of gene sequences and simple visual detection of products. Nat Protoc 3:877–882
72. Iseki H, Kawai S, Takahashi N et al (2010) Evaluation of a loop-mediated isothermal amplification method a tool for diagnosis of infection by the zoonotic simian malaria parasite *Plasmodium knowlesi*. J Clin Microbiol 348:2509–2514
73. Lau YL, Fong MY, Mahmud R et al (2011) Specific, sensitive and rapid detection of human *Plasmodium knowlesi* infection by loop-mediated isothermal amplification [LAMP] in blood samples. Malar J 10:197
74. Faith FA, Staines HM, Siner A et al (2013) Susceptibility of human *Plasmodium knowlesi* infections to anti-malarials. Malar J 12:425
75. Daneshvar C, Davis TM, Cox-Singh J, Rafa'ee MZ, Zakari SK, Divis PC, Singh B (2010) Clinical and parasitological response to oral chloroquine and primaquine in uncomplicated human *Plasmodium Knowlesi* infections. Malar J 9:238
76. Mj G, William T, Menon J et al (2016) Artesunate-mefloquine versus chloroquine for treatment for uncomplicated *Plasmodium knowlesi* malaria in Malaysia [ACT KNOW]: an open label, randomized controlled trial. Lancet Infect Dis 16:180–188
77. Rajahram GS, Barber BE, William T, Menon J, Anstey NM, Yeo TW (2012) Deaths due to *Plasmodium knowlesi* malaria in Sabah, Malaysia: association with reporting as *Plasmodium* malariae and delayed parenteral artesunate. Malar J 11:284
78. Neaffsey DE, Juraska M, Bedford T et al (2015) Genetic diversity and protective efficacy of the RTS,S/AS01 malaria vaccine. N Engl J Med 373:2025–2037
79. Hamid MM, Remarque EJ, Hassan IM et al (2011) Malaria infection by sporozoite challenge induces functional antibody titers against blood stage antigens after a DNA prime, poxvirus boost vaccination strategy in rhesus macaques. Malar J 10:29
80. Weiss WR, Jiang CG (2012) Protective. CD8+ T lymphocytes in primates immunized with malaria sporozoites. PLoS One 7:e31247

Chapter 10
Zoonotic Streptococci: A Focus on *Streptococcus suis*

10.1 Introduction

Human streptococcal pathogens are commonly recognized to produce a wide spectrum of clinical diseases, from asymptomatic colonization and mild skin infections to lethal septicemia and invasive diseases of various organs. However, animal-derived streptococci were only rarely reported to produce human diseases of similar clinical spectrum as human specific streptococci and may have been misidentified or not identified to species level and broadly labeled as β-hemolytic streptococci of human origin. In the past two decades, there have increased medical awareness of the importance and significance of animal streptococcal pathogens with the advent of serious outbreaks of pig-derived streptococcal infections in Asia. Within the recent decade, the scientific medical literature has been replete with hundreds of articles and studies on the porcine pathogen *Streptococcus suis*. For the past 50 years or more, there has been a global increase in the incidence of invasive human infections due to veterinary pathogens caused by group B, C, and G *Streptococcus*. Group B *Streptococcus, Streptococcus agalactiae,* was first identified as an animal pathogen, but over the decades, it has been recognized as mainly a human pathogen. Group C and G streptococci are still considered as mainly animal pathogens, but there are increasing reports of human infections.

10.2 Classification of Streptococci

There are more than 40 species of streptococci of human and animal origin divided into multiple groups that have had their taxonomy changed several times over the years [1]. The *Streptococcus* genus consists of gram-positive cocci in pairs or chains that are catalase negative. The initial classification was based on the hemolysis produced on blood agar culture media: α reduction of hemoglobin with a greenish zone around the colonies, β complete hemolysis of erythrocytes, and γ lack of visible hemolysis.

I.W. Fong, *Emerging Zoonoses*, Emerging Infectious Diseases of the 21st Century,
DOI 10.1007/978-3-319-50890-0_10

In the 1930s, the Lancefield classification was introduced to differentiate streptococci in groups based on the presence and type of surface antigen [carbohydrate or lipoteichoic acids] [2]. The Lancefield classification differentiates the β-hemolytic group well from A through W, but some streptococci with α or γ hemolysis [i.e., *Streptococcus pneumoniae*] do not encode Lancefield antigen. The major pathogenic streptococci belong to the "pyogenic" group of β-hemolytic *Streptococcus* A, B, C, and G. Based on 16S rRNA gene sequence and biochemical features, 138 *Streptococcus* strains were divided into seven species groups: "*pyogenes*," "*mitis*," "*anginosus*," "*bovis*," "*mutans*," "salivarius," and unknown [3]. Members of the *Streptococcus* genus belong to the phylum *Firmicutes* and are among the most diverse and significant zoonotic pathogens.

10.3 Zoonotic Streptococci

S. agalactiae [group B *Streptococcus*] is a major bovine pathogen that causes bovine mastitis and still causes a large economic impact on milk production; but it is largely recognized in humans to cause genitourinary tract infection or colonization in pregnant women and subsequent neonatal sepsis [4]. Infection in humans with *S. agalactiae* was first reported in the 1930s, and until the mid-1960s, human infection was infrequently reported [4]. Group B *Streptococcus* [*GBS*] colonizes the gastrointestinal and genitourinary tract of 30% of healthy adults, and most infections are of an endogenous source. Thus, this pathogen is now considered a primary human bacteria, but zoonotic transmission may still occur. Although genotyping data indicate that human- and bovine-derived GBS strains represent distinct populations, human colonization increases with cattle exposure, and this supports interspecies transmission [5]. Furthermore, *S. agalactiae* is a known fish pathogen, and seafood may represent another source of infection or colonization [6].

Groups C and G *Streptococcus* [GCS and GGS] are zoonotic pathogens of animal origin but can be carried by humans and cause similar diseases as human pyogenic streptococci, such as skin/soft tissue infection and pharyngitis [4]. Human infections with these zoonotic streptococci are often milk borne and can cause outbreaks complicated by post-streptococcal glomerulonephritis [7]. Streptococci of group C and G classification have changed over the last 40 years, and previously in 1974 only, four species of GCS were listed [*Bergey's Manual of Determinative Bacteriology 1974*]: *Streptococcus equisimilis, Streptococcus dysgalactiae, Streptococcus equi, and Streptococcus zooepidemicus.* In 1996, taxonomic changes were made to Lancefield streptococci groups C, G, and L. [8]. GCS include *S. equi* subsp. *equi, S. equi* subsp. *zooepidemicus*, *S. dysgalactiae* subsp. *dysgalactiae* [rarely as group L], and *S. dysgalactiae* subsp. *equisimilis* [GCS/GGS as contain group G antigen and rarely group L]. *Streptococcus canis* is a member of GGS and a close relative of *S. equisimilis*.

Before the 1970s, cases of invasive infection with GCS and GGS were rarely described, and during the 1980s–1990s, most cases were patients with underlying conditions such as malignancy and cardiovascular disease [4, 9]. Although the

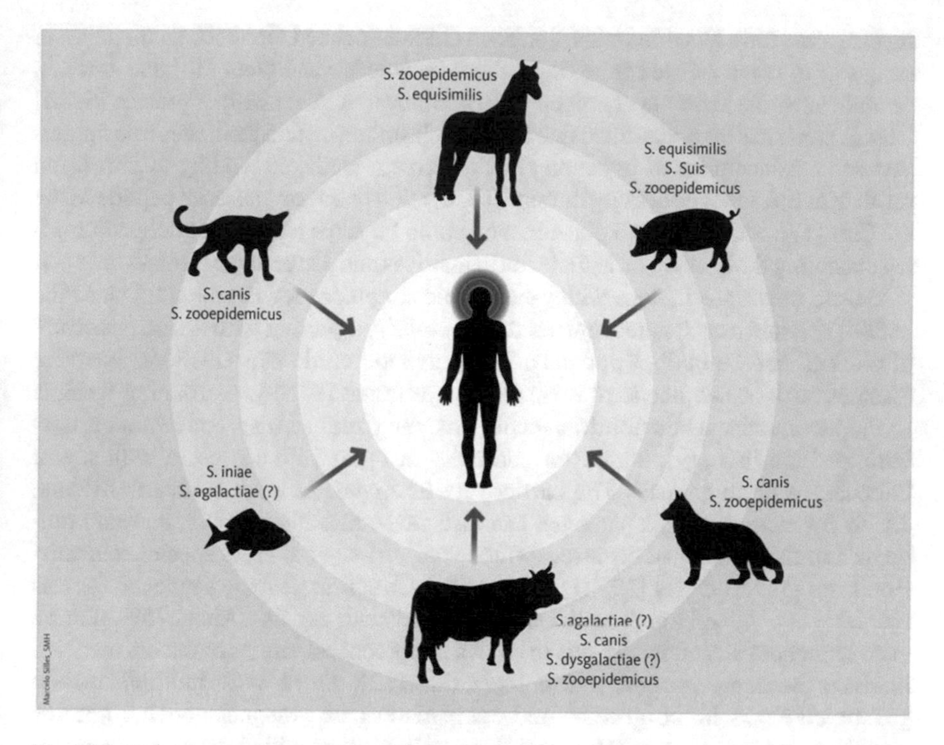

Fig. 10.1 Animal source of zoonotic streptococci

most frequent types of infection were bacteremia and endocarditis, a wide spectrum of clinical disease were reported including puerperal sepsis, skin/soft tissue infection, urinary tract infection, intra-abdominal infection, and spinal infection. In the last 25 years, there has been an increasing incidence of invasive GCS and GGS infections, and the burden of disease and fatalities may be approaching that of group A *Streptococcus* [10, 11]. The animal source of various zoonotic streptococci is outlined in Fig. 10.1.

10.4 *Streptococcus canis*

S. canis is a β-hemolytic, pyogenic, large-colony-forming GGS of animal origin that can be isolated from a wide range of mammals, but it is a resident microflora of domestic dogs and cats with colonization of the skin, mucosae of the genitourinary and gastrointestinal tract of these animals [12]. Phylogenetic analysis indicated that *S. canis* was a divergent taxon of the closely related species of *Streptococcus pyogenes* [GAS] and *S. equisimilis* and could acquire genetic material from the latter species [13]. The clinical spectrum and progression of diseases by *S. canis* share features with *S. pyogenes* infection, and this suggests similar virulence mechanisms.

Investigators have identified some common GAS-associated virulence factor genes in strains of *S. canis* isolated from dogs; these include *slo* and emm [14] and smeZ, a new allele of the GAS superantigen [15]. Moreover, a novel M-like protein [SCM] from *S. canis* has been identified that binds to plasminogen and facilitates transmigration and establishment of infection [16]. M-protein mediated binding to fibrinogen result in antiphagocytic activity through inactivation of the antibacterial peptide MIG/CXCL9 [17]. Surface-bound plasminogen could be activated by host derived urokinase enabling the bacteria to degrade fibrin matrices and disseminate through thrombi.

Severe infections in dogs with β-hemolytic streptococci were reported since the 1930s [18], and now it is recognized that *S. canis* can present with a wide spectrum of clinical disease in dogs and cats. These include mild skin infections to severe diseases such as streptococcal toxic shock syndrome [STSS], necrotizing fasciitis [NF], pneumonia, septic arthritis, meningitis, puerperal sepsis, septicemia of newborn, and mastitis [19]. Significant mortality of up to 30% can occur with severe infection in these animals. The carrier rate of *S.canis* in dogs is about 18% and 12.7% for cats [20]. This pathogen can also cause mastitis in cattle, and rare outbreak s in dairy herd had occurred, which was attributed to cross species transmission from domestic cats [19, 21]. Interestingly, the first genome sequence for this species was obtained from a milk isolate of a cow with mastitis. About 75% of these gene sequences were homologous to known streptococcal virulence factors involved in tissue invasion, evasion, and colonization [22]. There were multiple mobile genetic elements in the genome and comparison with two other bovine mastitis causing pathogens [*S. agalactiae* and *S. dysgalactiae*] provided strong evidence for lateral gene transfer [LGT], possibly contributing to host adaptation [22].

Human infections with *S. canis* are rarely reported, but this likely represents under-recognition, common for all animal pyogenic streptococci, as most laboratories continue to identify *Streptococcus* species just to the Lancefield group unless sent to a reference center. In many cases of human infection, the source of infection is unknown, but it is believed that animal contact or bites were the predisposition. In the last several years, there has been increasing detection of infected ulcers of dog owners with *S. canis* [23]. Septicemia has been reported following dog bite and close contact with pet canine [24, 25]. Two relatively recent cases of severe invasive disease have been described with dog contact without a bite. One case was a dog owner with septicemia with full recovery after treatment with ceftriaxone [26]. The other case is the first reported case of *Streptococcus canis* native valve endocarditis that was treated with antibiotics and bioprosthetic valve replacement [27].

10.5 *Streptococcus equi*, *Streptococcus zooepidemicus*, and *Streptococcus equisimilis*

S. zooepidemicus is closely related to *S. equi* and *S. equisimilis* serologically, and differentiation is done biochemically based on species-specific patterns of carbon source. *S. zooepidemicus* ferments lactose and sorbitol unlike *S. equi* and

S. equisimilis; but *S. equisimilis* utilize trehalose, and *S. equi* and *S. zooepidemicus* lack this ability [28]. These bacteria are also large-colony-forming, pyogenic, β-hemolytic streptococci. Phylogenetic analysis indicates that *S. equisimilis* is more closely related to GAS and forms a common clade with *S. canis* [29], whereas *S. equi* and *S. zooepidemicus* genomes are more than 98% identical [30]. It is believed that *S. equi* evolved from an ancestral form of *S. zooepidemicus* to become highly specialized virulent pathogen of horses as the only host [31]. It causes a major, highly contagious disease, strangles, of horses with upper respiratory tract infection, fever, and lymphadenitis with abscess formation of the head and neck that can rupture in the guttural pouch and restrict the airway [32]. Strangles remain the most common and important infectious disease of horses. Rare reports of *S. equi* producing human infections in subjects who frequently come in close contact with horses, such as an endovascular aortic prosthetic graft infection with *S. equi* [33], may in fact be due to *S. zooepidemicus* as the PCR used for diagnosis is a broad range 16S rDNA method that may not provide subspecies identification.

S. zooepidemicus is a commensal of the mucosae of the upper respiratory tract and lower genital tract of horses, and it is an opportunistic pathogen of a wide variety of mammals including dogs, cats, pigs, ruminants, rodents, monkeys, and seals [19]. It is a major pathogen in horses usually as a secondary infection after a primary viral respiratory infection or after stress and injuries. It can cause severe hemorrhagic pneumonia with dissemination to many organs from septicemia and form abscesses and as well cause endometritis with infertility and neonatal sepsis in fouls [19]. This animal microbe is also an emerging pathogen in dogs, with clinical spectrum of diseases similar to those in horses. Unlike horses, diseases in dogs frequently occur in outbreaks with high fatality rates, and animals can present with hemorrhagic pneumonia and septicemia [34].

In Asia, *S. suis* and *S. zooepidemicus* are the two major bacterial pathogens of swine. Large outbreaks of swine infection with *S. zooepidemicus* have occurred in China and Indonesia in the 1970s and 1990s, respectively [19]. Pigs may develop arthritis, diarrhea, pneumonia, meningitis, and endocarditis [35]. During a large outbreak in Indonesia, more than 300,000 pigs died within two weeks, as a result of a single highly virulent clone which also spread to monkeys with similar high fatality [36]. This pathogen can infect many ruminants such as cattle, sheep, goats, lama, alpaca, and camels [19]. Although mastitis is the most frequent manifestation, animals can develop severe deep tissue infections.

Human infections with *S. zooepidemicus* can occur from consumption of contaminated milk and milk products or close contact with animals, most commonly horses. Two outbreaks of human infection occurred in 2003 from eating cheese made from raw milk in Gran Canaria and Finland [37, 38]. Clinical manifestations included septicemia, pneumonia, septic arthritis, meningitis, and infected aortic aneurism with some fatalities. Similar to *S. pyogenes,* post-streptococcal nephritis have been reported after *S. zooepidemicus* infection. In a large outbreak in rural Brazil, from contaminated milk used to produce cheese, 253 cases of acute nephritis occurred leading to three fatalities and seven patients requiring dialysis [39]. Transmission of the pathogen from pigs to humans may occur from consumption of undercooked pork in Asia, as isolates from patients and pigs were very similar genotypically in the absence of direct animal contact [40]. *S. zooepidemicus* infections in a family cluster of severe clinical illness have recently been associated with exposure to guinea pigs [41].

Direct or indirect contact with companion animals especially horses can result in severe infection in humans. Within 6 months in 2011, three unrelated cases of severe, disseminated infection with *S. zooepidemicus* occurred in men working with horses in eastern Finland [42]. The spectrum of clinical diseases in humans transmitted by animal contact is the same as noted with outbreaks related to contaminated cheese. It is of interest to note that a few cases of *S. zooepidemicus* endovascular aortic grafts infection have been cured with conservative prolonged antibiotics [43]. At least 21 cases of meningitis due to *S. zooepidemicus* have been described, associated with animal contact or ingestion of unpasteurized dairy products, with hearing loss in 19% and complete recovery only in 38% [44].

S. equisimilis is also a large-colony, β-hemolytic *Streptococcus*; human pathogens belong to Lancefield group G, but veterinary pathogens belonged to Lancefield group C [10]. The organism can be found in the respiratory and reproductive tracts of horses and has been isolated from pigs. The clinical spectrum of diseases is similar to *S. pyogenes* from superficial skin, to deep, toxin-mediated diseases such as toxic shock syndrome. The organism has been associated with pharyngitis and post-streptococcal glomerulonephritis, and acute rheumatic fever has been described [45, 46]. *S. equisimilis* is considered a commensal organism in horses, but it has been increasingly isolated from horses with respiratory, reproductive, and other diseases. Although human and animal strains have been considered host specific, the sequence of the streptokinase gene of equine *S. equisimilis* has been found in two human cases of pneumonia in Japan [47]. Thus, indicating its zoonotic potential.

10.6 *Streptococcus iniae*

S. iniae has emerged in the past decade and a half to become one of the most serious aquatic pathogen globally, causing huge economic losses in farmed marine and freshwater finfish in warmer regions [48]. It was first described in 1976 from captive Amazon freshwater dolphin and is believed to have caused disease outbreaks in marine aquaculture in Japan for several decades [48]. *S. iniae* is an encapsulated β-hemolytic *Streptococcus* not assigned to any Lancefield group but is genetically closely related to group B streptococci from sequencing [19]. It is widely distributed geographically, mainly in North America, Middle East and the Asia-Pacific region. *S. iniae* is an invasive pathogen of various fish species and cause meningoencephalitis with brain and eye necrosis [49]. The bacteria colonize the scales and gills in high concentration and can be detected in the tank water a week post-infection. An outbreak of fatal septicemia by this pathogen has been described in Caribbean reef fish with necropsy findings of diffuse pericarditis, inflammation and necrosis of the pericardium, meninges, liver, kidneys, and gills [50].

This aquatic pathogen is considered an emerging zoonosis [International Conference on Emerging Infectious Diseases in 2000] although human infections have been sporadic and mainly identified in the United States, Canada, and throughout Asia. Infection occurs by preparation of contaminated fish by direct puncture wounds and primarily causes cellulitis and soft tissue infection [19]. However, severe complications such as septic arthritis, meningitis, osteomyelitis, and endocarditis have been reported [51, 52].

10.7 *Streptococcus suis*

10.7.1 General Background

S. suis is a neglected zoonotic pathogen of major global importance to the swine industry and to human health in Asia. It was first described by veterinarians in 1954, after outbreaks in piglets caused severe invasive diseases with septicemia, septic arthritis, and meningitis [53]. Cross infections in humans were initially reported in Denmark in 1968 and then subsequently in other European countries and Hong Kong [54–56]. Within the past decade, the number of human cases of *S. suis* reported in the literature has shown a logarithmic growth. In 2007, 409 cases were reported; by 2009 this had increased to >700 cases [57], and by 2012 a total of 1584 cases had been reported [58]. The global importance of this zoonotic pathogen has gained international scientific attention in recent years, as reflected by the number of publications with over 500 articles published in less than 10 years [Pubmed database], and the occurrence of the First International Workshop on *S. suis* in August 2013, Beijing, attended by 80 researchers from Asia, Europe, and North America.

10.7.2 Epidemiology of S. suis

Pigs and wild boars are the natural reservoirs of *S. suis*, and these animals can be asymptomatic carriers in the upper respiratory tract, intestines, and genital tract [19]. Healthy carriers are the main source of infection to other swine and to people. Weaning piglets are heavy colonizers with tonsillar carriage of up to 100%, which may persist after antibiotic treatment probably because of biofilm colonies [19, 59]. Pigs of any age can be infected, but the risk decreases with age after weaning. Source of infection between animals can be from direct contact, contaminated feces, dust, water, and feed, and vectors such as flies and mice can play a role [19]. Outbreaks in swine usually occur with introduction of a carrier of a virulent strain among the herd, and pigs can carry multiple strains. Animals are most susceptible to infection under stressful conditions and with preceding or concurrent viral infection. Vertical transmission can occur in piglets born to sows with genital infection or colonization [60].

Infection in pigs with *S. suis* is worldwide and has been documented in all continents except Africa, from North America [Canada and the United States] to South America [Brazil], Europe [Denmark, France, the Netherlands, Great Britain, Norway, Spain, and Germany], Asia [China including Hong Kong, Thailand, Vietnam, and Japan], Australia, and New Zealand [58]. *S. suis* can be found in other animals as intestinal commensal such as ruminants, cats, dogs, deer, horses, and rabbits [59, 61]. In developed countries such as in Europe, human infections occur sporadically from pig contact in countries with extensive swine production, even in countries such as Italy and Greece where endemic pig infections are not established [62, 63]. In Asia where human infections are much more prevalent, the major route of transmission is from consumption of undercooked pork or during the preparation and close animal contact in regions with intensive pig rearing.

10.7.3 Microbiology of S. suis

S. suis grows readily on commercial media and produces α-hemolysis on blood agar; hence, it is often misdiagnosed as viridans streptococci or *Streptococcus pneumoniae,* especially when recovered from the cerebrospinal fluid [CSF]. In a prospective study from Thailand, 70% of invasive viridans streptococci from patients were subsequently identified as *S. suis* [64]. In another series from Thailand, 62.5% of *S. suis* isolated from patients were initially reported to be viridans streptococci [65] and 20% of clinical isolates in a report from the Netherlands [55]. This gram-positive coccus usually appears on gram stain singly, frequently in pairs, or occasionally in short chains; it grows well in aerobic environment, but the growth is enhanced by microaerophilic conditions [66]. Initially *S. suis* was considered a Lancefield group D *Streptococcus*, but DNA hybridization demonstrated that it was not closely related to group D streptococci. Four biochemical tests can give a presumptive identification of *S. suis*: no growth in 6.5% NaCl agar, a negative Voges-Proskauer test, and production of acid in trehalose or salicin negative [66].

There are 35 known serotypes, but a large number of isolates especially from healthy pigs are untypable [67]. Infections in swine are usually caused by a limited number of virulent strains, serotypes 1–9 and 14 [68]. Serotype 2 is the most important *S. suis* strain responsible for majority of infections in pigs and humans [67, 69]. The organism is thickly encapsulated, and serotyping is based on use of antisera against 35 capsular polysaccharide antigens [serotypes 1–34 and serotype 1/2]. However, this is tedious and expensive for surveillance studies in animals. Real-time [RT] PCR has been used to diagnose patients with meningitis especially in cases pretreated with antibiotics [70], but mainly for serotype 2 or for limited serotypes. A novel PCR method for identification of all the currently recognized serotypes for epidemiological studies has recently been developed. This method targets the recombination/repair protein [recN] gene of *S. suis*, designated recN PCR, and appears to give reliable results [71]. Another group of investigators have also developed a multiplex PCR that identify 33 serotypes of *S. suis* [72].

Genotyping of strains of *S. suis* from different serotypes that cause disease in pigs and humans in various geographic regions of the world has been used for epidemiological surveillance. These strain variants have been categorized by genetic sequence type [ST] by multilocus sequence typing using genetic variation of seven housekeeping genes [*cpn60, dpr, recA, aroA, thrA, gki, and mutS]* [73]. Serotype 2 ST1 is responsible for most diseases in pigs and humans in most countries of Asia and Europe, and serotype 2 ST7 is responsible for major outbreaks in China [57, 58]. Combination of sequence type and virulence-associated genes with corresponding protein levels have been used to compare phenotypic characteristics [mainly for serotype 2]. These include *sly* gene that encodes suilysin, *mrp* gene that encodes muramidase-related protein, *epf* gene that encodes extracellular factor, and different pili [74]. Serotype 2 ST1 and ST7 strains from diseased pigs and humans all over the world and China, respectively, express the common virulence markers encoded by the genes *sly, mrp*, and *epf* [75].

10.7.4 Virulence and Pathogenesis of S. suis

Over the last 5 years, major progress has been made in elucidating the virulence factors and understanding the pathogenesis of *S. suis* infection in pigs and humans. Bacterial adhesion to the host epithelium is usually a prerequisite for infection. There is evidence that an immunogenic cell wall protein of *S. suis,* designated Sad P [streptococcal adhesion P], recognizes galactosyl-alpha-4-galactose-oligosaccharides and binds its natural glycolipid host receptor [76]. However, adhesion of *S. suis* to epithelial cells is multifactorial and involves bacterial surface proteins and extracellular matrix interaction [77]. The mechanism by which virulent strains of the bacteria breach the upper respiratory epithelium of the upper respiratory tract in pigs or the intestinal endothelium of humans to cause invasion is not well understood. There is recent evidence that a novel fibronectin-binding protein of serotype 2 contributes to epithelial cell invasion and in vivo dissemination [78]. Many microbial proteins involved in cell transport, biological regulation, signal transduction, and stress responses are associated with biofilm formation which allows *S. suis* to become persistent colonizer and resist clearance by the immune system [79, 80].

Other mechanisms used by the bacteria to evade the host mucosal immunity and allow epithelial invasion include the capsular polysaccharide, which is relatively resistant to phagocytosis by dendritic cells [81], and secretion of IgA protease. It has been reported that *S. suis* produces a protease capable of cleaving human IgA1 [82] and abrogates the adaptive mucosal immunity. Conversely, the bacterial capsule may decrease adhesion to epithelial cells and paradoxically abate invasion; however, the bacteria can counter this flaw by downregulation of genes involved in capsule biosynthesis to facilitate adhesion [77]. The capsular polysaccharide inhibits phagocytosis through destabilization of lipid microdomains and prevents lactosylceramide-dependent recognition [83]. Suilysin, a hemolysin with cholesterol-binding cytolytic activity, can disrupt epithelial cells to allow invasion

[77] and can impair opsonophagocytosis of invading bacteria [84]. It was recently shown that an epidemic strain of *S. suis* serotype 2, associated with severe disease and high fatality in pigs and humans, had enhanced secretion of suilysin compared to a non-epidemic strain [85]. Moreover, infection of mice with a mutant epidemic strain deficient in suilysin had increased survival and decreased level of bacteremia. This suggests that suilysin contributes to penetration and invasion of the bacteria, particularly for virulent strains.

It is currently believed that the initial site of mucosal invasion of *S. suis* occurs in the natural porcine host mainly in the upper respiratory tract and probably in the gastrointestinal tract [GIT], while in humans it is primarily via the GIT in endemic countries of Asia and probably through breaks in the skin in sporadic cases from developed countries. Based on in vitro and in vivo models, Ferrando and Schultsz [86] have proposed a model of zoonotic *S. suis* interaction with the intestinal mucosa shared by the porcine and human host. Adhesion of the bacteria to the host enterocytes is facilitated by adhesins, pili, and IgA1 protease, and variation in the capsule structure by environmental changes promotes adhesion by increasing bacterial surface polysaccharide to specific enterocyte receptors. Active translocation across epithelial cells then occurs via paracellular and/or transcellular routes and suilysin secretion damage the mucosal barrier to allow deeper tissue invasion. Translocated bacteria might survive inside or attached to dendritic cells [through anti-phagocytic properties of the capsule] to allow dissemination and invasive disease [86].

Certain genotypes of *S. suis* serotype 2, epidemic strain sequence type 7, had been associated with outbreaks of streptococcal toxic shock syndrome [STSS] in China. Investigation of the pathogenic mechanisms of the STSS produced by these strains has implicated a novel pathogenicity island as the genetic determinant, designated 89K, with universal properties of pathogenicity islands [87]. Enhanced mitogenic capacity of these strains allowed stronger stimulation of T-cells, naïve T-cells, and peripheral blood mononuclear cell proliferation than other strains [88]. Further studies in mice comparing the epidemic strain ST7 with a classical highly pathogenic strain and an intermediate pathogenic strain had shown that the epidemic strain produced massive amounts of proinflammatory cytokines that led to STSS, and this was related to acquisition of additional genomic pathogenicity islands compared to the classic highly pathogenic strain [89].

Meningitis is a major complication of invasive *S. suis* infection in humans and pigs, and studies have been performed to elucidate the mechanisms. Bacteria reach the brain via the blood stream from the GIT or upper respiratory tract. The choroid plexus epithelium is the structural basis of the blood-cerebrospinal fluid or blood-brain barrier. In meningitis, the blood-brain barrier is compromised as a result of inflammation of the choroid plexus, resulting in cell death and necrosis [90]. Pathological analysis of the brain of experimentally infected pig revealed bacteria localized within the cytoplasm of neutrophils and macrophages in the meninges, plus high expression of suilysin in the cytoplasm [91]. In the murine model, there is evidence that the bacteria induce inflammation by interaction of microglial cells via stimulation of Toll-like receptor 2, as well as phosphotyrosine, protein kinase C, and different mitogen-activated kinase signaling events [92]. There is also evidence that *S. suis* serotype 2 adheres and invades human astrocytes more avidly than meningeal cells [93]. Another study found that the virulent factor of the bacteria, enolase, enhances blood-brain barrier permeability by inducing interleukin [IL]-8 upregulation in the inflammatory cytokine cascade [94]. A critical host response to bacterial meningitis is the influx of neutrophils from the blood into the brain and across the blood-cerebrospinal fluid barrier. In an ex vivo cell model using porcine choroid plexus epithelial cells, *S. suis* causes influx of neutrophils via stimulation of TNF-α and increased expression of cell adhesion molecules ICAM-1 and VCAM-1 [95]. Furthermore, neutrophils preferentially migrated across the blood-cerebrospinal fluid barrier via the transcellular route, which was dependent on CD11b/CD18 expression.

10.7.5 Clinical Aspects of S. suis *Infection*

10.7.5.1 Human Epidemiology

S. suis zoonotic infections in humans are limited to pig-rearing countries of the world, but the mode of transmission varies between developed and developing countries. In European pig-rearing countries and North America, human infections are primarily from occupational exposure to pigs or processing of pork, with an incidence of 3/100,000 and about 1500-fold greater than the population not exposed to the pig industry [55]. On rare occasion, infection had occurred from contact with wild boar or processing of the carcass, including serotype 2 presenting with STSS in a hunter [96]. Transmission of infection is from contamination via the skin or conjunctivae, but oral ingestion is feasible. The lack of reported cases of *S. suis* infection associated with ingestion of undercooked pork in Europe is difficult to explain as this is seen with other porcine zoonosis such as hepatitis E. This may suggest that a high inoculum is needed to produce infection via the GIT, or the organism is more susceptible to heat than hepatitis E. Unrecognized or subclinical infection with *S. suis* in swine workers in developed countries is fairly common as reflected by the development of antibodies in 9.6% of swine exposed subjects [97]. Most of human cases from developed countries are from Europe, mainly the Netherlands, United Kingdom, France, and Spain and rarely from Austria, Belgium, Croatia, Denmark, Germany, Greece, Italy, Ireland, Poland, Portugal, Serbia, and Sweden [98]. Europe accounts for 8.5% of all reported cases [second after Asia], but only a few sporadic cases have been in North America despite being the continent with the highest *S. suis*-infected diseased pigs. This is attributed to lower rate of the virulent serotype 2 in infected swine in Canada and the United States. Serotype 2 and sequence type 1 [ST1] strains are responsible for most clinical disease in pigs and humans worldwide, and ST7 was responsible for outbreaks in China. In North America, serotype 2 ST25 and ST28 are the predominant strains in diseased pigs and are less virulent than ST1 and ST7. Sporadic cases [0.8% of all reported cases] have been identified in Australia, New Zealand, and South America [Argentina, Chile, and French Guiana] [98].

Outbreaks of *S. suis* human infections are seen only in low-income countries with extensive pig rearing. Infection is transmitted by direct contact with infected pigs or pork and consumption of raw or undercooked pork or pig products. Most cases of human infection occur in Asia, which account for 90% of all cases globally, especially Vietnam, Thailand, and China where 83.6% of all cases arise [98]. Human cases were also reported from Cambodia, Hong Kong, Japan, Laos, Singapore, South Korea, and Taiwan. Although cases have not been reported from the Philippines, infection in humans probably occurs and may be unrecognized. A retuning traveler in the United States developed *S. suis* meningitis after returning from the Philippines, where he consumed raw pork [98]. In Asia farmers' dwellings are in close proximity with pigs, and it is common practice to slaughter sick pigs at home for consumption [57]. Besides occupational exposure, risk factors for infection include eating "high-risk" dishes such as undercooked pig blood and pig intestine and contact with pork in the presence of skin injuries [99]. Pork is the most common meat source in Vietnam and is usually obtained fresh from wet markets [57], and in Hong Kong, *S. suis* has been isolated from 6.1% of fresh pork meat from half of the wet markets studied [100] but in 11% of pork products from wet/retail markets in Vietnam [98]. In southern Vietnam, 41% of slaughterhouse pigs were found to carry *S. suis* [101]. Although *S. suis* colonization of the throat was not been found in employees of slaughterhouse in Brazil, it was isolated from 25% of environmental cultures [102]. However, nasopharyngeal carriage of the bacteria has been described in 5.3% of workers in the swine industry in Germany [57], but has not been studied in Asia.

A total of 1642 cases of human infection worldwide had been reported as of December 2013 [98], but this likely is a gross underestimate of the actual burden of human infections globally, as this infection is not reportable in many countries and problems of reporting bias. Serotype 2 represented 74.7% of the reported cases, followed by serotype 14 with 2%, and sporadic cases had been reported with serotypes 4, 5, 16, 21, and 24 [98] and more recently with serotype 31 [103]. In Thailand serotype 14 is the second commonest and occurred in 25.8% of the human cases [104]. In a systematic review of the human experience, the risk factors identified were occupational exposure, overall 38.1% [but 83.8% for industrialized countries]; contact with pigs or pork, 33.9% for the meta-analysis; consumption of pork, pooled estimate 37.3% but 55.8% in Thailand; skin injury, one-fourth of single cases; and alcohol in one-third of single cases [58]. Although in vitro alcohol promotes the intestinal translocation of *S. suis* [105], a case control study from Vietnam did not confirm alcohol consumption as an independent risk factor [99].

10.7.5.2 Clinical Features of *S. suis* Infection

The incubation period of *S. suis* infection in humans varies considerably probably associated with the route of transmission. In a Chinese outbreak, the incubation period ranged from 3 h to 14 days, median 2.2 days [106], but it is likely that cases with very short incubation period were infected via skin wounds with rapid dissemination hematogenously. Other studies had reported longer incubation period of 60 h to 1 week, where inoculation of the bacteria may have occurred from oral ingestion to a large degree [107].

The clinical manifestations of *S. suis* are very diverse with the majority of cases occurring in healthy, predominantly male [76%] adults, mean age of 51 years, and extremely rare in children [58]. Pooled data of clinical studies including single case reports indicate that meningitis is the most common clinical disease of human infections with *S. suis* [68%], followed by sepsis [25%], septic arthritis [12.9%], endocarditis [12.4%], toxic shock syndrome [25.7%], and endophthalmitis [4.6%] [58]. However, the clinical disease manifestations and rates were largely related to country of publication. For instance, STSS occurred at high rates in two outbreaks in China [64% and almost 30%] and in Thailand [37.7%] but in only 2.9% of case reports [58], whereas studies in Vietnam and Thailand had reported very high rates of meningitis. The reason for these geographic differences in clinical manifestations is unclear but may be related to sequence type or genetic virulence factors, host factors, and mode of transmission. For instance, a study in Thailand found that meningitis [58.9% of the cases] was associated with serotype 2 ST1 strains and nonmeningitis [41.1%, and including sepsis in 35.4%] was associated with ST104 strains, which is unique to Thailand [108]. Also the high rates of STSS and sepsis in the two outbreaks in China appear to be associated with ST7 strains.

In a recent systematic review and meta-analysis of *S. suis* meningitis, 913 cases were identified in 24 studies from 1980 to 2015 [109]. The mean age and sex were similar to the overall human experience [58], and the main risk factor was exposure to pig or pork in 61% of cases, and 20% of patients had skin injury with exposure. Other predisposing factors include alcoholism [19%], diabetes [5%], splenectomy [1%], cancer [6%], and immunosuppressive drugs [0.3%]. While headache, fever, and neck stiffness were very common [93–97%], the classic triad of fever, neck stiffness, and altered level of consciousness occurred only in 9% of cases. Penicillin and ceftriaxone were used for treatment, and no resistance to these agents was reported. Remarkably, the case fatality rate was only 2.9% which is much lower than the pooled case fatality of 12.8% for all *S. suis*-infected patients reported in the general systematic review in 2014 [58]. Similar difference between meningitis and sepsis case fatality rates has been reported for invasive meningococcal disease [110]. The low case fatality rates of *S. suis* and meningococcal meningitis compared to pneumococcal meningitis [20–30%] may be related to age as both the former conditions' mean age is in young middle-aged range, whereas the latter condition mean age is in older age above 60–65 years, often with comorbidities. In a study from Thailand, the risk factors for mortality were septic shock and elevated alanine transaminase [111]. Hearing loss was a common complication occurring in 53% of patients, and dexamethasone treatment was associated with reduced hearing loss in one randomized controlled study [107, 109]. Dexamethasone did not reduce mortality in *S. suis* meningitis probably because the case fatality was very low. The hearing loss can occur early at presentation or as a sequelae of the meningitis and has been attributable to supportive labyrinthitis or inflammation of the eight nerve during meningitis [112].

10.7.5.3 Diagnosis and Antimicrobial Resistance

Diagnosis of *S. suis* infection is usually made by blood and/or cerebrospinal fluid culture on standard commercial media. Molecular methods were used in some large studies [11studies] or in cases treated with antibiotics before obtaining cultures [58]. Misidentification of organism by conventional biochemical tests or commercial identification systems was fairly common. In one study, misidentification of *S. suis* was more common with the Phoenix Identification system than the VITEK 2 GPI Card system [113].

S. suis is almost universally susceptible to penicillin and cephalosporin, and resistance was only reported in two patient isolates [58]. Thus, penicillin or cephalosporin should be considered drugs of choice. Studies on antimicrobial susceptibility profile of large number of isolates have been mainly performed from pig isolates, but this should reflect the pattern in human isolates as well. The antimicrobial susceptibility data indicate that the antibiotic resistance rate was highest for tetracycline [99.1%], followed by macrolides [68–68%], clindamycin [67.9%], trimethoprim/sulfamethoxazole [16%], levofloxacin [2.8%], chloramphenicol [1.9%], and ceftriaxone [0.9%] [114]. Similar results were also found from another study in China but limited to serotype 2 strains from pigs [115]. These results indicate that for patients with history of severe penicillin allergy, the best alternative treatment would be vancomycin [although not tested in these studies, it should be highly effective] or chloramphenicol for severe infections or meningitis, but for moderate skin/soft tissue infections, levofloxacin is a reasonable choice.

10.7.5.4 Prevention and Potential Vaccines for *S. Suis*

An effective vaccine against *S. suis* is unavailable but urgently needed by the global swine industry and for the at-risk population in Asia. In humans in Asia, the most effective preventative measures would be communication campaigns and education on local risk of infection, such as unsafe handling and consumption of raw or undercooked pork.

For pig farmers and the swine industry in general, alternative treatment and prophylactic agents are needed to reduce antibiotic use and mitigate the spiraling trend of widespread antimicrobial resistance in pig-derived *S. suis*. A novel approach to reduce carriage of the bacteria in the nasopharynx of pigs is by disrupting biofilms formed by the pathogen. A bacteriophage lysine, designated LySMP, had been shown to disrupt biofilm structure and inactivate *S. suis* cells in vitro [116]. Other alternatives include natural substances such as bacteriocins, protein substances released by certain bacteria that can kill other unrelated strains of bacteria. Nisin is the only approved bacteriocin for food preservation, and treatment with this agent can cause bacterial lysis from cell membrane breakdown [68, part 2]. Nisin has been reported to be synergistic with antibiotics against *S. suis* [117]. Strains of *S. suis* isolated from healthy pig carriers produce new bacteriocins belonging to the lantibiotic class [67], but it is unclear if these suicins would have any effect on pathogenic strains.

Intensive research has been ongoing since 1990 to find an effective vaccine for *S. suis* in pigs. From 1990 to 2015, 54 studies on the subject had been reported, and an excellent comprehensive review by M. Segura was recently published [118]. Most of the studies on experimental vaccines had been performed for elucidating protection for serotype 2, but there are multiple other virulent serotypes causing disease outbreaks in swine and sporadic cases in humans. A major challenge is to develop an effective and safe vaccine to provide cross-protection for multiple serotypes and multiple sequence types. Inactivated whole-cell bacteria vaccines and heat-killed or formalin-inactivated bacterins, in experimental and field trials, had produced inconsistent results and limited protection against mortality in piglets [118]. Some studies showed protective immunity against homologous challenge but no protection against heterologous strains. Others found bacterins reduced mortality but not morbidity; also protection in mice may not produce protection in pigs. Vaccination of pregnant sows to prevent infection in newborn piglets had mixed results, and although opsonizing antibodies could be increased with vaccination in the serum and colostrum of sows, immunity in newborn piglets lasted only 6 weeks [119]. Live-attenuated vaccines with pathogenic and nonpathogenic bacteria, including nonencapsulated strains, have been studied with mixed results. Some vaccines required multiple doses [by intramuscular injections] with high inoculum of bacteria [107] but a single dose would be more suitable. A single intranasal immunization with an attenuated virulent strain of *S. suis* serotype 2 elicited high humoral immune response, but failed to protect against homologous strain or serotype 9 strain [120]. A naturally avirulent serotype 2 strain from the tonsils of a pig appears promising, as two doses of vaccine produced complete protection against challenge in a swine model with a highly virulent Chinese ST7 strain [121].

Attempt has been made in the last several years to find a novel immunogenic protein conserved among most serotypes that could be used as a "universal" subunit vaccine through immunoproteonomic-based research. Although mouse models are often used in these studies, protection in pigs is usually lower than that observed in mice [118], but they are still useful for screening. Inclusion of putative virulence-associated proteins EF, MRP, and SLY in a subunit vaccine may produce high antibodies but restricted protection in pigs, and their benefit in cross-protection of other serotypes has been limited [118]. Probably the most promising molecule is a 110-kDa protein labeled "surface antigen one" [Sao], which can elicit cross-reactive antibodies against 28 of 33 serotypes of *S. suis* and 25 of 26 ST of serotype 2 [122]. Other prospective subunit vaccine candidates include four membrane-associated proteins recently shown to elicit strong humoral antibody response and protect mice from lethal challenge against *S. suis* [123]. However, studies should be performed in pigs and determine cross-protection for other serotypes.

Theoretically a multivalent surface capsular vaccine should be effective, as antibodies against the capsular polysaccharide [CPS] increases killing of the bacteria by phagocytosis, and monoclonal antibody specific for *S. suis* serotype 2 CPS protects mice against the bacterial challenge [124]. However, CPS is a poor immunogen, and antibody response was impaired and lacked isotype switching in *S. suis*-infected pigs [125].

10.7.6 Future Directions

A global strategy is needed to prevent and rapidly diagnose zoonotic streptococcal infections in livestock, pets, and humans. Much research is still needed to elucidate the evolution of human pathogenic strains and the mechanisms of host-specific adaptation. There are still large gaps in the knowledge of the geographic epidemiology of these zoonoses even for *S. suis*. For instance, there is sparse data on the prevalence of *S. suis* infection in animals and humans in many pig-rearing countries such as in South America [i.e., Brazil, Mexico], Eastern Europe, Russia, and the Philippines. Studies on other streptococci considered strictly animal pathogens need to be explored for zoonotic potential. For instance, *S. dysgalactiae* subsp. *dysgalactiae* that are considered exclusive animal pathogens [found in cattle] have recently been associated with limb cellulitis, prosthetic joint infection, and infective endocarditis in humans [126]. Also the transmission routes and zoonotic potential of *Streptococcus gallolyticus* subsp. *gallolyticus* needs clarification. Formerly known as *S. bovis* biotype 1, the bacteria are a commensal of the GIT in various animals, pigeons, birds, and cattle and are found in 2.5% to 15% of healthy animals [127]. It can also be a pathogen in animals and can cause mastitis, septicemia, and meningitis, and in humans, it had been reported to cause endocarditis and neonatal sepsis [127].

Further vaccine studies need to evaluate combination of promising proteins for subunit vaccine, such as Sao and the membrane-associated proteins in pigs. It should also be feasible to develop and test a multivalent conjugated CPS vaccine for *S. suis*, which may be more effective in inducing protective antibodies. In a recent report, a *S. suis* type 2 CPS glycoconjugate vaccine was shown to induce potent IgM and isotype-switched IgGs in mice and pigs and provided protection against lethal challenge in vivo [128]. Hence, this would be a suitable candidate for field trials in pigs and especially piglets.

References

1. Kohler W (2007) The present state of species within the genera *Streptococcus* and *Enterococcus*. Int J Med Microbiol 297:133–150
2. Lancefield RC (1933) A serological differentiation of human and other groups of hemolytic streptococci. J Exp Med 57:571–595
3. Gao XY, Zhi ZY, Li HW, Klenk HP, Li WJ (2014) Comparative genomics of the bacterial genus *Streptococcus* illuminates evolutionary implications of species groups. PLoS One 9:e101229
4. Sitkiewicz I, Hryniewicz W (2010) Pyogenic streptococci danger of re-emerging pathogens. Pol J Microbiol 59:219–226
5. Manning SD, Springman AC, Million AD et al (2010) Association of Group B *Streptococcus* colonization and bovine exposure: a prospective cross-sectional cohort study. PLoS One 5:e8795
6. Liu G, Zhang W, Lu C (2013) Comparative genomics analysis of *Streptococcus agalactiae* reveals that isolates from cultured tilapia in China are closely related to the human strain A909. BMC Genomics 14:775

7. Bordes-Benitez A, Sanchez-Onoro M, Suarez-Bordon P et al (2006) Outbreak of *Streptococcus equi* subsp *zooepidemicus* infection on the island of Gran Canaria associated with the consumption of inadequately pasteurized cheese. Eur J Clin Microbiol Infect Dis 25:242–246
8. Vandamme P, Pot B, Falsen E, Kersters K, Devriese LA (1996) Taxonomic study of Lancefield streptococcal groups C, G, and L [Streptococcus dysgalactiae] and proposal of *S. dysgalacv-tiae* subsp. *equisimilis* subsp. nov. Int J Syst Bacteriol 46:774–781
9. Bradley SF, Gordon JJ, Baumgartner DD, Marasco WA, Kauffman CA (1991) Group C streptococcal bacteremia: analysis of 88 cases. Rev Infect Dis 13:270–280
10. Brandt CM, Spellerberg B (2009) Human infections due to *Streptococcus dysgalactiae* subspecies *equisimlis*. Clin Infect Dis 49:766–772
11. Ikebe T, Oguro Y, Ogata K et al (2002) Surveillance of severe invasive group G streptococcal infections in Japan during 2002-2008. Jpn J Infect Dis 63:372–375
12. Devriese LA, Hommez J, Klipper-Balz R, Schleifer K-H (1986) *Streptococcus canis* sp. nov.: a species of group G Streptococci from animals. Int J Syst Bacteriol 36:422–425
13. Pinho MD, Matos SC, Pomba C et al (2013) Multilocus sequence analysis of *Streptococcus canis* confirms the zoonotic origin of human infections and reveals genetic exchange with *Streptococcus dysgalactae* subsp. *equisimilis*. J Clin Microbiol 51:1099–1109
14. DeWinter LM, Low DE, Prescott JF (1999) Virulence of *Streptococcus canis* from canine streptococcal toxic shock syndrome and necrotizing fasciitis. Vet Microbiol 70:95–110
15. Igwe EI, Shewmaker PL, Facklam RR, Farley MM, van Beneden C, Beall B (2003) Identification of superantigen genes specM, ssa and smeZ in invasive strains of beta-hemolytic group C and G streptococci recovered from humans. FEMS Microbiol Lett 229:259–264
16. Fulde M, Rohde M, Hitznmann A, Preissner KT, Nitsche-Schmitz DP, Nerlich A, Chhatwal GS, Bergmann S (2011) SCM, a novel M-like protein from *Streptococcus canis,* binds [mini-] plasminogen with high affinity and facilitates bacterial transmigration. Biochem J 434:523–525
17. Egesstein A, Frick IM, Olin AI, Bjorck L (2011) Binding of albumin promotes bacterial survival at the epithelial surface. J Biol Chem 286:2469–2476
18. Hare T, Frye RM (1938) Preliminary observation of an infection in dogs by beta-hemolytic streptococci. Vet Rec 50:213–218
19. Fulde M, Valentin-Weigand P (2013) Epidemiology and pathogenicity of zoonotic streptococci. Curr Top Microbiol Immunol 368:49–81
20. Lyskova P, Vydrazalova M, Mazurova J (2007) Identification and antimicrobial susceptibility of bacteria and yeasts isolated from healthy dogs and dogs with otitis externa. J Vet Med A Physiol Pathol Clin Med 54:559–563
21. TiKofsky LL, Zadoks RN (2005) Cross-infection between cats and cows: origin and control of *Streptococcus canis* mastitis in a dairy herd. J Dairy Sci 88:55–62
22. Richards VP, Zadoks RN, Pavinski Bitar PD et al (2012) Genome characterization and population genetic structure of the zoonotic pathogen, *Streptococcus canis*. BMC Microbiol 12:293
23. Lam MM, Claridge JE III, Young EJ, Mizuki S (2007) The other group G Streptococcus: increased detection of *Streptococcus canis* ulcer infection in dog owners. J Clin Microbiol 45:2327–2329
24. Bert F, Lambert-Zechovsky N (1997) Septicemia caused by *Streptococcus canis* in a human. J Clin Microbiol 35:777–779
25. Takeda N, Kikuchi K, Asano R et al (2001) Recurrent septicemia caused by *Streptococcus canis* after a dog bite. Scand J Infect Dis 33:927–928
26. Ohtaki H, Ohkusu K, Ohta H et al (2013) A case of sepsis caused by *Streptococcus canis* in a dog owner: a first case report of sepsis without dog bite in Japan. J Infect Chemother 19:1206–1209
27. Amsallem M, Lung B, Bnouleti C et al (2014) First reported human case of native mitral infective endocarditis caused by *Streptococcus canis*. Can J Cardiol 30:1462.e1–1462.e2
28. Bannister MF, Benson CF, Sweeney CR (1985) Rapid species identification of group C streptococci isolated from horses. J Clin Microbiol 21:524–526
29. Jensen A, Kilian M (2012) Delineation of *Streptococcus dysgalactae*, its subspecies, and its clinical and phylogenetic relationship to *Streptococcus pyogenes*. J Clin Microbiol 50:113–126

30. Timoney JF (2004) The pathogenic equine streptococci. Vet Res 35:397–4009
31. Holden MT, Heather Z, Pailott R et al (2009) Genomic evidence for the evolution of *Streptococcus equi*: host restriction, increased virulence, and genetic exchange with human pathogens. PLoS Pathog 5:e1000346
32. Walker AS, Jolley KA (2007) Getting a grip on strangles: recent progress towards improved diagnostics and vaccines. Vet J 173:492–501
33. Parmar J, Winterbottom A, Cooke F, Lever AM, Gaunt M (2013) Endovascular aortic stent graft infection with *Streptococcus equi*: the first documented case. Vascular 21:14–16
34. Priestnall S, Erles K (2011) *Streptococcus zooepidemicus*: an emerging canine pathogen. Vet J 188:142–148
35. Feng Y, Zhang H, Ma Y, Gao GF (2010) Uncovering newly emerging variants of *Streptococcus suis*, an important zoonotic agent. Trends Microbiol 18:124–131
36. Soedarmanto I, Pasaribu FH, Wibawan IW, Lammler C (1996) Identification and molecular characterization of serological group C streptococci isolated from diseased pigs and monkeys in Indonesia. J Clin Microbiol 34:2201–2204
37. Bordes-Benitez A, Sanchez-Onoro M, Suarez-Maroto A et al (2006) Outbreak of *Streptococcus equi* subsp. *zooepidemicus* on the island of Gran Canaria associated with the consumption of inadequately pasteurized cheese. Eur J Clin Microbiol Infect Dis 25:242–246
38. Kuusi M, Lahti E, Virolainen A, Hatakka M et al (2006) An outbreak of *Streptococcus equi* subspecies *zooepidemicus* associated with consumption of fresh goat cheese. BMC Infect Dis 6:36
39. Balter S, Benin A, Pinto SW et al (2000) Epidemic nephritis in Nova Serrana, Brazil. Lancet 355:1776–1780
40. Yuen KY, Seto WH, Choi CH, Ng W, Ho SW, Chau PY (1990) *Streptococcus zooepidemicus* [Lancefield group C] septicemia in Hong Kong. J Infect 21:241–250
41. Guszynski K, Young A, Levine SJ et al (2015) *Streptococcus equi* subsp. *zooepidemicus* infections associated with guinea pigs. Emerg Infect Dis 21:156–158
42. Pelkonen S, Lindahl SB, Suomala P et al (2013) Transmission of *Streptococcus equ*i subspecies *zooepidemicus* infection from horses to humans. Emerg Infect Dis 19:1041–1048
43. Alttreuther M, Lange C, Ho M, Hannula R (2013) Aortic graft infection and mycotic aneurism with *Streptococcus equi zoooepidemicus*: two cases with favorable outcome of antibiotic treatment. Vascular 21:6–9
44. Eyre DW, Kebnkre JS, Bowler IC, McBride SJ (2010) *Streptococcus equi* subspecies *zooepidemicus* meningitis a case report and review of the literature. Eur J Clin Microbiol Infect Dis 29:1459–1463
45. Reid HF, Bassett DC, Poon-King T, Zabriskie JB, Read SE (1985) Group G streptococci in healthy school-children and in patients with glomerulonephritis in Trinidad. J Hyg [London] 94:61–68
46. Haidan A, Talay SR, Rohde M et al (2000) Pharyngeal carriage of group C and G streptococci and acute rheumatic fever in an aboriginal population. Lancet 356:1167–1169
47. Preziuso S, Pinho MD, Attili AR et al (2014) PCR differentiation between *Streptococcus dysgalactiae* subsp. *equisimilis* strains isolated from humans and horses. Comp Immunol Microbiol Infect Dis 37:169–172
48. Agnew W, Barnes AC (2007) *Streptococcus iniae*: an aquatic pathogen of global veterinary significance and a challenging candidate for reliable vaccination. Vet Microbiol 122:1–15
49. Baums CG, Hermeyer K, Leimbach S et al (2013) Establishment of a model of *Streptococcus iniae* meningoencephalitis in Nile tilapia [*Oreochromis niloticus*]. J Comp Pathol 149:94–102
50. Keirstead ND, Brake JW, Griffin MJ, Halliday-Simmonds I, Thrall MA, Soto E (2014) Fatal septicemia caused by the zoonotic bacterium *Streptococcus iniae* during an outbreak in Caribbean reef fish. Vet Pathol 51:1035–1041
51. Weinstein MR, Litt M, KLertesz DA et al (1997) Invasive infection due to a fish pathogen, *Streptococcus iniae*. *S. iniae* study group. N Engl Med J Med 337:3016–3024

52. Sun JR, Yan JC, Yeh CY, Lee SY, Lu JJ (2007) Invasive infection with *Streptococcus iniae* in Taiwan. J Med Microbiol 56:1246–1249
53. Field HI, Buntain D, Done JT (1954) Studies on pig mortality. I. Streptococcal meningitis and arthritis. Vet Rec 66:453–455
54. Perch B, Kristjansen P, Skadhauge K (1968) Group R streptococci pathogenic for man: two cases of meningitis and one fatal case of sepsis. Acta Pathol Miocrobiol Scand 74:69–76
55. Arends JP, Zanen HC (1988) Meningitis caused by *Streptococcus suis* in humans. Rev Infect Dis 10:131–137
56. McLendon BF, Bron AJ, Mitchell CJ (1978) *Streptococcus suis* type II [group R] as a cause of endophthalmitis. Br J Ophthalmol 26:723–731
57. Wertheim HFL, Nghia HDT, Taylor W, Schultsz C (2009) *Streptococcus suis:* an emerging human pathogen. Emerg Infect Dis 48:617–625
58. Huong VTL, Ha N, Huy NT et al (2014) Epidemiology, clinical manifestations, and outcomes of *Streptococcus suis* infection in humans. Emerg Infect Dis 20:1105–1114
59. Staats JJ, Feder I, Okwumabua O, Chengappa MM (1997) *Streptococcus suis:* past and present. Vet Res Commun 21:381–407
60. Amass SF, SanMiguel P, Clark LK (1997) Demonstration of vertical transmission of *Streptococcus suis* in swine by genomic fingerprinting. J Clin Microbiol 35:1595–1596
61. Sanchez del Rey V, Fernandez-Garayzabal JF, Briones V et al (2013) Genetic analysis of *Streptococcus sui* from wild rabbits. Vet Microbiol 165:483–486
62. Camporese A, Tizianel G, Bruschetta G, Cruchiatti B, Pomes A (2007) Human meningitis caused by *Streptococcus suis*: the first reported case from north-east Italy. Infez Med 15:111–114
63. Chatzopoulou M, Voulgaridou I, Papalas D, Vasilliou P, Tsiakalou M (2015) Third case of *Streptococcus suis* infection in Greece. Case Rep Infect Dis 2015:505834
64. Fongcom A, Prusksakorn S, Netsirisawan P, Pngprasert R, Onisibud P (2009) *Streptococcus suis* infection: a prospective study in northern Thailand. Southeast Asian J Trop Med Public Health 40:511–517
65. Donsakul K, Dejthevvporn C, Witoonpanich R (2003) *Streptococcus suis* infection: clinical features and diagnostic pitfalls. Southeast Asian J Trop Med Publ Health 34:154–158
66. Higgins R, Gottschalk M (1990) An update on *Streptococcus suis* identification. J Vet Diagn Investig 2:249–252
67. Segura M, Zheng H, de Greeff A et al (2014) Latest developments on *Streptococcus suis*: an emerging zoonotic pathogen: part 1. Future Microbiol 9:441–444 part 2: 587–591
68. Gottschalk M, Segura M, Xu J (2007) *Streptococcus suis* infections in humans: the Chinese experience and the situation in North America. Anim Health Res Rev 8:29–45
69. Lun ZR, Wang QP, Chen XG, Li AX, Zhu XQ (2007) *Streptococcus suis*: an emerging zoonotic pathogen. Lancet Infect Dis 7:201–209
70. Nga TV, Nghia HD, Tu le TP et al (2011) Real-time PCR for detection of *Streptococcus suis* serotype 2 in cerebrospinal fluid of human patients with meningitis. Diagn Microbiol Infect Dis 70:461–467
71. Ishida S, Tien le HT, Osawa R et al (2014) Development of an appropriate PCR system for the reclassification of *Streptococcus suis*. J Microbiol Methods 107:66–70
72. Liu Z, Zheng H, Gottschalk M et al (2013) Development of multiplex PCR assays for the identification of the 33 serotypes of *Streptococcus suis*. PLoS One 8:e72070
73. King SJ, Leigh JA, Heath PJ et al (2002) Development of a multilocus sequence typing scheme for the pig pathogen *Streptococcus suis*: identification of virulent clones and potential capsular serotype exchange. J Clin Microbiol 40:3671–3680
74. Fittipaldi N, Xu J, Lacouture S et al (2011) Lineage and virulence of *Streptococcus suis* serotype 2 isolates from North America. Emerg Infect Dis 17(12):2239–2244
75. Zhu W, Wu C, Sun X et al (2013) Characterization of *Streptococcus suis* serotype 2 isolates from China. Vet Microbiol 166:527–534

76. Kouki A, Haataja S, Loimaranta V, Pulianinen AT, Nilsson UJ, Finne J (2011) Identification of a novel streptococcal adhesion P [SadP] protein recognizing galactosyl-alpha 1-4-galactose-containing glycoconjugate: convergent evolution of bacterial pathogens to binding of the same host receptor. J Biol Chem 286:38854–38864
77. Fittipaldi N, Segura M, Grenier D, Gottschalk M (2012) Virulence factors involved in the pathogenesis of the infection caused by the swine pathogen and zoonotic agent *Streptococcus suis*. Future Microbiol 7:259–279
78. Li W, Wan Y, Tao Z, Chen H, Zhou R (2013) A novel fibronectin-binding protein of *Streptococcus suis* serotype 2 contributes to epithelial cell invasion and in vivo dissemination. Vet Microbiol 162:186–194
79. Zhao YL, Zhou YH, Chen JQ et al (2015) Quantitative proteonomic analysis of sub-MIC erythromycin inhibiting biofilm formation of *S. suis* in vitro. J Proteomics 116:1–14
80. Dawei G, Liping W, Chengping L (2012) In vitro biofilm forming potential of *Streptococcus suis* isolated from human and swine in China. Braz J Microbiol 43:993–1004
81. Lecours MP, Segura M, Lachance C et al (2011) Characterization of porcine dendritic cell response to *Streptococcus suis*. Vet Res 42:72
82. Zhang A, Mu X, Chen B et al (2010) Identification and characterization of IgA1 protease from *Streptococcus suis*. Vet Microbiol 140:171–175
83. Houde M, Gottschalk M, Gagnon F, Van Calsteren MR, Segura M (2012) *Streptococcus suis* capsular polysaccharide inhibits phagocytosis through destabilization of lipid microdomain and prevents lactosylceramide-dependent recognition. Infect Immun 80:506–517
84. Lecours MP, Gottschalk M, Houde M, Lemire P, Fittipaldi N, Segura M (2011) Critical role for *Streptococcus suis* cell wall modifications and suilysin in resistance to complement-dependent killing by dendritic cells. J Infect Dis 204:919–929
85. He Z, Pian Y, Ren Z et al (2014) Increased production of suilysin contributes to invasive infection of the *Streptococcus suis* strain 05ZYH33. Mol Med Rep 10:2819–2826
86. Ferrando ML, Schultsz C (2016) A hypothetical model of host-pathogen interaction of *Streptococcus suis* in the gastrointestinal tract. Gut Microbes 7:154–162
87. Chen C, Tang J, Dong W et al (2007) A glimpse of streptococcal toxic shock syndrome from comparative genomics of *S. suis* 2 Chinese isolates. PLoS One 2:e315
88. Zheng H, Ye C, Segura M, Gottschalk M, Xu J (2008) Mitogenic effect contributes to increased virulence of *Streptococcus suis* sequence type 7 to cause streptococcal toxic shocke-like syndrome. Clin Exp Immunol 153:385–391
89. Ye C, Zheng H, Zheng J et al (2009) Clinical, experimental, and genomic differences between intermediate pathogenic, highly pathogenic, and epidemic *Streptococcus suis*. J Infect Dis 199:97–107
90. Tenenbaum T, Essmann F, Adam R et al (2006) Cell death, caspase activation, and HMGB1 release of porcine choroid plexus epithelial cells during *Streptococcus suis* infection in vitro. Brain Res 1100:1–12
91. Zheng P, Zhao YX, Zhang AD, Kang C, Chen HC, Jin ML (2009) Pathological analysis of the brain from *Streptococcus suis* type 2 experimentally infected pigs. Vet Pathol 46:531–535
92. Dominguez-Punaro Mde L, Segura M, Conteras I et al (2010) In vitro characterization of the microglial inflammatory response to *Streptococcus suis*, an important emerging zoonotic agent of meningitis. Infect Immun 78:5074–5085
93. Auger JP, Christodoulides M, Segura M, Xu J, Gottschalk M (2015) Interactions of *Streptococcus suis* serotype 2 with human meningeal cells and astrocytes. BMC Res Notes 8:607
94. Sun Y, Li N, Zhang J et al (2016) Enolase of *Streptococcus suis* serotype 2 enhances blood-brain barrier permeability by inducing IL-8 release. Inflammation 39:718–726
95. Wewer C, Seibert A, Wolburg H et al (2011) Transcellular migration of neutrophil granulocytes through the blood-cerebrospinal fluid barrier after infection with *Streptococcus suis*. J Neuroinflammation 8:51
96. Eisenberg T, Hudemann C, Hossain HM et al (2015) Characterization of five zoonotic *Streptococcus suis* strains from Germany, including one isolate from a recent fatal case of streptococcal toxic shock-like syndrome in a hunter. J Clin Microbiol 53:3012–3015

97. Smith TC, Capuano AW, Boese B, Myers KP, Gray GC (2008) Exposure to *Streptococcus suis* among US swine workers. Emerg Infect Dis 14:1925–1927
98. Goyette-Desjardins G, Auger J-P, Xu J, Segura M, Gottschalk M (2014) *Streptococcus suis,* an important pig pathogen and emerging zoonotic agent an update on the worldwide distribution based on serotyping and sequencing typing. Emerg Microbes Infect 3:e45. doi:10.1038/emi.2014.45
99. Nghia HD, le TP T, Wolbers M et al (2011) Risk factors of *Streptococcus suis* infection in Vietnam. A case-control study. PLoS One 6:e17604
100. Ip M, Fung KS, Chi F et al (2007) *Streptococcus suis* in Hong Kong. Diagn Microbiol Infect Dis 57:15–20
101. Ngo TH, Tran TB, Tran TT et al (2011) Slaughterhouse pigs are a major reservoir of *Streptococcus suis* serotype 2 capable of causing human infection in southern Vietnam. PLoS One 6:e17943
102. Soares TC, Gottschalk M, Lacourture S et al (2015) *Streptococcus suis* in employees and the environment of swine slaughterhouse in Sao Paolo, Brazil: Occurrence, risk factors, serotype distribution, and antimicrobial susceptibility. Can J Vet Res 79:279–284
103. Hatrongjit R, Kerdsin A, Gottschalk M et al (2015) First human case report of sepsis due to infection with *Streptococcus suis* serotype 31 in Thailand. BMC Infect Dis 15:392
104. Takeuchi D, Kerdsin A, Pienpringam A et al (2012) Population-based study of *Streptococcus suis* infection in humans in Phayao Province in northern Thailand. PLoS One 7:e31265
105. Nakayama T, Takeuchi D, Matsumura T, Akeda Y, Fujinaga Y, Oishi K (2013) Alcohol consumption promotes the intestinal translocation of *Streptococcus suis* infection. Microb Pathog 65:14–20
106. Yu H, Jing H, Chen Z et al (2006) Human *Streptococcus suis* outbreak, Sichuan, China. Emerg Infect Dis 12:912–914
107. Mai NT, Hoa NT, Nga TV et al (2008) *Streptococcus suis* meningitis in adults in Vietnam. Clin Infect Dis 46:659–667
108. Kerdsin A, Dejsirilert S, Puangpatra P et al (2011) Genetic profile of *Streptococcus suis* serotype 2 and clinical features of clinical infection in humans, Thailand. Emerg Infect Dis 17:835–842
109. van Samkar A, Brouwer MC, Schultsz C, van der Ende A, van de Beek D (2015) *Streptococcus suis* meningitis: a systematic review and meta-analysis. PLoS Negl Trop Dis 9:e0004191
110. Heckenberg SG, de Gans J, Brouwer MC et al (2008) Clinical features, outcome, and meningococcal genotype in 258 adults with meningococcal meningitis: a prospective cohort study. Medicine (Baltimore) 87:185–192
111. Wangsomboonsiri W, Luksananun T, Saksornchai S, Ketwong K, Sungkanuparph S (2008) *Streptococcus suis* infection and risk factors for mortality. J Infect 57:392–396
112. Dominguez-Punaro MC, Koedel U, Hoegen T, Demel C, Klein M, Gottschalk M (2012) Severe cochlear inflammation and vestibular syndrome in an experimental model of *Streptococcus suis* infection in mice. Eur J Clin Infect Dis 1:2391–2400
113. Tsai HY, Liao CH, Liu CY, Huang YT, Teng LJ, Hsueh PR (2012) *Streptococcus suis* infection in Taiwan, 2000-2011. Diagn Microbiol Infect Dis 74:75–77
114. Chen L, Song Y, Wei Z, He H, Zhang A, Jin M (2013) Antimicrobial susceptibility, tetracycline and erythromycin resistance genes, and multilocus sequence typing of *Streptococcus suis* isolates from diseased pigs in China. J Vet Med Sci 75:583–587
115. Zhang C, Zhang Z, Song L et al (2015) Antimicrobial resistance profile and genotype of *Streptococcus suis* capsular type 2 isolated from clinical carrier sows and diseased pigs in China. Biomed Res Int 2015:284303
116. Meng X, Shi Y, Ji W et al (2011) Application of a bacteriophage lysine to disrupt biofilms formed by the animal pathogen *Streptococcus suis*. Appl Environ Microbiol 77:8272–8279
117. Lebel G, Piche F, Frenette M, Gottschalk M, Grenier D (2013) Antimicrobial activity of nisin against the swine pathogen *Streptococcus suis* and its synergisyic interaction with antibiotics. Peptides 50:19–23
118. Segura M (2015) *Streptococcus suis* vaccines: candidate antigens and progress. Expert Rev Vaccines 14:1587–1608

119. Baums CG, Bruggemann C, Kock C et al (2010) Immunogenicity of an autogenous *Streptococcus suis* bacterin in preparturient sows and their piglets after weaning. Clin Vaccine Immunol 17:1589–1597
120. Kock C, Beineke A, Seirz M et al (2009) Intranasal immunization with a live *Streptococcus suis* isogenic of mutant elicited suilysin neutralization titers but failed to induce opsonizing antibodies and protection. Vet Immunol Immunopathol 132:135–145
121. Yao X, Li M, Wang J et al (2015) Isolation and characterization of a native avirulent strain of *Streptococcus suis* serotype 2: a perspective for vaccine development. Sci Rep 5:9835
122. Li Y, Martinez G, Gottschalk M et al (2006) Identification of a surface protein of *Streptococcus suis* and evaluation of its immunogenic and protective capacity in pigs. Infect Immun 74:305–312
123. Zhou Y, Wang Y, Deng L et al (2015) Evaluation of the protective efficacy of four novel identified membrane associated proteins of *Streptococcus suis* serotype 2. Vaccine 33:2254–2260
124. Charland N, Jacques M, Lacouture S et al (1997) Characterization and protective activity of a monoclonal antibody against a capsular epitope shared by *Streptococcus suis* serotypes 1, 2 and ½. Microbiology 143:3607–3614
125. Calzas C, Lemiere P, Auray G, Gerdts V, Gottschalk M, Segura M (2015) Antibody response specific to the capsular polysaccharide is impaired in *Streptococcus suis* serotype 2-infected animals. Infect Immun 83:441–453
126. Roma-Rodrigues C, Alves-Barroco C, Raposo LR et al (2016) Infection of human keratinocytes by *Streptococcus dysgalactiae* subspecies *dysgalactiae* isolated from milk of the bovine udder. Microbes Infect 18:290–293
127. Dumke J, Hinse D, Vollmer T, Knabbe C, Dreier J (2014) Development and application of a multilocus sequence typing scheme for *Streptococcus gallolyticus* subsp. *gallolyticus*. J Clin Microbiol 52:2472–2478
128. Goyette-Desjardins G, Clazas C, Shiao TC et al (2016) Protection against *Streptococcus suis* serotype 2 infection using capsular polysaccharide glycoconjugate vaccine. Infect Immun 84:2059–2075

Chapter 11
New and Emerging Parasitic Zoonoses

11.1 Introduction

The global burden of zoonotic parasites is huge but not well characterized and the annual human incidence or prevalence worldwide [not been calculated] is likely in the hundreds of millions. These parasites are widespread in wildlife, domestic animals used for food, and pets. Transmission of zoonotic parasites can be from the food chain, contamination of water/food through fecal/soil contamination, and insect vectors. Zoonotic parasites are not limited to tropical and subtropical regions but are also prevalent in the temperate hemisphere, and no country in the world is spared. Globalization of these uncommon conditions are facilitated by ease of international travel, large migration of refugees fleeing regional conflicts and wars, and change in dynamic interactions between arthropods and their hosts, including humans. These latter factors include climate change, urbanization and deforestation to change demographics in developed and developing countries, impact of economic crisis, and increased movement of people and animals following major catastrophes [1].

This chapter will review existing parasitic zoonosis that rarely affected humans but has been emerging as a problem, such as sarcocystosis, and parasites that were considered only animal pathogens but have been increasingly recognized to infect humans, i.e., *Baylisascaris procyonis*, *Dirofilaria* species, *Onchocerca lupi*, and *Trypanosoma* species. See Table 11.1 for a summary of these conditions.

11.2 Baylisascariasis

Baylisascariasis is a zoonotic infection caused by the raccoon roundworm, *Baylisascaris procyonis* [2]. Raccoons [*Procyon lotor*] are the definitive host and the adult worms live in the small intestine, where females pass embryonated eggs in the feces. The life cycle is maintained in young raccoons by the ingestion of embryonated

I.W. Fong, *Emerging Zoonoses*, Emerging Infectious Diseases of the 21st Century,
DOI 10.1007/978-3-319-50890-0_11

Table 11.1 Summary of new and emerging parasitic zoonoses

Parasite	Final host	Distribution	Transmission	Disease	Diagnosis	Treatment
1. *B. procyonis* [baylisascariasis]	Raccoon	North and Central America	Fecal-oral	CNS	ELISA	Albendazole
2. *D. immitis*	Canines	Worldwide	Mosquito	Lung lesion	Histology	Surgery
D. repens [dirofilariasis]	same	Europe, Asia, Africa	same	Subcut. nodule	same	same
3. *O. lupi* [ocular onchocerciasis]	Canines	Europe, US	Blackflies	Eye nodule, spinal cord	Histology	Surgery, iver., doxycycline
4. *T. evansi*	Camels	Africa, Asia, Latin Am.	Biting insects	Fever	Blood smear	Suramin
T. lewisi [zoonotic trypanosomiasis]	Rodents	Worldwide, Asia	Fleas	Fever	same	Suramin, melar.
5. *S. hominis*/*suihominis*	Cattle, pigs	Europe, Asia, Latin Am.	Beef/pork	Enteritis	Stool microscopy	None
S. nesbitti/others [sarcocystosis]	Mammals, reptiles	Malaysia, Asia	Fecal-oral	Myositis	Muscle biopsy	Co-trimoxazole/steroid

Am. America, *B Baylisascaris*, *D Dirofilaria*, *O Onchocerca*, *iver.* ivermectin, *melar.* melarsoprol, *S* sarcocystis, *T Trypanosoma*

eggs from the ground [or contaminated feed] or through predation of small paratenic hosts [mice, birds, etc.] [3]. A wide variety of vertebrate hosts can become infected with the larval form to become intermediate hosts [150 species], including humans, rodents, lagomorphs, carnivores, and birds [3]. Embryonated eggs ingested by paratenic hosts result in the release of third-stage larvae [300 μm] that penetrate the intestinal wall and migrate via the blood stream to various organs and tissues, including the brain and eyes [3, 4]. *Baylisascaris* belong to the family Ascarididae, phylum Nematoda, and one of several genera, i.e., *Ascaris*, *Parascaris*, *Toxascaris*, and *Toxocara* [3]. There are several *Baylisascaris* species that can cause visceral larva migrans in animals and humans, including species infecting skunks, badgers, and other animals, but *B. procyonis* is the most prevalent and pathogenic [3, 5].

Raccoons defecate in certain areas [base of large trees, tree stumps, woodpiles, pile of debris, barn loft] to form a communal latrine that is often invaded by other animals seeking feed [birds and rodents] and accidentally by children. Infected animals can shed 20,000 eggs/g of feces [2] and *B. procyonis* eggs can accumulate in the latrines and remain viable for extended periods [6]. Children may accidentally ingest large number of *B. procyonis* eggs while playing in the areas used as raccoon latrines. Raccoons are abundant in the urban environment and their latrines [from 1 to 6] have been found in 51% of suburban backyards, with 23% found to contain *B. procyonis* eggs [7]. Factors associated with raccoon latrines included proximity to forests or natural areas, food sources such as pet food, garbage, and bird feed.

11.2.1 Epidemiology of Human Infection with B. procyonis

The extent and prevalence of human baylisascariasis are unknown and it is likely that only the severe infections are recognized and reported. Raccoons are native to North and Central America, extending from Canada to Panama, and the range of *B. procyonis* expands along with its raccoon host. Most of the cases of human baylisascariasis are reported from areas with the highest prevalence of the parasite in raccoons. The regions surrounding the Great Lakes account for 68% of the reported human *B. procyonis* infections, including the midwestern [36%] and northeastern [24%] regions of the USA and Ontario Province, Canada [8%], and 28% from the western regions of the USA and Canada [British Columbia] [8].

Raccoons were initially introduced into Germany and the former Soviet Union for hunting and the fur trade, but the animals have spread across Europe to colonize 20 European countries [9]. Germany has over 100,000 raccoons in the wild, with a prevalence of 71% infection with *B. procyonis* [3]; and baylisascariasis neuroretinitis has been described in a native German [10]. Since 1977 more than 20,000 raccoons were introduced into Japan as pets, many of which escaped and flourished in the wild, and Japan has now reported visceral larva migrans with *B. procyonis* in macaques from a zoo and wild animals [11, 12]. Importation of raccoons also occurred in the past in China, and they have spread within the country; as a result infected raccoons have been found throughout the wildlife parks [13]. Human cases of baylisascariasis have at present only been reported in the USA, Canada, Germany, and Austria, but animal invasive larva migrans are known in Europe and Japan [8]. However, many more cases of infections in expanded areas are predicted in the future wherever raccoons are present. The aggressive invasive nature of raccoons with adaptation to human food and habitat has created opportunities for infections closer to homes, as raccoon feces can be found on porches, decks, gardens, sandboxes, and children play areas. Domestic animal and exotic pets can be infected with *B. procyonis* and potentially expand environmental contamination beyond raccoon latrines. Infection of wild animals maybe associated with extinction of vulnerable species, as believed to have occurred with Allegheny woodrat [*Neotoma magister*] from baylisascariasis [14].

11.2.2 Pathogenesis of Baylisascariasis

The pathologic features of invasive *B. procyonis* are mainly secondary to mechanical injury from migrating larvae [which grow in size during this process] and inflammatory reaction to secreted or excreted antigens [8]. During migration the mechanical damage results in characteristic necrotic linear tracts in the tissues, i.e., in the central nervous system [CNS] called malacic tracks [15]. Hematogenous dissemination of the larvae has been documented in numerous organs and tissues and in a large variety of animals [3, 15]. The tracts of the migrating larvae consist of hemorrhagic necrosis and edema, with eosinophils and other leucocyte infiltration. Tissue injury is also caused by the inflammatory response to excretory-secretory antigens such as migratory enzymes, shed surface proteins, and metabolic waste products which stimulate the host immune response. Molecules generated by the inflammatory response, especially eosinophil products from degranulation of reactive cells in lesions, are toxic to tissue [16]. Also high levels of eosinophil-derived neurotoxin and major basic protein were found in the cerebrospinal fluid [CSF] of disseminated baylisascariasis in children, which may be related to the neuropathology and clinical signs [17]. Granulomas of the CNS can be found in *B. procyonis* infection, especially in those with long-standing or chronic infections, or low-grade infections with a few larvae with mild or no symptoms [15]. The larvae may become walled off in localized granulomas and with time undergo post-inflammatory degeneration. There is recent evidence to suggest that parasitic infection of the brain [exemplified by toxocariasis] can lead to development and progression of neuropsychiatric and neurodegenerative disorders [18]. This may also apply to baylisascariasis as the mechanisms of brain injury are probably similar and CNS involvement can lead to seizures, cognitive deficits, and brain atrophy [8].

11.2.3 Clinical Aspects of Baylisascariasis

Invasive infection with *B. procyonis* can manifest as three clinical syndromes: neural larva migrans [NLM], visceral larva migrans [VLM], and ocular larva migrans [OLM] [8]. Toxocara species, especially *T. canis*, are the most important and frequent cause of VLM and OLM but does not usually present with overt CNS disease. Baylisascariasis VLM is typically related to heavy or repeated infection, with migration of the larvae through the viscera [15] and this is similar to toxocariasis. Migration of the larvae to the CNS results in NLM and the clinical manifestation depends on the larval load and areas damaged by the migration [8, 15]. Severe neurological disease is associated with ingestion of a large number of embryonated eggs, typically seen in small children with pica or from consumption of contaminated material with raccoon feces [8, 19]. The majority of *B. procyonis* infections are from a few larvae and neurologically patients are intact. In contrast, OLM can result from a single larva migrating to the eye [15, 20].

Recent tabulation by Graeff-Teixeira et al. [8] indicates that there has been at least 25 human cases of NLM and VLM due to *B. procyonis*, and the first human infection [with infantile hemiplegia] was reported in 1975 [21]. However, subclinical infections are more frequent than clinical disease, as reflected by the results of a few epidemiological seroprevalence surveys, which showed that 30 of 389 [7.7%] asymptomatic children and 7 of 43 [16.3%] raccoon trappers were seropositive for *B. procyonis* infection [8]. Also the majority of mild infections are probably undiagnosed or misdiagnosed. Visceral larva migrans secondary to *B.* procyonis, compared to toxocariasis as previously defined in 1969 [22], have a more rapid aggressive course and more common neurological involvement [8]. In the typical *T. canis* infection, the larvae migrate extensively and last for years walled off in the tissues of its host. In the classic VLM due to *Toxocara*, the signs and symptoms are related to the organs being invaded, such as the intestines, lungs, liver, heart, and adjacent tissues. Thus, patients may present with abdominal pain, pneumonitis, and hepatomegaly with persistent high eosinophils and hypergammaglobulinemia [22, 23]. Whereas, in *B. procyonis*, the larvae migrate through the liver and lungs rapidly to the CNS, and the primary presentation is usually with neurological manifestation [8]. However, patients can present initially with cough and low-grade fever before developing CNS symptoms/signs and rarely mild pulmonary infiltrates on presentation with neurological signs [24]. Some patients may have mild hepatomegaly early with respiratory symptoms and eosinophilia preceding CNS disease.

In animal models of *B. procyonis* infection in mice and nonhuman primates, transient respiratory symptoms develop 2–5 days after inoculation and larvae can be found in the lungs with migratory lesions [25, 26]. Early migratory lesions in the lungs account for the respiratory symptoms and usually precede CNS signs, similar to the course seen in some patients. NLM is related to the invasion of the CNS but only a few larvae [5–7%] usually enter the CNS [4, 27]. The clinical manifestations of NLM depend on the larval load in the CNS, area of involvement and inflammatory reaction, and extent of tissue damage and may vary from mild to severe disease.

The brain is usually most severely affected, although cervical and thoracic cord lesions can be present [8]. The incubation period and acuteness of disease are largely related to the larval load and vary from 1 to 21 days with a mean of 7 days. Low burden of larvae may take several weeks for disease manifestation [8].

Children with heavy burden of infection may present with sudden onset of irritability, lethargy, ataxia, loss of fine motor skills, inability to stand or walk independently, weakness, impaired vision, seizures, incontinence, and partial paralysis [8, 19, 28]. Neck stiffness, opisthotonos, stupor, and coma may be present leading to death. Blindness may be the result of lesions in the visual cortex or neuroretinitis. Low levels of larval burden may present with subacute to chronic manifestations with nonspecific signs with or without eosinophilia more typical for toxocariasis. Gradual progression with abnormal behavior, motor weakness, incoordination, visual and speech disturbance, and subsequently decreased level of consciousness and meningismus may occur. Fever was uncommonly present in only 24% of patients, and subtle CNS disease is also present in natural infection of nonhuman primates [3, 8]. In most patients with NLM, there are high eosinophils in the blood and cerebrospinal fluid [CSF], ranging from 4 to 68%, and magnetic resonance imaging of the brain may reveal diffuse, white matter lesions with periventricular and cerebellar distribution consistent with disseminated encephalomyelitis [24]. Other change on imaging may include cerebral edema, generalized cerebral atrophy in chronic cases with ventricular dilatation, and spinal cord lesion with enhancement.

Ocular larva migrans [OLM] is caused by migration of ascarids to the tissues of the eye by *Toxocara* spp. and *Baylisascaris* [15]. OLM usually occurs without any other manifestation of VLM and NLM, from dissemination of a single larva to the eye. Symptoms usually occur with sudden decreased vision unilaterally with photophobia and may progress to blindness. The disease is characterized by chorioretinitis, vitritis, intraocular granulomas, diffuse unilateral subacute neuroretinitis [DUSN], and occasionally panophthalmitis and retinal detachment [29, 30]. Features of DUSN include papillitis, vitritis, recurrent crops of gray-white retinal lesions, and loss of vision and with progression can result in optic nerve atrophy, narrowing of vessels, and epithelial degeneration. Some patients presenting with NLM may have concomitant invasion of the eye with typical signs of DUSN. While VLM and NLM tend to affect infants <2 years of age, isolated OLM usually present in older children and adults [8, 31].

11.2.4 Diagnosis of Baylisascariasis

The diagnosis of baylisascariasis can be suspected based on the clinical features, exposure risk to raccoon feces, and usually the presence of eosinophilia. The major differential diagnosis includes toxocariasis, but the initial presentation may mimic viral upper respiratory infection and aseptic meningoencephalitis. Eosinophilic pleocytosis of the CSF should raise suspicion of baylisascariasis and serology can confirm the diagnosis. Enzyme-linked immunosorbent assay [ELISA] and immunoblotting assay for IgG antibodies from serum and CSF are very sensitive and also can be performed on vitreous fluid [32, 33]. Excretory-secretory antigens of *B. procyonis* are used in the ELISA and Western blot assays, with selection of the most specific components in the 30–45-kDa fractions [34]. OLM or DUSN can be made clinically by the characteristic ocular findings and the direct visualization of intraocular parasites, typically large larvae [1500–2000 μm in length compared to 350–450 μm in *Toxocara* spp.] which maybe mobile [30]. Larvae may also be found on tissue biopsy [i.e., brain] and the diagnosis can be confirmed by immunofluorescence staining of frozen sections for third-stage larvae [15]. Real-time PCR has also been used to identify *Baylisascaris* in fixed tissues [35].

11.2.5 Treatment of Baylisascariasis

Clinical suspicion is essential for early and rapid diagnosis with immediate treatment in order to limit the tissue damage and improve the outcome. Patients with neurological signs already have significant CNS damage and may worsen quickly. In cases with heavy infection, the prognosis is poor even with treatment [8]. The aim of effective treatment is to eradicate the larvae and control the inflammatory response. Albendazole is considered the agent of choice for human baylisascariasis, at 20–40 mg/kg or 400 mg twice daily for 3–4 weeks but a minimum of 10 days [8]. In mice experimentally infected with *B. procyonis*, albendazole at 25–50 mg/kg of weight given 1–10 days post-infection resulted in 100% protection against NLM compared to death in all untreated animals [36, 37]. Albendazole is well absorbed with oral administration, produces good tissue concentration, and penetrates the blood-brain barrier [19]. While ivermectin is also larvicidal, it is not effective for NLM as it does not cross the blood-brain barrier and clinical failures with recovery of the larvae from brain has been reported in animals and humans [8, 38].

Concomitant corticosteroid is usually used and is recommenced to prevent worsening symptoms/signs from the inflammation associated with death of the larvae during albendazole therapy. Although this is of unproven benefit and experimental data are inconclusive [27], there is evidence to support the use of steroids in other parasitic CNS infections such as *toxocariasis*, eosinophilic meningitis from angiostrongyliasis, and neurocysticercosis, at the time of antihelminthic therapy [23, 39, 40]. Early and aggressive therapy is recommended based on clinical and epidemiological data, even before serological confirmation to limit further neurological damage in suspected cases of NLM [8]. Preventive therapy with albendazole for 10 days is also recommended for suspected ingestion of material from raccoon latrine, to prevent CNS invasion and avoid serious brain damage [41]. This is reasonable since albendazole is safe and well tolerated.

Localized OLM/DUSN without evidence of VLM or NLM can be treated with laser photocoagulation with corticosteroid, when larvae are visualized on fundoscopy away from critical areas [8]. This procedure has been shown to kill the larvae in the eye and is best performed in early disease [29, 42]. Treatment with albendazole is not considered necessary unless there is associated VLM or NLM.

11.2.6 Control and Prevention of Baylisascariasis

Measures to control and prevent baylisascariasis have involved steps to reduce environmental contamination of *B. procyonis* eggs and larvae, and education of the local communities of the risk and dangers of infection and means of avoidance. It appears that these measures have been taken mainly after recognition of infection in children of suburban/urban areas, rather than proactively wherever raccoons are prevalent. Understanding the epidemiology of *B. procyonis* in raccoons is important to develop strategies to control the spread of this zoonosis to humans. The prevalence of infection in wild animals can be assessed by necropsies of captured raccoons and examination of their intestines for parasites. In one such study from Ontario [Canada], 38% of raccoons were found to be infected with the roundworm from March to October, and young raccoons had greater number of parasites than adults [43]. Environmental contamination of raccoon roundworm eggs/larvae has been assessed by examination of material from raccoon latrines by microscopy of fecal flotation solution [7]. Risk of transmission to humans and animals can be estimated from the number of raccoon latrines, size and density of parasites per latrine, or scat in the backyards of neighborhoods [44].

Effective control of raccoon roundworms is best accomplished by treatment, removal, and translocation of infected raccoons in the area. Anthelmintic baiting of raccoons is an effective means of lowering the risk of transmission. This can lower the eggs in the environment and the prevalence of larvae in intermediate animal hosts [45]. Infected pets should also be treated and raccoon latrines decontaminated. Materials from latrines are best removed and burned, as the eggs are relatively resistant to household bleach. Embryonated eggs can be rendered nonviable to temperature of 62°C for less than a minute; thus, pouring hot or boiling water over contaminated material should be effective for decontamination [46].

More public education of the population in raccoon-infested communities is warranted, about the disease and mode of transmission. Children should be taught to avoid raccoon latrines and hand sanitization after playing in their backyards. Other simple measures include removal of bird feeders and protection of garbage and pet food from exposure to raccoons. No vaccine is available and not likely to be developed as the disease in humans is still limited in extent.

11.3 Dirofilariasis Emerging in Humans

Dirofilariasis is a zoonotic nematode of domestic and wild carnivores that can be transmitted to humans by infected mosquitoes. Canine filariasis or dog heartworm was first described in Europe in the seventeenth century [47]. *Dirofilaria* is a genus of the class Nematoda of the superfamily Filarioidea, and members have very long filiform bodies, with striated cuticles [48]. The genus consists of two subgenera: *Dirofilaria* [*Dirofilaria*] with one species, the canine heartworm [*D. immitis*], distributed worldwide, and *Dirofilaria* [*Nochtiella*], with over 20 species distributed regionally of which the most important is *D.* [*Nochtiella*] *repens* [49]. *D. immitis* produces canine and feline cardiopulmonary dirofilariasis, and *D. repens* is responsible for canine and feline subcutaneous dirofilariasis. In humans *D. immitis* causes pulmonary dirofilariasis and *D. repens* produces subcutaneous and ocular dirofilariasis [49, 50].

11.3.1 Biology of Dirofilaria

Both *D. immitis* and *D. repens* also belong to the Onchocercidae family and are transmitted by mosquitoes, primarily by peridomestic *Culex* and *Aedes* species but also by *Anopheles* and *Mansonia* species [51]. *Dirofilaria* species can infect numerous mammalian species but are best adapted to wild canine species and domestic dogs, which function as the main reservoir or definitive hosts, and wild felines, domestic cats, and humans are less suitable hosts and are intermediate or incidental hosts [52]. Female mosquitoes transmit the infective L3 larvae during blood feeding, which subsequently enters the circulation. The larvae undergo molting to produce the preadult worms that reach the pulmonary artery and canine heart about 70–85 days after inoculation. Maturity is achieved about 120 days post-inoculation and the adult *D. immitis* female measures 250–300 mm in length and the males 120–200 mm [53]. Microfilariae are produced by the female worms about 6–9 months post-infection that live in the blood stream and are 290–330 μm in length. Adults can live for over 7 years and the microfilariae up to 2 years [54]. Some infected dogs do not produce microfilariae, possibly from host immune response or aging female worms. In cats the adult *D. immitis* worms are shorter than those in dogs, have a shorter life expectancy [about 2 years], and generally do not produce microfilariae.

D. repens adult worms reside in the subcutaneous tissues of the definitive hosts, but may be found in the muscle fasciae and the abdominal cavity [53]. They mature 6–9 months post-infection and the adult worms are smaller than those of *D. immitis*: females 100–170 mm and males 50–70 mm long. The microfilariae also reside in the blood stream and also can be found in feline hosts which may act as reservoir for this species and are about 350–385 μm in length [53].

There is evidence that an arthropod endosymbiont bacteria, *Wolbachia*, present in other species of filariae, such as *Onchocerca volvulus* [55], also reside intracellularly in *Dirofilaria* species in a symbiotic relationship [56]. Moreover, studies suggest that *Wolbachia* bacteria are involved in molting and embryogenesis of filariae [57]. Recent trials in patients infected with *O. volvulus* showed that therapy with doxycycline aimed at *Wolbachia* bacteria eradication resulted in effective clearing of microfilariae in the skin in patients with persistent microfilaridermia and enhanced killing of adult worms on antiparasitic agent [58].

11.3.2 Dirofilariasis in Animals

D. immitis infect wild and domestic canines and felines in tropical and temperate regions throughout the world, and *D. repens* is confined to countries of the Old World, Europe, Asia, and Africa [49]. However, in the last 15 years, despite efforts to control infection in dogs, canine dirofilariasis has been spreading into new areas previously considered free of the disease. This has resulted in increased recognition of dirofilariasis in both felines and humans in endemic and neighboring regions. Expansion of dirofilariasis in the world has been attributed to combination of factors: increased recognition, climate change, and increased range of specific vectors of *Dirofilaria* species in some regions [59]. Expansion of dirofilariasis to central and northern Europe [previously virgin to these nematodes] may be due to the introduction of the Asian mosquito [*Aedes albopictus*] into Europe several years ago and increased vector competence of *Culex pipiens* [49].

The immune response of the host and pathogenesis of the disease in animals appear to be complex, as it involves reactions to the worm antigens as well as to the *Wolbachia* symbiont. In general, immune response to filarial infection invokes both Th1 and Th2 reactions. Studies in mice indicate that *Wolbachia* induce preferentially a Th1 response during the filarial infection, and the nematode antigens drive the Th2 response [60]. There is evidence that dead adult worms and microfilariae release symbiotic *Wolbachia* bacteria in the blood, which interact with host tissues. IgG antibodies specific for Wolbachia surface protein can be detected in infected dogs, cats, and humans with dirofilariasis [59]. Inflammation results from the adult worms dying driven by the proinflammatory response to the symbiotic bacteria surface protein. During *D. immitis* infection, there is inflammation and vascular endothelial disorganization [59]. There is also evidence that *D. immitis* secretes substances that impair vascular endothelial function [via the prostaglandin pathway], and this partly explains the inability of infected dogs to adapt to physical exertion [61].

Heartworm disease in dogs, cats, and ferrets is caused by *D. immitis* worms. The initial damage occurs in the pulmonary arteries with endothelial dysfunction and narrowing from endarteritis that can lead to pulmonary hypertension, right heart strain and congestive heart failure, and chronic progression with thromboembolism and may be fatal [59, 62]. *D. immitis* can also cause membranous glomerulonephritis with renal dysfunction and occasionally vena cava syndrome especially in small dogs. Occasionally, some dogs with occult infection develop eosinophilic pneumonia with respiratory distress, and dirofilariae sometimes result in aberrant lesions in the brain, liver, eye, and peritoneum [59].

Infection in cats with *D. immitis* primarily involves the lungs and the animal may be asymptomatic but can develop eosinophilic pneumonitis. Hyperacute infection can result in severe respiratory insufficiency and gastrointestinal, cardiovascular, and neurological signs, with dyspnea, cough, vomiting, and diarrhea [62]. Acute infection in the cat can lead to chronic dirofilariasis or become asymptomatic but later develop into chronic disease. Chronic dirofilariasis is mainly associated with respiratory and gastrointestinal symptoms with gradual cachexia [62].

Infection with *D. repens* is associated with subcutaneous nodules and conjunctivitis in dogs. Infection with this species is often subclinical and never become detected. Diverse dermatological manifestations can occur including pruritus; multifocal nodular dermatitis, which is usually present on the face; and prurigo papularis dermatitis [63]. Extradermal involvement may include conjunctivitis [46%], fever, vomiting, lethargy, and lymphadenopathy. Massive infection with adult worms and microfilariae rarely results in invasion of the spleen, liver, kidney, lungs, heart, brain, and vitreous of the eye [59].

11.3.3 Human Dirofilariasis

Human dirofilariasis can occur wherever the infection is endemic in dogs and cats, but zoonotic diseases are most common in the southern USA and the coastal Mediterranean regions of Europe. Canine heartworm infection has been expanding in North America since the 1950s from hyperendemic regions in the south [coastal area of the Mississippi river] to northern states. Presently, dog heartworm is present in 50 states of the USA and the risk of human infection has increased [64]. This has been attributed to greater movements of heavily infected dogs for hunting, breeding, and dog shows and increased awareness by veterinarians [65]. Climate change may also be a contributory factor, as following the Hurricane Katrina in August 2005 heavily infected dogs [prevalence 34–51%] from Louisiana were exported to northern states and Canada [66].

Both *D. immitis* and *D. repens* infections were mainly found in the southern regions of Europe, with the greatest endemic area in the Po River Valley in Italy, until the latter half of the twentieth century [67]. However, the risk of infection has been spreading after introduction of the Pet Travel Scheme in 2000, allowing easier movements of companion animals throughout the European Union [68]. Besides increased movement of dogs, climate change is considered to have played a major role in the spread of zoonotic infections, including dirofilariasis, in Europe. Vector-borne pathogens are sensitive to climate and there is evidence that climate change can influence their increased intensity and incidence [69]. Climate change [temperature, precipitation, and humidity] determines the abundance of mosquitoes and mosquito-borne diseases and may influence the development and transmission of *Dirofilaria* [70]. *Aedes albopictus* was transported into Italy in 1990 and has spread throughout Europe as far north as the Netherlands [71], and this mosquito may have contributed to the spread of dirofilariasis.

Transmission of *Dirofilaria* to humans and frequency depend on the prevalence and burden of disease in wild and domestic canines with microfilariae and the appropriate mosquito vectors. In North America *D. immitis* is the primary cause of dirofilariasis or human pulmonary dirofilariasis [HPD]. On rare occasions other species can cause human disease from mosquito or blackfly bites associated with other zoonotic reservoirs [raccoons, bears, porcupines, and wild felids], to produce subcutaneous nodules or ocular disease: *D.* [*Nochtiella*] *tennis*, *D.* [*Nochtiella*] *ursi*,

D. [*Nochtiella*] *subdermata*, and *D.* [*Nochtiella*] *striata* [72]. HPD most commonly are found as incidental findings on chest radiograph as a "coin lesions" and result in further investigations to exclude a malignant tumor. The lesion result from migration of fourth-stage larva to small pulmonary artery, causing inflammation and initial endarteritis followed by organizing granuloma, with worm structures at varying stages of decomposition [59]. Histology of the lung nodules usually reveals cellular infiltration with eosinophils, lymphocytes, plasma cells, and histiocytic reaction around adjacent blood vessels. Most commonly single lesions are found in HPD, but multiple lesions can be present and usually located in the subpleural regions of the lungs. A small proportion of patients may present with mild symptoms of cough, low-grade fever, malaise, chest pain, and hemoptysis, and wheezing may be present on auscultation [73]. The radiographic features of HPD on chest films consist of well-circumscribed, non-calcified peripheral nodules with a mean diameter of 1.9 cm, and calcified lesions were rare [74]. Management of cases of HPD usually requires a lung biopsy by thoracotomy or by video-assisted thoracic surgery in order to exclude malignancy.

Subcutaneous and ocular dirofilariasis caused by *D. repens* is more prevalent than HPD caused by *D. immitis* globally. By 2012, about 1782 cases of human dirofilariasis had been reported of which 372 were pulmonary disease reported mainly from North America [72]. The remaining cases [1410] were subcutaneous or ocular dirofilariasis, mainly caused by *D. repens* in Europe, including 24 in returning travelers [72]. Subcutaneous nodules usually occur at the site of mosquito bite and gradually grow over weeks or months, as the larvae develop to preadult and adult worms [59]. The subcutaneous nodules are firm with elastic consistency and associated with erythema. In contrast to HPD, subcutaneous dirofilariasis is more frequent in women than men and most commonly present in middle aged adults. However, in Sri Lanka, 33.6% of the cases were reported in children under age 10 years [59].

Ocular dirofilariasis relative frequency appears to be increasing in recent years. It has been estimated that 30–35% of *D. repens* infection present with involvement around the eyes, orbital zone, eyelids, and subconjunctival and intravitreous tissues [67]. Patients occasionally present with decreased vision, floaters, or loss of sight. Permanent complications may develop in 10% of cases, including retinal detachment, glaucoma, crystalline lens, vitreous humor opacity, and loss of visual acuity [75]. Orbital dirofilariasis may present with eye discomfort, blepharedema, and palpebral ptosis.

Although in general, *D. immitis* is associated with pulmonary nodules and *D. repens* with subcutaneous nodules or ocular dirofilariasis, both species can be found in other sites occasionally. Worms of *D. immitis* can be found in cranial, hepatic, intraocular, mesenteric adipose, and conjunctival tissues and in testicular arteries; and *D. repens* have been found in the lungs, scrotum, penis, spermatic cord, epididymis, and female breasts [59].

Immunological and molecular diagnostic tests are not widely available for pulmonary and subcutaneous/ocular dirofilariasis except in research laboratories and include indirect hemagglutination, enzyme-linked immunosorbent assay [ELISA], and DNA detection by PCR. However, these methods are not highly sensitive or very specific for human diagnosis, and there is a wide range of other nematodes capable of producing cross-reactive antibodies [76]. The diagnosis and management of subcutaneous and ocular dirofilariasis are similar to that of HPD: excision of nodules and careful extraction of intact adult worms to avoid release of antigens to produce further inflammatory reaction. No pharmacotherapy is recommended.

11.3.4 Control of Dirofilariasis

Prevention of human dirofilariasis can be achieved through control of canine infection and vector control, which has proven difficult for many mosquito-borne diseases. Further epidemiological surveys are needed on a regular basis to define the at-risk areas, prevalence, and disease burden in wild and domestic canines. Improved serological methods are needed to identify infections in dogs to allow institution of treatment to prevent transmission and for humans in order to avoid invasive diagnostic procedures for benign "coin lesions." Dogs can be checked with commercial kit that is able to detect *D. immitis* circulating antigen only [67]. However, real-time PCR coupled to high-resolution melting analysis [HRMA] can be used to identify *D. immitis* from *D. repens* in canine blood [77]. It has been suggested that combination or sequential use of drugs to improve canine chemoprophylaxis should be used, such as ivermectin, moxidectin, selamectin, milbemycin, and melarsomine hydrochloride [72]. A combination of ivermectin [microfilaricide] and doxycycline [an adulticide targeting the *Wolbachia* endosymbiont] has been used in dogs to control *D. immitis* and *D. repens* [78, 79].

11.4 *Onchocerca lupi*

Onchocerca lupi [Spirurida, Onchocercidae] is another vector-borne filarial worm first described in wolves in 1967 [80]. It is a canine ocular onchocerciasis that causes disease of the connective tissue of the sclera of dogs, wolves, and cats [81]. *O. lupi* is an atypical *Onchocerca* species that is separated from other *Onchocerca* spp. early in evolution but still retains the *Wolbachia* endosymbiont [82]. The adult worm resides in the conjunctival tissue of the sclera or in the retrobulbar region, and microfilariae are found in the skin, especially in the head and interscapular region [83]. Microfilariae are more common in the ears and nose, with greater concentration in the afternoon followed by night and morning, which may correlate to the feeding habits of the natural vector [83]. Although the life cycle has not been well established, it is believed to be similar to *Onchocerca volvulus*. The microfilariae are believed to be ingested by blackflies [Simuliidae] and undergo several stages of development, and in the infectious stage larvae migrate to the mouth parts of the blackflies and enter the host skin during a blood meal, for adult stage of development [81]. Although blackflies are considered the presumptive vectors, smidges have been considered as possible vectors [81].

11.4.1 Epidemiology of Zoonotic Ocular Onchocerciasis

Since the recognition of *O. lupi* canine ocular onchocerciasis in 2001 in Hungary [84], the disease has been increasingly recognized from dogs across Europe, including Greece, Portugal, Germany, and Switzerland [81], and more recently in Romania and Spain [85, 86]. Although canine ocular onchocerciasis was first reported in the USA in 1991 [87], infection due to *O. lupi* was not documented in cats and dogs in the USA until 2011–2013 [88, 89]. The earlier canine ocular onchocerciasis was attributed to species from other animal hosts [cattle, horses, or wild ungulates] but may have been unrecognized *O. lupi*. There is scant data on the epidemiology of *O. lupi* in domestic or wild canines. In an epidemiology survey of dogs from areas in Greece and Portugal with reported cases, 8% of 107 dogs were found to be infected with *O. lupi* from skin samples taken in the frontal area of the head [90]. Dogs that were positive for *O. lupi* from one area had ocular signs ranging from conjunctival swelling and mucopurulent discharge to blindness; but animals from the other area were asymptomatic. The number of symptomatic ocular canine *O. lupi* infection confirmed that infection in Greece was 158 as of 2015 [81]. Although *O. lupi* was first recognized in a wolf from Georgia, there is no data on the prevalence in wild canines.

Human onchocerciasis secondary to *O. lupi* was first confirmed in 2011 in a patient from Turkey [91], but the infection was suspected in a man in 2002 [92]. As of 2015, 18 human infections were reported from Europe, Tunisia, Iran, and the USA [81], and more recently a case series of six patients was reported from the USA of which two were previously reported [93]. In retrospect, it has been speculated that *O. lupi* might have been responsible for a human ocular infection in the Crimean Peninsula in 1965 [before discovery of *O. lupi* as a species], attributed to other *Onchocerca* species [94].

11.4.2 O. lupi *Clinical Aspects in Animals*

Onchocerciasis lupi can present as acute or chronic ocular infection in dogs of any age but most commonly in adult animals [81]. Acute infection of the canine eye is associated with variable degree of conjunctivitis with conjunctival congestion, redness, periorbital swelling, exophthalmia, blepharitis, excessive lacrimation, and serous or purulent discharge and can progress to corneal stromal edema, corneal ulcers, and anterior/posterior uveitis. In some cases parts of the worm may be visible on the surface or under the conjunctiva [95, 96].

In chronic canine infection, there is usually localized swelling or pea-size nodule of the subconjunctival tissue, consisting of granulomatous reaction surrounding the parasite. Cyst-like nodules may penetrate the retrobulbar space, orbital fascia, eyelid, conjunctiva, nictitating membrane, and sclera with exophthalmia and signs of inflammation. Anterior uveitis and glaucoma can occur and rarely the worms may invade the anterior chamber [97]. The diagnosis of canine *Onchocerciasis lupi* was based on proper eye examination, skin snips, and periocular tissue biopsies and occasionally by the presence of characteristic threadlike worms. Skin snips from dog's ear tip were found to be the most sensitive diagnostic method [83]. Imaging has been used to localize nodules or cyst containing adult worms [98]. Specific identification of nematodes has been based on the morphology and molecular analysis. Detection of *O. lupi* antigens in sera of dogs by specific ELISA may be a promising diagnostic test if the sensitivity can be improved, as only 3 of 6 dogs with confirmed infection had positive serology in a preliminary study [99].

Treatment of animals [dogs and cats] consists of surgical excision of nodules, cysts or visible worms [even from the anterior chamber], antihelminthic drugs [microfilaricide ivermectin and adulticide melarsomine in dogs], and topical or systemic steroid for 2–3 days for control of inflammation and as well for uveitis and orbital disease [95, 96].

11.4.3 Human Onchocerciasis Lupi

As of August 2016, there have been a total of 24 human cases of human zoonotic onchocerciasis reported [81, 93], but some of the earlier cases were not confirmed to be due to *O. lupi*. Clinical presentation in most patients was for symptoms concerning the eye, with 16 patients showing evidence of ocular onchocerciasis. The majority of patients manifested subconjunctival nodules and/or conjunctival inflammation and swelling, but two cases presented with iritis or anterior chamber involvement. In the case series from the USA, however, only 1 of 6 patients had ocular or periorbital involvement [93]. Three of the patients presented with mass encroaching on the spinal cord with intradural/extradural lesions of the upper cervical spine, and two patients had subdermal nodule or inflammatory mass on the forearm and the scalp. The cases in the USA were identified in Arizona [$N = 3$], New Mexico [$N = 2$], and Texas [$N = 1$]. The marked difference in the manifestation of infection between the cases described in Europe/Middle East versus those in North America suggest that there may be different subtypes of the parasite in separate geographic areas.

Diagnosis of zoonotic onchocerciasis was based on surgical excision of the lesions and morphological appearance of the parasite, and some of the recent cases were confirmed to be *O. lupi* by molecular phylogenetic analysis or PCR [81, 93]. Fifteen patients were treated with surgical excision alone with no antihelminthic agents, but one patient required corticosteroid for generalized urticarial rash with eosinophilia after extraction of the worm, a Mazzotti reaction [81]. Ivermectin was used in six cases and doxycycline was added in four patients to kill and sterilize female adult worms [93]. There are no properly conducted treatment trials for *O. lupi* infection; thus, treatment has been based on the experience with other filarial parasites. A single dose of ivermectin150 μg/kg is recommended if microfilariae or gravid worms are identified on histopathology and 4–6 weeks course of doxycycline [4 mg/kg/day] or 100 mg twice daily for all patients to kill any adult worm not found at surgical resection [93].

Ivermectin has been proven to treat *O. volvulus* infection and a single dose has been shown to be effective to kill the microfilariae and suppress the appearance of new ones for several months [100]. Doxycycline is more effective in killing and sterilizing the adult worm of *O. volvulus* and is of proven benefit in randomized, controlled trials [101].

11.4.4 Control of O. lupi *Infection*

Onchocerciasis lupi is a new emerging zoonosis as evidenced by its expansion in Europe, Middle East [Turkey, Tunisia, and Iran], and southern parts of North America, and no studies have yet been published on control strategies. It is likely that the condition has been unrecognized by veterinarians and medical practitioners and is more frequent than estimated. Moreover, there is evidence that previous cases diagnosed as dirofilariasis were probably ocular onchocerciasis due to *O. lupi* [81]. Hence, education of health workers, veterinarians, and ophthalmologists on the recognition and diagnosis is necessary to determine the true extent and prevalence of *Onchocerciasis lupi*. Further research is needed to define the potential hosts and vectors of the parasite and to understand the reservoir and seasonality. Widely available simple diagnostic tests such as serology are needed for animal and human applications in order to determine the extent of the burden of infection. Studies should also be done in Africa and Asia, where blackflies are prevalent and other *Onchocerca* species exist, but there has been no report of *Onchocerciasis lupi*. Potential control measures in areas where canine and human zoonotic onchocerciasis exist would be treatment of infected dogs and cats with single dose of ivermectin and a course of doxycycline, similar to methods used for dirofilariasis.

11.5 *Trypanosoma evansi*, *T. lewisi*, and Others: New Zoonotic Trypanosomiasis

Trypanosomes are flagellate protozoa found in the blood and sometimes tissues of mammals, including humans, of the order Kinetoplastida, family of the Trypanosomatidae, and genus *Trypanosoma*. They are transmitted by biting insects to mammals to complete their biological cycle. There are several members of the genus, but the most well recognized, responsible for a large burden of disease globally, are *Trypanosoma brucei* species [*T. brucei* subspecies *gambiense* and *T. brucei* subspecies *rhodesiense*] responsible for sleeping sickness in sub-Saharan Africa and *Trypanosoma cruzi* which causes Chagas disease in Latin America [102]. Animal *Trypanosoma* species such as *T.b. brucei*, *T. congolense*, *T. vivax*, *T. lewisi*, and *T. evansi* have rarely been reported to cause human disease [103], but the latter two trypanosomes may be emerging.

These parasites are distributed worldwide and can infect a variety of wild and domestic animals with humans as incidental hosts. African livestock trypanosomes [*T. congolese*, *T. vivax*, *T. b. brucei*], called "nagana," have large reservoir in wild animals, where it produces mild disease, but causes huge loss of cattle yearly from severe, often fatal disease, and the geographic distribution is evolving to livestock of Asia and Latin America [103]. Animal trypanosomes are transmitted cyclically by biting insects such as the tsetse fly. *T. vivax* infects primarily bovines and less commonly horses, is transmitted by tabanids and stomoxes in Africa and Latin America, and to date has not been found in Asia and Europe. *T. congolese* affect mainly cattle in Central Africa with relatively low parasitemia, which may restrict its transmissibility and expansion [103]. *T. lewisi* is host restricted primarily to rodents [*Rattus* spp.] worldwide and is transmitted by rat fleas [103].

11.5.1 Animal Infection with T. evansi

Trypanosoma evansi was the first pathogenic mammalian trypanosome to be described in1880, in the blood of Indian equines and dromedaries [104]. It is a salivarian trypanosome, originating from Africa, and is believed to have been derived from *T. brucei* by a mutation of the kinetoplast mitochondrial DNA, resulting in the lack of transmissibility by the tsetse fly [105]. The primary host is the camel, but it can be found in dromedaries, horses, bovines, other livestock, and a large range of other mammals [sheep, goats, pigs, dogs, buffaloes, deer, etc.]. The parasite is transmitted by biting insects, tabanid [horsefly], and stomoxes [genus of fly that bites livestock and humans]. The parasite is present in all countries with camels and may have spread from northern Africa thousands of years ago by this principal host, from North Africa toward the Middle East, Turkey, Spain, north to Russia, then east to India, all of Southeast Asia, down to Indonesia and the Philippines, and by conquistadores into Latin America [106]. In Latin America, it occurs in new range of domestic and wild hosts, including the main reservoirs [capybaras] and other biological vectors [vampire bats].

T. evansi produces a disease in animals called "surra" derived from the Indi meaning rotten, as the result of its chronic evolution [106]. Surra is a major disease of camels, equines, and dogs, which is frequently fatal without treatment, and presents with anemia, weight loss, abortion, and death [106]. While *T. evansi* has a large host range with most mammals being susceptible, their degree of susceptibility is widely variable depending on species and geographic regions. In Africa and the Middle East, the parasite mainly affects camels, but it is also found in horses, donkeys and mules, and dogs and cats from eating raw meat of slaughtered animals. In Asia *T. evansi* is a common parasite found in bovine animals, especially in water buffaloes; and it is economically important in the Philippines due to affliction of water buffaloes, cattle, pigs, goats, and horses [107]. Asian bovines/cattle appear to be more susceptible to *T. evansi* infection than those in Africa and Latin America, with more severe illness and higher degree of parasitemia [106]. The clinical manifestation and course varies with the animal affected. In camels infection can be acute, chronic, and subclinical with healthy carriers, and chronic disease can lead to emaciation and death. In horses, it produces a more acute and progressively fatal disease with weakness of legs nervous system involvement. While in bovines the parasite results in mild regular illness and acts as efficient reservoir [106].

11.5.2 New Zoonotic Trypanosome Infections in Humans

There are only a few cases of human *T. evansi* infection reported, but it is estimated that zoonotic infections frequently occur in rural areas of tropical and subtropical countries, with limited resources, that often failed to be diagnosed. This has led to the Vietnam initiative to study new emerging zoonotic infections [108]. A review of new or atypical zoonotic trypanosomiasis [Chagas disease and sleeping sickness being classical forms] in 2013 enumerated only 19 cases, of which five were attributed to *T. evansi* [103]. Three of the cases were from India and one each from Sri Lanka and Egypt; and the diagnosis were made by morphology of the parasite on blood smear in four of the patients and by PCR in one. The first suspected case of *T. evansi* infection was reported in India in 1977, following an accidental injection with a syringe containing *T. evansi*-infected blood [109]. In 1999, another patient with suspected of *T. evansi* infection with headaches and high fever with high levels of parasitemia was reported from Sri Lanka, but the case was not well documented [110]. Similarly, a patient with suspected infection, based on morphology on blood smear, from Kolkata, India, died 2 days after hospital admission [110]. A cattle farmer from Egypt was diagnosed with *T. evansi* trypanosomiasis based on morphology of the parasite, but details of drug treatment were not provided [111]. The first well-documented case of zoonotic *T. evansi* infection was reported in 2005 from Maharashtra State in India [112]. The patient was a cattle farmer with febrile episodes associated with fluctuating parasitemia, confirmed by morphology of the parasites, serology, and molecular tests. Treatment with suramin resulted in cure of the patient. It was later shown that this patient lacked apolipoprotein L-1 [APOL-1] that predisposed to this infection [113]. Normally, humans are resistant to infection to animal trypanosomes such as *T. b. brucei* and *T. evansi* by the trypanolytic activity of APOL-1 [114]. However, in certain conditions, *T. evansi*, *T. congolense*, and *T. vivax* can be resistant to human plasma [115]. Recently, another case of human *T. evansi* was reported from Vietnam with high parasitemia [>50,000 parasites/μL] confirmed by PCR and without APOL-1 deficiency [116]. The 38-year-old female contracted the infection via a wound while butchering raw beef, with 18 days of fever, headaches, and arthralgia, and was treated successfully with suramin after initial improvement and relapse with a course of amphotericin B. Investigation revealed widespread burden of the parasites in local cattle [41%] and buffaloes [67%].

T. lewisi- or *T. lewisi*-like parasites have been the most commonly reported atypical human trypanosomes, with a total of eight cases [103]. The first case was reported from Malaysia in 1933, five cases from India, and one each from Thailand and Gambia [the only case outside of Asia]. The species diagnosis was based on morphology of the parasites on blood smear in four patients, especially the earlier reports, and four by PCR. Humans infected with *T. lewisi* usually lived in poor dwellings infested with rats and appears to be transmitted by rat fleas to humans. Half of the reported cases had mild febrile illness with transient parasitemia and spontaneous cure with no specific treatment. However, infants can become very sick [117] and a fatal case was reported from India [118]. Treatment has included melarsoprol, suramin, and pentamidine. *T. lewisi* has been detected in 14.3–18.0% of different species of rodents in Thailand, and sequence from one sample showed 96.4% similarity to that of an infected infant [117].

T. b. brucei, a cause of nagana in cattle of Africa, was able to infect only one of seven volunteers by the bite of infected tsetse fly and produced transient infection for 3 weeks [119]. Humans are naturally resistant to these species of trypanosomes. However, a patient with sleeping sickness in western Ethiopia was found to be infected with *T. b. brucei* rather than *T. gambiense*, confirmed by blood incubation test [120]. *T. b. brucei* is indistinguishable on blood smear morphology from *T. gambiense* and *T. rhodesiense*, which are the classic causes of chronic or acute sleeping sickness, respectively [103]. A single case of *T. vivax* from Ghana was reported in 1917 and not since then, similarly only one case of *T. congolense* in Cote d'Ivoire, has been reported in a patient coinfected with another *T. brucei* spp. [103].

Based on current data from case reports over the years, these atypical zoonotic trypanosomes are not a major threat to human health, but infections due to *T. evansi* and *T. lewisi* in humans should be monitored by public health officials in countries where the parasites are found in animals, especially in livestock and rodents, respectively.

11.6 Sarcocystosis

Sarcocystosis, a universal zoonotic parasitic disease, is caused by small intracellular apicomplexan/coccidian protozoa of the genus *Sarcocystis* [Eucoccidiorida: Sarcocystidae]. *Sarcocystis* species are ubiquitous in the environment and are present worldwide. There are >150 species in the genus, and they require two hosts, intermediate and definitive, to complete the life cycle; herbivores or preys are the primary intermediate hosts and carnivores are the definitive hosts [121]. Omnivores such as primates can be intermediate or definitive hosts. The intermediate hosts carry the sarcocystis with infectious zoites in their muscles, and definitive hosts excrete the oocysts or sporocysts into the environment. Carnivores or definitive hosts usually do not manifest much symptoms or only mild disease of the noninvasive intestinal infection, whereas intermediate hosts [including humans] usually show severe symptoms of the invasive disease. *Sarcocystis* was first described in deer muscles in Switzerland in 1843 [121] and in humans in 1975 in Malaysia and was considered a rare disease in humans until a large cluster of disease was reported in 2011–2012 from Tioman Island, Pahang State, with worldwide distribution from affected travelers [122, 123].

11.6.1 Biology of Sarcocystosis

Sarcocystosis is one of the most widespread infection in wild and domestic animals in the world. The invasive muscle disease is found in a wide variety of intermediate hosts, i.e., mammals [74%], birds [14%], reptiles [10%], and rarely fish [0.5%] [124]. The most common combinations of intermediate/definitive hosts are mammals/mammals, mammals/reptiles, reptiles/reptiles, and mammals/birds [124]. Some species are of economic importance to the agriculture industry as they may produce severe disease in affected animals, cattle, sheep, goats, pigs, buffaloes, and horses and can infect cats and dogs [125].

The life cycle is based on findings in cattle and starts with ingestion by the intermediate host of oocysts or sporocysts in feces from the definitive or final host. Sporozoites [released from sporocysts] migrate through the gut wall to enter the circulation to be dispersed in the body [126]. The parasite undergoes four cycles of asexual development [schizogony] in the vascular endothelium to produce merozoites, and the last generation develops in skeletal, cardiac, and smooth muscles and occasionally the central nervous system, as sarcocystis. Sarcocystis can be found in muscles of limbs, tongue, esophagus, diaphragm, heart, and rarely brain and spinal cord. In definitive or final hosts, the sexual stage of development occurs. Carnivores [or omnivores] eat the meat of infected intermediate host containing mature sarcocystis, and digestion of the wall allows release of bradyzoites which infect the epithelial cells of small intestinal villi intracellularly into microgamont [male] or macrogamont [female] stage [126]. The microgamont fuses and fertilizes the macrogamont and develops into oocysts or sporocysts with sporozoites that shed in excreted feces.

Humans can be final or intermediate hosts. Humans can be final hosts for only two known *Sarcocystis* species: *S. hominis*, from eating undercooked beef from cattle or water buffaloes, or *S. suihominis* from eating infected pork from pigs or wild boars, the intermediate hosts. Tissues of many domestic and wild mammals, birds, and reptiles are eaten by humans all over the world that may be infected with sarcocystis and infect humans. Based on differences observed in the morphology of sarcocystis found in human tissues, it is possible that humans may be intermediate hosts for seven or more species [127]. Humans can become intermediate hosts by ingesting food or water contaminated from feces of definitive hosts. Molecular sequence studies indicate that humans can be accidental intermediate host for *S. nesbitti*, which could be transmitted from reptile excreta as snakes appear to be the definitive hosts [128, 129]. Different species of nonhuman primates can also be infected by this *Sarcocystis* spp. [126].

11.6.2 Clinical Sarcocystosis in Humans

Intestinal sarcocystosis in humans that results from ingestion of undercooked meat of the intermediate hosts [mainly beef and pork] is present worldwide, but it has not been identified in Africa and the Middle East, but infections are probably unrecognized [126]. Intestinal infections with *Sarcocystis* are most commonly reported from Europe [the Netherlands, Germany, France, Poland, Slovakia, and Spain], also in Asia [China, Laos, Tibet, and Thailand], South America [Argentina and Brazil], and Australia. Intestinal infections previously attributed to *Isospora hominis* represented infection with multiple species of *Sarcocystis*, including *S. hominis* and *S. suihominis* [126]. Studies in volunteers infected with consumption of raw beef or pork showed incubation or prepatent period of 9–17 days and patent or duration of shedding from 8 to 120 days [126]. However, the incubation period and duration of symptoms or shedding of parasites probably are related to infectious dose, virulence of the strain, and the host natural innate immunity. For instance, in a study from Germany, volunteers who fed raw pork heavily infected with *S. suihominis* developed symptoms of acute gastroenteritis 8 h later which persisted for 48 h [130]. Many human infections are mild or asymptomatic, but acute, severe, or chronic enteritis can occur with nausea, vomiting, and diarrhea. Diagnosis of intestinal sarcocystosis can be made by identification of oocysts or sporocysts in the stool by microscopy, but species cannot be distinguished by morphology. Intestinal infections are usually self-limited and there is no known effective therapy and recurrent infection is possible.

Humans can develop extraintestinal sarcocystosis, primarily manifested as muscular sarcocystosis, by consumption of oocysts or sporocysts likely from different species of *Sarcocystis* but confirmed only for *S. nesbitti*, thus, humans becoming aberrant intermediate hosts. Initial cases of invasive human sarcocystosis were described from Malaysia [1975–1992] as incidental findings of surgical biopsy samples or at autopsy [131]. The first case reported in 1975, however, had hoarseness of voice with fibromuscular nodule on the larynx [122]. The actual frequency of invasive sarcocystosis in humans is unknown, but over 100 cases have been reported worldwide, of which most are from Southeast Asia, especially Malaysia, but also cases have been recognized in India, Central and South America, Africa, Europe, and the USA and single case from China, Egypt, and Australia [126]. However, several of these cases are retuning travelers from Malaysia or other parts of Southeast Asia. In Malaysia a high prevalence of muscle sarcocystosis was found from autopsy studies, with the presence of cysts in about 21% of tongues [132].

Many, if not the majority, of patients with muscle sarcocystosis are subclinical or had transient mild illnesses with nonspecific symptoms. Symptomatic disease has been best documented in outbreaks of infection arising from Malaysia. A small cluster of cases was initially reported in 1993 and involved seven US Air Force Team members deployed near a jungle village and exposed to contaminated water and food [133]. Six of the seven developed acute symptoms of eosinophilic myositis within 3–8 weeks after the high-risk exposure with symptoms of intermittent fever, myalgias, arthralgia, headache, bronchospasm, lymphadenopathy, weight loss, rashes, and muscle wasting. One of the subjects remained symptomatic intermittently for several years and muscle biopsy 3 months after onset of illness revealed sarcocystis; a year later this patient developed probable *Sarcocystis*-related myocarditis.

The largest single outbreak of muscle sarcocystosis involved college students and teachers attending a retreat on Pangkor Island, Malaysia, in 2012 [128].

Within 26 days after leaving the island, 89 of 92 persons attending the retreat developed symptoms of fever [94%], myalgia [91%], headache [87%], and cough [40%]. Eight patients developed facial swelling for 4–6 weeks with changes of inflammatory edema seen on magnetic resonance imaging [MRI] [134]. This outbreak was attributed to *S. nesbitti* possibly from snake feces contamination of food or water [129]. Four of the symptomatic patients had muscle biopsies which showed sarcocystis in three and positive PCR in one. Between 2011 and 2012, two outbreaks of infection occurred in 99 tourists visiting Tioman Island off the coast of Malaysia, with all but two from Europe [123]. One cluster of cases occurred 2 weeks after returning from the island and the other cluster occurred 6 weeks after. Myalgia, fatigue, fever, and headaches were the most frequent symptoms, and eosinophilia and elevated creatine kinase [CK] were present by the fifth week after the exposure [135]. Sixty-two patients were considered probable *Sarcocystis* myositis and six cases were confirmed by muscle biopsy. A longitudinal study was performed from 39 travelers returning to Germany in 2011–2014 from Tioman Island, some of whom were reported in the first cluster of cases [136]. The median duration of symptoms was 2.2 months [range 0–23 months] and 17% of patients had symptoms greater than 6 months, and two patients had diminished but persistent symptoms at 13 and 23 months.

Although most patients with *Sarcocystis* myositis had skeletal muscle involvement, myocarditis has been reported but more frequently in older reviews. In a previous review of 40 cases of human sarcocystosis in 1979, 11 patients at autopsy had cardiac sarcocystis [127]. Among the 68 cases of sarcocystosis from Tioman Island, only one patient had mild myocarditis which subsequently resolved, but ten patients had mildly elevated CK MB fraction [cardiac muscle CK isoenzymes] [135]. Only one patient with muscle sarcocystosis has been reported with glomerulonephritis [137], but causation was not established. However, in cattle with acute *S. cruzi* infection, schizonts of the parasite can be found in the glomeruli.

11.6.3 Diagnosis, Treatment, and Prevention of Human Sarcocystosis

Invasive sarcocystosis were mainly diagnosed with muscle biopsy with histology showing sarcocystis in tissues with or without inflammatory reaction. Patients commonly have evidence of eosinophilia, elevated transaminases, and CK. A major differential diagnosis is trichinosis which has different appearing cysts on histology and can be excluded by serology for *Trichinella spiralis*. There is no severe or protracted myositis, and treatments have included non-steroidal anti-inflammatory drugs, albendazole with or without oral steroids, and co-trimoxazole [126]. In vitro susceptibility assays using cell cultures have shown cidal activity of trimethoprim and pyrimethamine, with enhanced activity in combination with sulfonamides [138]. Co-trimoxazole with or without prednisone has been used in several patients with improvement and may be a reasonable choice in severe or protracted symptomatic patients. Randomized therapeutic trials are probably not feasible due to the few numbers of sporadic symptomatic cases.

Approach to prevention and control should include measures to decrease infection in livestock and pets as well as humans. In domestic animals, interrupting the life cycle of the parasite is a considered a key factor, such as avoiding feeding raw meat of slaughtered/dead animals to carnivores and preventing the contamination of water and feed of livestock with animal feces [125]. Treatment of experimental infected animals with some antiparasitic agents [amprolium and maduramicin] can reduce fecal excretion of sporocysts and reduce the transmission risk to other animals [139]. To prevent intestinal sarcocystosis in humans, meat should be properly cooked of frozen to kill the bradyzoites in the sarcocystis. Prevention of invasive or muscle sarcocystosis requires avoidance of contaminated water or food. Possible contaminated drinking water is best decontaminated by boiling or filtering to remove the sporocysts. Food can be contaminated during production, preparation, and distribution and best cooked or reheated even at 100 °C for 5 min [140].

11.7 Future Directions

A comprehensive global network through the auspices of the WHO is needed to track and monitor these novel emerging parasitic zoonoses. More accounting and precise epidemiological data are needed from existing countries with these parasites and from countries with infections in animals but not yet reported in humans. Further research is needed to explore biology, pathogenesis, and transmissibility to humans.

References

1. Colwell DD, Dantas-Torres F, Otranto D (2011) Vector-borne parasitic zoonoses: emerging scenarios and new perspectives. Vet Parasitol 182:14–21
2. Sorvillo F, Ash LR, Berlin OGW, Yatabe J, Degiorgio C, Morse SA (2002) *Baylisascaris procyonis*: an emerging helminthic zoonosis. Emerg Infect Dis 8:355–359
3. Kazacos KR (2001) *Baylisascaris procyonis* and related species. In: Samuel WM, Kocan AA, Pybus MJ (eds) Parasitic diseases of wild mammals, 2nd edn. Iowa State University Press, Ames, pp 301–341
4. Tiner JD (1953) The migration, distribution in the brain, and growth of ascarid larvae in rodents. J Infect Dis 92:105–113
5. Bauer C (2012) Baylisascariosis infections of animals and humans with 'unusual roundworms'. Vet Parasitol 193:404–412
6. Page LK, Swihart RK, Kazacos KR (1999) Implications of raccoon latrines in the epizootiology of baylisascariasis. J Wildl Dis 35:474–480
7. Page LK, Anchor C, Luy E et al (2009) Backyard raccoon latrines and risk for *Baylisascaris procyonis* transmission to humans. Emerg Infect Dis 15:1530–1531
8. Graeff-Texeira C, Morassutti AL, Kazakos KR (2016) Update on Baylisascariasis, a highly pathogenic zoonotic infection. Clin Microbiol Rev 29:375–399

9. Beltrain-Beck B, Garcia FJ, Gortazar C (2012) Raccoons in Europe: disease hazards due to the establishment of an invasive species. Eur J Wildl Res 58:5–15
10. Kuhle M, Knorr HLJ, Medenblik-Frysch S, Weber A, Bauer C, Naumann GOH (1993) Diffuse unilateral subacute neuroretinitis syndrome in a German most likely caused by the raccoon roundworm, *Baylisascaris procyonis*. Graefes Arch Clin Exp Ophthalmol 231:48–51
11. Sato H, Une Y, Kawakami S, Sato E, Kamiya H, Akao N, Furuoka H (2005) Fatal *Baylisascaris* larva migrans in a colony of Japanese macaques kept in a safari-style zoo in Japan. J Parasitol 91:716–719
12. Furuoka H, Sato H, Kubo M, Owaki S, Kobayashi Y, Matsui T, Kamiya H (2003) Neuropathological observation of rabbits [*Oryctolagus cuniculus*] affected with raccoon roundworm [*Baylisascaris procyonis*] larva migrans in Japan. J Vet Med Sci 65:695–699
13. Xie Y, Zhou X, Li M et al (2014) Zoonotic *Baylisascaris procyonis* roundworms in raccoons, China. Emerg Infect Dis 20:2170–2172
14. Page KL (2013) Parasites and the conservation of small populations: the case of *Baylisascaris procyonis*. Int J Parasites Wildl 2:203–210
15. Kazacos KR (1997) Visceral, ocular and neural larva migrans. In: Connor DH, Chandler FW, Schwartz DA, Manz HJ, Lack EE (eds) Pathology of infectious diseases, vol 2. Appleton & Lange, Stamford, CT, pp 1459–1473
16. Hamann KJ, Kephart GM, Kazacos KR, Gleich JG (1989) Immunofluorescent localization of eosinophil granule major basic protein in fatal human cases of *Baylisascaris procyonis* infection. Am J Trop Med Hyg 40:291–297
17. Mortel CL, Kazacos KR, Butterfield JH, Kita H, Watterson J, Gleich GJ (2001) Eosinophil-associated inflammation and elaboration of eosinophil–derived proteins in 2 children with raccoon roundworm [*Baylisascaris procyonis*] encephalitis. Pediatrics 108:e93
18. Fan CK, Holland CV, Loxton K, Barghout U (2015) Cerebral toxocariasis: silent progression to neurodegenerative disorders? Clin Microbiol Rev 28:663–686
19. Wise ME, Sorvillo FJ, Shafir SC, Asch LR, Berlin OG (2005) Severe and fatal central nervous system disease in humans caused by *Balylisascaris procyonis*, the common roundworm of raccooons: a review of current literature. Microbes Infect 7:317–323
20. Saffra NA, Perlman JE, Desai RU et al (2010) *Baylisascaris procyonis* induced diffuse unilateral subacute neuroretinitis in New York City. J Neuroparasitol 1:N100401. doi:10.4303/jnp/n100401
21. Anderson DC, Greenwood R, Fishman M, Kagan IF (1975) Acute infantile hemiplegia with cerebrospinal fluid eosinophilic pleocytosis: an unusual case of visceral larva migrans. J Pediatr 86:247–249
22. Beaver PC (1969) The nature of visceral larva migrans. J Parasitol 55:3–12
23. Magnaval J-F, Glickman LT, Dorchies P, Morassin B (2001) Highlights of human toxocariasis. Korean J Parasitol 39:1–11
24. Rowley HA, Uht RM, Kazacos KR, Sakanari J, Wheaton WV, Barkovich AJ, Bollen AW (2000) Radiological-pathological findings in raccoon roundworm [*Bayliscaris procyonis*] encephalitis. Am J Neuroradiol 21:415–420
25. Kazacos KR, Wirtz WL, Burger PP, Christmas CS (1981) Raccoon ascarid larvae as a cause of fatal central nervous system disease in subhuman primates. J Am Vet Med Assoc 179:1089–1094
26. Sato H, Matsuo K, Osanai A, Kamiya H, Akao N, Owaki S, Furuoka H (2004) Larva migrans by *Baylisascaris transfuga*: fatal neurological diseases in Mongolian birds, but not in mice. J Parasitol 90:774–781
27. Kazacos KR, Jelicks LA, Tanowitrz HB. *Baylisascaris* larva migrans. In: Garcia HH, Tanowitz HB, Del Brutto OH (eds) Handbook of clinical neurology. Neuroparasitology and tropical neurology, vol. 114 (3rd series), pp 251–262. Elsevier BV, Amsterdam 2013
28. Murray WJ, Kazacos KR (2004) Raccoon roundworm encephalitis. Clin Infect Dis 39:1484–1492
29. Silvalingam A, Goldberg RE, Ausberger J, Frank P (1991) Diffuse unilateral subacute neuroretinitis. Arch Ophthalmol 109:1028

30. Brasil OF, Lewis H, Lowder CY (2006) Migration of *Baylisascaris procyonis* into the vitreous. Br J Ophthalmol 90:1203–1204
31. Goldberg MA, Kazacos KR, Boyce WM, Ali E, Katz B (1993) Diffuse unilateral subacute neuroretinitis. Morphometric, serologic, and epidemiologic support for *Baylisascaris* as a causative agent. Ophthalmology 100:1695–1701
32. Dangoudoubiyam S, Vemulapalli R, Ndao M, Kazacos KR (2011) Recombinant antigen-based enzyme linked immunosorbent assay for diagnosis of *Baylisascaris procyonis* larva migrans. Clin Vaccine Immunol 18:1650–1655
33. Roscoe LN, Santamaria C, Handali S, Dangoudoubiyan S, Kazacos KR, Wilkins PP, Ndar M (2013) Interlaboratory optimization and evaluation of a serological assay for diagnosis of human baylisascariasis. Clin Vaccine Immunol 20:1758–1763
34. Boyce WM, Asai DJ, Wilder JK, Kazacos KR (1989) Physiochemical characterization and monoclonal and polyclonal antibody recognition of *Baylisascaris procyonis* larval excretory-secretory antigens. J Parasitol 75:540–548
35. Dangoudoubiyam S, Vemulapalli R, Kazacos KR (2009) PCR assays for detection of *Baylisascaris procyonis* eggs and larvae. J Parasitol 95:571–577
36. Miyashita M (1993) Prevalence of *Baylisascaris procyonis* in raccoons in Japan and experimental infections of the worm to laboratory animals. J Urban Living Hlth Assoc 37:137–151 [In Japanese with English summary]
37. Garrison RD 1996. Evaluation of anthelmintic and corticosteroid treatment in protecting mice [*Mus musculus*] from neural larva migrans due *to Baylisascaris procyonis*. M.S. thesis, Purdue University, West Lafayette, IN
38. Fu Y, Nie HM, Niu LL et al (2011) Comparative efficacy of ivermectin and levimasole for reduction of migrating and encapsulated larvae of *Baylisascaris procyonis* in mice. Korean J Parasitol 49:145–151
39. Sawanyawisuth K (2008) Treatment of angiostrongyliasis. Trans R Soc Trop Med Hyg 102:990–996
40. Garcia HH, Nash TE, Del Brutto OH (2014) Clinical symptoms, diagnosis, and treatment of neurocysticercosis. Lancet Neurol 13:1202–1215
41. Kazacos KR (2000) Protecting children from helminthic zoonoses. Contemp Pediatr 17:1–24
42. Williams GA, Aaberg TM, Dudley SS (1988) Perimacular photocoagulation of presumed *Baylisascaris procyonis* in diffuse unilateral subacute neuroretinitis. In: Gitter KA, Schatz H, Yamnnuzzi LA, HR MD (eds) Laser photocoagulation of retinal disease. Pacific Medical Press, San Francisco, pp 275–280
43. Jardine CM, Pearl DL, Puskas K, Campbell DG, Shirose L, Peregrine AS (2014) The impact of land use, season, age, and sex on the prevalence and intensity of *Baylisascaris procyonis* infection in raccoons [*Procyon lotor*] from Ontario, Canada. J Wildl Dis 50:784–791
44. Smyser TJ, Page LK, Rhodes OE (2010) Optimization of raccoon latrine surveys for quantifying exposure to *Baylisascaris procyonis*. J Wildl Dis 46:929–933
45. Page K, Beasely JC, Olson ZH et al (2011) Reducing *Baylisascaris procyonis* roundworm larvae in raccoon latriens. Emerg Infect Dis 17:90–93
46. Shafer SC, Wang W, Sorvillo FJ, Wise ME, Moore L, Sorvillo T, Eberhard MT (2007) Thermal death point of *Baylisascaris procyonis* eggs. Emerg Infect Dis 13:172–173
47. Birago F (1626) Trattato cinegetico, ovvero della caccia. Sfondrato V, Milan, p. 77
48. Anderson DM, Keith J, Novak PD (2000) Dorland's illustrated medical dictionary, vol 509, 29th edn. WB Saunders Company, Philadelphia
49. Simon F, Morchon R, Gonzales-Miguel J, Marcos-Atxutegi C, Siles-Lucas M (2009) What is new about animal and human dirofilariasis? Trends Parasitol 25:404–409
50. Pampiglione S, Canestri Trotti G, Rivasi F (1995) Human dirofilariasis due to *Dirofilaria* [*Nochtiella*] *repens*: a review of world literature. Parasitolgia 37:149–193
51. Cancrini G, Kramer L (2001) Insect vectors of *Dirofilaria* spp. In: Simon F, Genchi C (eds) Hearworm infection in human and animals. Ediciones Universidad de Salamanca, Salamanca, pp 63–82

52. Bareriga OO (1982) Dirofilariasis. In: Steele JH, Schultz MG (eds) Handbook series in zoonoses, vol 2. CRC, Boca Raton, pp 93–110
53. Manfredi MT, Di Cerbo A, Genchi M (2007) Biology of filarial worms parastising dogs and cats. In: Genchi C, Rinaldi L, Cringoli G (eds) *Dirofilaria immitis* and *D. repens* in dog and cat and human infections. Mappe parasitologiche. Universita degli Studi di Naploi Federico II, Naples, pp 39–47
54. Venco L, Genchi C, Simon F (2011) La filariasis cardiopulmonary [*Dirofilaria immitis*] en el perro. In: Simon F, Genchi C, Venco I, Montoya MN (eds) La filariasis en lase species domesticas y en hombre. Meriql Laboratorios, Barcelona, pp 19–60
55. Kozec WJ, Marroquin HF (1977) Intracytoplasmic bacteria in *Onchocerca volvulus*. Am J Trop Med Hyg 26:663–778
56. Marcus-Atxutegi C, Gabrielli S, Kramer LH, Cancrini G, Simon F (2003) Antibody response against *Dirofilaria imitis* and the *Wolbachia* endosymbiont in naturally infected canine and human hosts. In: Mas-Coma S (ed) Proceeding of IX European Multicolloquium Parasitol [EMOP IX]. Editorial Medimond, Piaoro, pp 297–302
57. Bandi C, Dunn AM, Hurst GD, Rigaud T (2001) Inherited microorganisms, sex-specific virulence and reproductive parasitism. Trends Parasitol 17:88–94
58. Debrah AY, Specht S, Klarmann-Schultz U et al (2015) Doxycycline leads to sterility and enhanced killing of female *Onchocerca volvulus* worms in an area with persistent microfilaridermia after repeated ivermectin treatment: a randomized, placebo-controlled, double-blind trial. Clin Infect Dis 61:517–526
59. Simon F, Siles-Lucas M, Morchon R, Gonzalez-Miguel J, Mellado I, Carreton E, Montoya-Alonso JA (2012) Human and animal dirofilariasis: the emergence of a zoonotic mosaic. Clin Microbiol Rev 25:507–544
60. Marcos-Atxutegi C, Kramer LH, Fernandez I, Simoncini L, Genchi M, Simon F (2003) Th1 response in BALB/c mice immunized with *Dirofilaria immitis* soluble antigens: a possible role for *Wolbachia*? Vet Parasitol 112:117–130
61. Kaiser L, Spickard RC, Sparks HV Jr, Williams JF (1989) *Dirofilaria immitis*: alteration of endotrhelial-dependent relaxation in the in vivo canine femoral artery. Exp Parasitol 69:9–15
62. Venco L, Genchi C, Simon F (2011) La filariasis cardiopulmonary [*Dirofilria immitis*] en el perro. In: Simon F, Genchi C, Venco L, Montoyo MN (eds) La filariqasis en las especvioes domesticas y en el hombre. Merial Laboratorios, Barcelona, pp 19–60
63. Scarzi M (1995) Cutaneous dirofilariasis in dogs. Obviettivi Doc Vet 16:11–15
64. Nelson CT, McCall JW, Rubin SB et al (2005) Guidelines for the diagnosis, prevention and management of heartworm [*Dirofilaria immitis*] infection in dogs. Vet Parasitol 133:255–266
65. Roncalli RA (1998) Tracing the history of heartworms: a 400 year perspective. In: Seward RL (ed) Recent advances in heartworm disease. American Heartworm Society, Batavia, IL, pp 1–14
66. Levy JK, Edinboro CH, Gloptfelty CS et al (2007) Seroprevalence of *Dirofilaria immitis*, feline leukemia virus, and ferline immunodeficiency virus infection among dogs and cats exported from the Gulf Coast hurricane disaster area. JAVMA 231:218–225
67. Genchi C, Kramer LH, Rivasi F (2011) Dirofilarial infections in Europe. Vector-Borne Zoonotic Dis 11:1307–1317
68. Trotz-Williams LA, Trees AJ (2003) Systematic review of the distribution of the major vector-borne parasitic infections in dogs and cats in Europe. Vet Rec 152:97–105
69. Purse BV, Mellor PS, Rogers DJ et al (2005) Climate change and the recent emergence of bluetongue in Europe. Nat Rev Microbiol 53:171–181
70. Genchi C, Rinaldi L, Mortarino M, Genchi M, Cringoli G (1634) Climate and *Dirofilaria* infection in Europe. Vet Parasitol 2005:286–292
71. Benedict MQ, Levine RS, Hawley WA, Lounibos LP (2007) Spread of the tiger: global risk of invasion by the mosquito *Aedes albopictus*. Vector Borne Zoonot Dis 7:76–85
72. Diaz JH (2015) Increasing risks of human dirofilariasis in travelers. J Transl Med 22:116–123

73. Muro A, Cordero M (2001) Clinical aspects and diagnosis of human dirofilariasis. In: Simon F, Genchi C (eds) Heartworm infections in humans and animals. Ediciones Universidad de Salamanca, Salamanca, pp 191–202
74. Shah MK (1999) Human pulmonary dirofilariasis: review of the literature. South Med J 92:276–279
75. Avidiukhina TI, Lysenko AI, Supriaga VG, Postnova VF (1996) Dirofilariasis of the vision organ: registry and analysis of 50 cases in the Russian Federation and in countries of the United Independent States. Vestn oftalmol 112:35–39
76. Glickman L, Grieve R, Schantz P (1986) Serological diagnosis of zoonotic pulmonary dirofilariasis. Am J Med 80:161–164
77. Albonico F, Loiacono M, Gioia G et al (2014) Rapid differentiation of *Dirofilaria immitis* and *Dirofilaria repens* in canine peripheral blood by real-time PCR coupled to high resolution melting analysis. Vet Parasitol 200:128–132
78. Grandi G, Quintavalla C, Mavropoulou A et al (2010) A combination of doxycycline and ivermectin is adulticidal in dogs with naturally acquired heartworm disease [*Dirofilaria immitis*]. Vet Parasitol 169:347–351
79. Giannelli A, Ramos RA, Trversa D et al (2013) Treatment of *Dirofilaria repens* microfilariaemia with a combination of doxycycline hyclate and ivermectin. Vet Parasitol 197:702–704
80. Rodonaja TE (1967) A new species of nematode, *Onchocerca lupi* n. sp., from *Canis lupus cubanensis*. Bull Acad Sci Georgia SSR 45:715–719
81. Gracio AJ, Richter J, Komnenou AT, Gracio MA (2015) Onchocerciasis caused by *Onchocerca lupi*: an emerging zoonotic infection. Syst Rev Parasitol Res 114:2401–2413
82. Egyed Z, Sreter T, Szel Z, Nyiro G, Marialigeti K, Varga L (2002) Molecular phylogeny analysis of *Onchocerca lupi* and its *Wolbachia* endosymbiont. Vet Parasitol 108:153–161
83. Oltranto D, Dantas-Torres F, Gianelli A et al (2013) Cutaneous distribution and circadian rhythm of *Onchocerca lupi* microfilariae in dogs. PLoS Negl Trop Dis 7:e2585
84. Szell Z, Erdelyi L, Sretwer T, Albert M, Varga I (2001) Canine ocular onchocerciasis in Hungary. Vet Parasitol 97:243–249
85. Tudor P, Turcitu M, Mateescu C et al (2016) Zoonotic ocular onchocerciasis caused by *Onchocerca lupi* in dogs in Romania. Parasitol Res 115:859–862
86. Miro G, Montoya A, Checa R, Galvez R, Minguez JJ, Marino V, Otranto D (2016) First detection of *Onchocerca lupi* infection in dogs in southern Spain. Parasit Vectors 9:290. doi:10.1186/s13071-016-1587-1
87. Orihel TC, Ash LR, Holshuh HJ, Satenelli S (1991) Onchocerciasis in a California dog. Am J Trop Med Hyg 44:513–517
88. Al L, Daniels JB, Dix M, Labelle P (2011) *Onchocerca lupi* causing ocular disease in two cats. Vet Ophthalmol 14(Suppl. 1):105–110
89. Labaeele AL, Maddox CW, Daniels JB et al (2013) Canine ocular onchocerciasis in the United States is associated with *Onchocerca lupi*. Vet Parasitol 193:297–301
90. Otranto D, Dantas-Torres F, Giannelli A, Latrofa MS, Papadopoulos E, Cardosa L, Cortes H (2013) Zoonotic *Onchocerca lupi* infection in dogs, Greece and Portugal, 2011–2012. Emerg Infect Dis 19:2000–2003
91. Otranto D, Testini G, Gurlu VP, Yakar K, Dantas-Torres F, Bain O (2011) First evidence human zoonotic infection by *Onchocerca lupi* [Spirurida, Onchoceridae]. Am J Torp Med Hyg 84:55–58
92. Sreter T, Szell Z, Egyed Z, Varga I (2002) Subconjunctival zoonotic onchocerciasis in a man: aberrant infection with *Onchocerca lupi*? Ann Trop Med Parasitol 96:497–502
93. Cantey PT, Weeks J, Edwards M et al (2016) The emergence of zoonotic *Onchocerca lupi* infection in the United States a case series. Clin Infect Dis 62:778–783
94. Azarova NS, Miretsky OY, Soniry MD (1965) The first instance of detection of Nematode *Onchocerca* Diesing 1841 in a person in the URSS, Moscow. Med Parasit Parasit Dis 34:156–158
95. Komnenou A, Eberhard M, Kaedrymidou E, Tsaile E, Dessiris A (2002) Subconjunctival filariasis due to *Onchocerca sp.* in dogs. Vet Ophthalmol 5:119–126

96. Komnenou A, Koutinas AF (2007) Ocular manifestations of some canine infectious and parasitic diseases commonly encountered in the Mediterranean. Euro J Comp Anim Pract 17:271–279
97. Komnenou AT, Thomas ALN, Papadopoulos E, Koutinas AF (2016) Intraocular localization of *Onchocerca lupi* adult worm in a dog with anterior uveitis: a case report. Vet Ophthalmol 19:245–249
98. Franchini D, Giannelli A, Di Paula G et al (2014) Image diagnosis of zoonotic onchocerciasis by *Onchocerca lupi*. Vet Parasitol 203:91–95
99. Giannelli A, Cantacessi C, Graves P, Becker L, Campbell BE, Dantas-Torres F, Otranto D (2014) A preliminary investigation of serological tools for the detection of *Onchocerca lupi* infection in dogs. Parasitol Res 113:1989–1991
100. Basanez MG, Pion SD, Boakes E, Filipe JA, Churcher TS, Boussinesq M (2008) Effect of a single-dose ivermectin on *Onchocerca volvulus*: a systematic review and meta-analysis. Lancet Infect Dis 8:310–322
101. Hoerauf A, Specht S, Buttner M et al (2008) *Wolbachia* endobacteria depletion by doxycycline as antifilarial therapy has macrofilaricidal activity in onchocerciasis: a randomized placebo-controlled study. Med Microbiol Immunol 197:295–311
102. Barrett MP, Burchmore RJS, Stich A et al (2003) The trypanosomiases. Lancet 362:1469–1480
103. Truc P, Buscher P, Cuny G et al (2013) Atypical human infections by animal trypanosomes. PLoS Negl Trop Dis 7:e2256
104. Hoare CA (1972) The trypanosomes of mammals: a zoological monograph. Blackwell Scientific Publication, Oxford
105. Lai D-H, Hashimini H, Lun Z-R, Ayala FJ, Lukes J (2008) Adaptation of *Trypanosoma brucei* to gradual loss of kinetoplast *DNA*: *Trypanosoma equiperdum* and *Trypanosoma evansi* are petite mutants *of T. brucei*. Proc Natl Acad Sci U S A 105:1999–2004
106. Desquesnes M, Holzmuller P, Lai D-H, Dargantes A, Lun Z-R, Jittaplapong S (2013) *Trypanosoma evensi* and surra: a review and perspectives on origin, history, distribution, taxonomy, morphology, hosts, and pathogenic effects. BioMed Res Int 2013:321237
107. Dargantes AP, Mecado RT, Dobson RJ, Reid SA (2009) Estimating the impact *of Trypanosoma evansi* infection [surra] on buffalo population dynamics in southern Philippines using data from cross-sectional surveys. Int J Parasitol 39:1109–1114
108. Rabaa MA, Tue NT, Phuc TM et al (2015) The Vietnam Initiative on Zoonotic Infections [VIZIONS]: a strategic approach to studying emerging zoonotic infectious diseases. EcoHealth 12:726–735
109. Gill BS (1977) Trypanosomes and trypanosomiases of Indian livestock. New Delhi: Indian Council of Agriculture Research Report, p 36
110. Touratier L, Das S. Confirmation et suspicion de trypanosomose humaine a *Trypanosoma evansi* dans des zones a forte prevalence de Surra. Premiere rencontre algero-francaise de parasitology. Alger, Algeria, 15–16 November 2006
111. Haridy FM, El-Metwally MT, Khalil HH (2011) Morsy. *Trypanosoma evansi* in dromedary camel: with a case report of zoonosis in greater Cairo, Egypt. J Egypt Soc Parasitol 41:65–76
112. Joshi PP, Shegokar V, Powar S et al (2005) Human trypanosomiasis caused by *Trypanosoma evansi* in India: the first case report. Am J Trop Med Hyg 73:491–495
113. Vanhollebeke B, Truc P, Poelvoorde P et al (2006) Human *Trypanosoma evansi* infection linked to a lack of Apolipoprotein L-1. N Engl J Med 355:2752–2756
114. Vanhamme L, Paturiaux-Hanocq F, Poelvoorde P et al (2003) Apolipoprotein L-1 is the trypanosome lytic factor of human serum. Nature 422:83–87
115. Hawking F (1978) The resistance of *Trypanosoma congolense*, *T. vivax* and *T. evansi* to human plasma. Trans R Soc Trop Med Hyg 72:405–407
116. Chau NVV, Chau LB, Desquesnes M et al (2016) A clinical and epidemiological investigation of the first reported human infection with the zoonotic parasite *Trypanosoma evansi* in Southeast Asia. Clin Infect Dis 62:1002–1008
117. Sarataphan N, Vongpakorn M, Nuansrichay B et al (2007) Diagnosis of a *Trypanosoma lewisi*-like [*Herpetosoma*] infection in a sick infant from Thailand. J Med Microbiol 56:1118–1121

118. Doke PP, Kar A (2011) A fatal case of *Trypanosoma lewisi* in Maharashtra, India. Ann Trop Med Public Hlth 4:91–95
119. Van Hoff L, Henrad C, Pweel E (1948) Observations sur le *Trypanosome brucei* produissant des infections naturelles dans une region infeste de *Glossina palpalis* en absence de *G. morsitans*. Liber Jubilaris J. Rodhain: Societe Belge de Medecine Tropicale Brussels, p 359
120. Abebe M, Bulto T, Endesha T, Nigatu W (1988) Further studies on the *Trypanosoma brucei* group trypanosome, isolated from a patient infected in Anger-Didessa Valley, West Ethiopia, using the blood incubation infectivity test [BIIT]. Acta Trop 45:185–186
121. Dubey JP, Speer CA, Fayer R (1989) Sarcocystis of animals and man. CRC, Boca Raton, FL
122. Kutty MK, Dissanaike AS (1975) A case of human *Sarcocystis* infection in west Malaysia. Trans R Soc Trop Med Hyg 69:503–504
123. Esposito DH, Freedman DO, Neumayr A, Parola P (2012) Ongoing outbreak of an acute muscular *Sarcocystis*-like illness among travelers returning from Tioman Island, Malaysia, 2011–2012. Eurosurveillance 8:17
124. Prakas P, Butkausas D (2012) Protozoan parasites from genus *Sarcocystis* and their investigation in Lithuania. Ekologija 58:45–58
125. Chhabra MB, Samantaray S (2013) *Sarcocystis* and sarcocystosis in India: status and emerging perspectives. J Parasit Dis 37:1–10
126. Fayer R, Esposito DH, Dubey JP (2015) Human infection with *Sarcocystis* species. Clin Microbiol Rev 28:295–311
127. Beaver PC, Gadgil RK, Morera P (1979) *Sarcocystis* in man: a review and report of five cases. Am J Trop Med Hyg 28:819–844
128. AbuBaker S, Teoh BT, Sam SS et al (2013) Outbreak of human infection with *Sacocystis nesbitti*, Malaysia, 2012. Emerg Infect Dis 19:1989–1991
129. Lau YL, Chang PY, Tan CT, Fong MY, Mahmud R, Wong KT (2014) *Sarcocystis nesbitti* in human skeletal muscle: possible transmission from snakes. Am J Trop Med Hyg 90: 361–364
130. Heydorn AO (1977) Sarkosporidieninfiziertes Fleish als mogliche Krankhheitsursache fur den Menschen. Arch Leb 28:27–31
131. Pathmanathan R, Kan SP (1992) Three cases of human *Sarcocystis* infection with a review of human muscular sarcocystosis in Malaysia. Trop Geogr Med 44:102–108
132. Wong KT, Pathmanathan R (1992) High prevalence of human muscle sarcocystosis in southeast Asia. Trans R Soc Trop Med Hyg 86:631–632
133. Arness MK, Brown JD, Dubey JP, Neafie RC, Gransom DE (1999) An outbreak of acute eosinophilic myositis attributed to human Sarcocystis parasitism. Am J Trop Med Hyg 1:548–553
134. Italiano CM, Wong KT, AbuBaker S et al (2014) *Sarcocystis nesbitti* causes acute relapsing febrile myositis with high attack rate: description of a large outbreak of muscular sarcocstosis in Pangkor Island, Malaysia, 2012. PLoS Negl Trop Dis 8:e2876
135. Esposito DH, Stich A, Epelboin L et al (2014) Acute muscular sarcopcystosis: an international investigation among ill travelers returning from Tioman Island, Malaysia, 2011–2012. Clin Infect Dis 59:1401–1410
136. Slesak G, Schafer J, Langeheinecke A, Tappe D (2015) Prolonged clinical course of muscular sarcocytosis and effectiveness of cotrimoxazole among travelers to Tioman Island, Malaysia, 2011–2014. Clin Infect Dis 60:329
137. Balakrishna JP, Chacko G, Manipadam MT, Ramya I (2013) Glomerulopathy in a patient with sarcocystis infestation. Indian J Pathol Microbiol 56:285–287
138. Lindsay DS, Dubey JP (1999) Determination of the activity of pyrimethamine, trimethoprim, sulfonamides, and combination of pyrimethamine against Sarcocystis neurona in cell cultures. Vet Parasitol 82:205–210
139. Srinivasa Rao K, Md H (2002) Efficacy of amprolium and maduramycin against sarcocystosis in experimentally infected pups. J Parasit Dis 26:111–113
140. Saleque A, Juyal PD, Bhatia BB (1990) Effect of temperature on the infectivity of *Sarcocystis miescheriana* cysts in pork. Vet Parasitol 36:343–346

ERRATUM

Emerging Zoonoses

I.W. Fong

I.W. Fong, *Emerging Zoonoses*, Emerging Infectious Diseases of the 21st Century,
DOI 10.1007/978-3-319-50890-0

DOI 10.1007/978-3-319-50890-0_12

On the series page, Dr. Fong's affiliation should read as:
Professor of Medicine, University of Toronto

Preface, Dedication, and Acknowledgements have been added to the Front Matter.

The updated original online version fort this book can be found at
http://dx.doi.org/10.1007/978-3-319-50890-0

I.W. Fong, *Emerging Zoonoses*, Emerging Infectious Diseases of the 21st Century,
DOI 10.1007/978-3-319-50890-0_12

Index

I.W. Fong, *Emerging Zoonoses*, Emerging Infectious Diseases of the 21st Century,
DOI 10.1007/978-3-319-50890-0

B

C

T

Printed in the United States
By Bookmasters